SOLUTIONS MANUAL TO ACCOMPANY

PHYSICAL CHEMISTRY

First Edition

Robert A. Alberty
Professor of Chemistry
Massachusetts Institute of Technology

Robert J. Silbey
Professor of Chemistry
Massachusetts Institute of Technology

John Wiley & Sons, Inc.
New York Chichester Brisbane Toronto Singapore

ISBN 0-471-55118-X

Printed in the United States of America
Printed and bound by the Courier Companies, Inc.

10 9 8 7 6 5 4 3 2 1

PREFACE

This manual gives solutions of the problems in the first set of the text R. A. Alberty and R. J. Silbey, PHYSICAL CHEMISTRY, John Wiley and Sons, Inc. New York, 1991, and it gives answers for the second set.

Working problems is an important part of learning physical chemistry. Not all knowledge of physical chemistry is quantitative, but much of it is. Since physical chemistry utilizes physics and mathematics to predict and interpret chemical phenomena, there are many opportunities to use quantitative methods.

The availability of hand-held electronic calculators has made it much easier to work physical chemistry problems and has made it possible to include more difficult problems.You may want to use a personal computer to solve some of the problems

The units of physical quantities are usually shown in solving problems in this manual. It is important to develop the habit of using units and cancelling them to obtain the units for the answer because this helps prevent errors.

In this book, as in your handwritten lecture notes, there is no distinction between italic (sloping) and roman (upright) type, but it is important to note that in the printed literature italic type is used for symbols for physical quantities and roman type is used for units.

We are indebted to many physical chemists who have recommended problems and who have suggested improvements in this SOLUTIONS MANUAL. We thank Vera Spanos for her careful typing of the solutions.

Robert A. Alberty
Robert J. Silbey

CONTENTS

PART I. THERMODYNAMICS

PART II. QUANTUM MECHANICS AND SPECTROSCOPY

PART III. STATISTICAL MECHANICS AND KINETICS

PART IV. MACROSCOPIC AND MICROSCOPIC STRUCTURES

Part 1

Thermodynamics

CHAPTER 1: Zeroth Law of Thermodynamics and Equations of State

1.1 The ideal gas law also represents the behavior of mixtures of gases at low pressures. The molar volume of the mixture is the volume of a mole of the mixture. The partial pressure of gas in a mixture is defined as $y_i P$, where y_i is its mole fraction, and P is the total pressure. Ten grams of N_2 is mixed with 5 g of O_2 and held at 25 °C at 0.750 bar. (a) What are the mole fractions of N_2 and O_2? (b) What are the partial pressures of N_2 and O_2? (c) What is the volume of the ideal mixture?

SOLUTION

(a) $$n_{N_2} = \frac{(10\ \text{g})}{(28.013\ \text{g mol}^{-1})} = 0.357\ \text{mol}$$

$$n_{O_2} = \frac{(5\ \text{g})}{(32.000\ \text{g mol}^{-1})} = 0.156\ \text{mol}$$

$$y_{N_2} = \frac{(0.357\ \text{mol})}{(0.513\ \text{mol})} = 0.696$$

$$y_{O_2} = \frac{(0.156\ \text{mol})}{(0.513\ \text{mol})} = 0.304$$

(b) $$P_{N_2} = (0.696)(0.750\ \text{bar}) = 0.522\ \text{bar}$$

$$P_{O_2} = (0.304)(0.750\ \text{bar}) = 0.228\ \text{bar}$$

(c) $$V = nRT/P$$

$$= \frac{(0.513\ \text{mol})(0.083145\ \text{L bar K}^{-1}\ \text{mol}^{-1})(298.15\ \text{K})}{0.750\ \text{bar}}$$

$$= 17.00\ \text{L}$$

1.2 A mixture of methane and ethane is contained in a glass bulb of 500 cm^3 capacity at 25 °C. The pressure is 1.25 bar, and the mass of gas in the bulb is 0.530 g. What is the average molar mass, and what is the mole fraction of methane?

SOLUTION

$$PV = nRT = \left(\frac{m}{M}\right) RT$$

$$M = \frac{mRT}{PV} = \frac{(0.530\ g)(0.08315\ L\ bar\ K^{-1}\ mol^{-1})(298\ K)}{(1.25\ bar)(0.500\ L)}$$

$$= 21.0\ g\ mol^{-1}$$

$$M = y_1M_1 + y_2M_2$$

$$21.0 = y_1\ 16.0 + (1 - y_1)30.0$$

$$14.0y_1 = 9.0$$

$$y_1 = \frac{9.0}{14.0} = 0.643$$

1.3 Nitrogen tetroxide is partially dissociated in the gas phase according to the reaction

$$N_2O_4(g) = 2NO_2\ (g)$$

A mass of 1.588 g of N_2O_4 is placed in a 500-cm^3 glass vessel at 1.0133 bar at 298 K and dissociates to an equilibrium mixture.

(a) What are the mole fractions of N_2O_4 and NO_2?

(b) What percentage of the N_2O_4 has dissociated?

Assume the gases are ideal.

SOLUTION

$m_{total} = 1.588\ g = m_{N_2O_4} + m_{NO_2}$

$$P_{total} = 1.0133\ bar = \frac{m_{N_2O_4}(0.08314\ L\ bar\ K^{-1}\ mol^{-1})\ (298\ K)}{(92\ g\ mol^{-1})(0.500\ L)}$$

$$+ \frac{m_{NO_2}(0.08314\ L\ bar\ K^{-1}\ mol^{-1})(298\ K)}{(46\ g\ mol^{-1})(0.500\ L)}$$

$= 0.5386\ m_{N_2O_4} + 1.0772\ m_{NO_2}$

$= 1.7106 - 0.5386\ m_{N_2O_4}$

Because of equation 1.

$m_{N_2O_4} = 1.295$ g or 0.01407 mol

$m_{NO_2} = 0.293$ g or 0.06370 mol

$n = 0.02044$ mol

(a) $y_{N_2O_4} = \dfrac{0.01407}{0.02044} = 0.6884$

$y_{NO_2} = 0.3116$

(b) $\%\ \text{undissociated} = \dfrac{(0.01407\ mol)(100)}{(1.588\ g)/(92\ g\ mol^{-1})} = 81.51\ \%$

$\%\ \text{dissociated} = 100 - 81.51 = 18.49\%$

1.4 Calculate the second virial coefficient of hydrogen at 0 °C from the fact that the molar volumes at 50.7, 101.3, 202.6, and 303.9

bar are 0.4634, 0.2386, 0.1271, and 0.09004 L mol^{-1}, respectively.

<u>SOLUTION</u>

$$\frac{P\bar{V}}{RT} = 1 + \frac{B}{\bar{V}} + \frac{C}{\bar{V}^2} + \cdots$$

P/bar	50.7	101.3	202.6	303.9
$\bar{V}$/L mol^{-1}	0.4634	0.2386	0.1271	0.09004
$P\bar{V}/RT$	1.035	1.064	1.134	1.205
$(1/\bar{V})$/mol L^{-1}	2.158	4.191	7.868	11.106

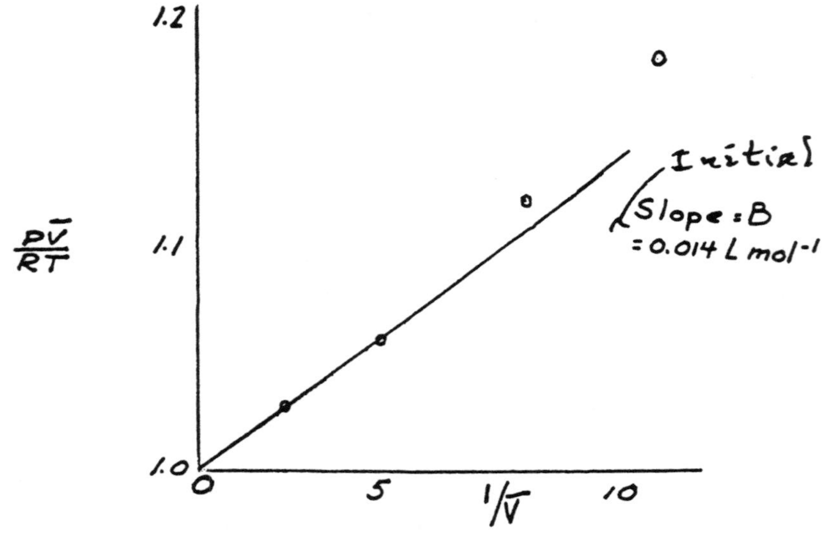

1.5 The critical temperature of carbon tetrachloride is 283.1 °C. The densities in g/cm^3 of the liquid ρ_l and vapor ρ_v at different temperatures are as follows:

t/°C	100	150	200	250	270	280
ρ_1	1.4343	1.3215	1.1888	0.9980	0.8666	0.7634
ρ_v	0.0103	0.0304	0.0742	0.1754	0.2710	0.3597

What is the critical molar volume of CCl_4? It is found that the mean of the densities of the liquid and vapor does not vary rapidly with temperature and can be represented by

$$\frac{\rho_1 + \rho_v}{2} = AT + B$$ where A and B are constants.

The extrapolated value of the average density at the critical temperature is the critical density. The molar volume $\bar{V}_c$ at the critical point is equal to the molar mass divided by the critical density.

SOLUTION

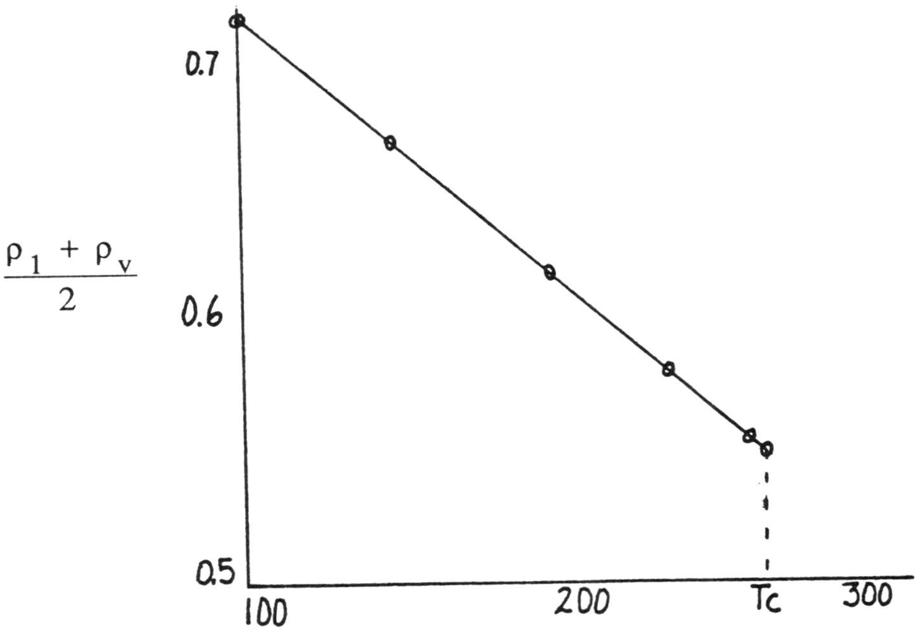

Extrapolating $\frac{\rho_1 + \rho_v}{2}$ to T_c we obtain

$\rho_c \quad = 0.557 \text{ g cm}^{-3}$

$$V_c = \frac{153.84 \text{ g mol}^{-1}}{0.557 \text{ g cm}^{-3}} = 276 \text{ cm}^3 \text{ mol}^{-1}$$

A hand-held calculator that will do linear expression can be used to obtain

$$\frac{\rho_l + \rho_v}{2} = 0.8110 - 8.95 \times 10^{-4} \, t$$

which is the equation that represents the experimental data with the smallest sum of squares of deviation. This yields the same critical density that is given above.

1.6 Show that for a gas of rigid spherical molecules, b in the van der Waals equation is four times the molecular volume times Avogadro's constant. If the molecular diameter of Ne is 0.258 nm (Table 16.1), approximately what value of b is expected.

SOLUTION

The molecular volume for a spherical molecule is

$$\frac{4}{3} \pi \left(\frac{d}{2}\right)^3 = \frac{\pi}{6} d^3$$

where d is the diameter. Since the center of a second spherical molecule cannot come within a distance d of the center of the first spherical molecule, the excluded volume per pair of molecules is

$\frac{4}{3}\pi d^3$

The constant b in van der Waals equation is the excluded volume per molecule times Avogadro's constant

$$b = \frac{2}{3}\pi d^3 N_A = 4\left(\frac{\pi}{6} d^3\right)N_A$$

For neon,

$$b = \left(\frac{2}{3}\right)\pi(6.02 \times 10^{23}\ \text{mol}^{-1})(0.258 \times 10^{-9}\ \text{m})^3(6.02 \times 10^{23}\ \text{mol}^{-1})$$

$$= 2.17 \times 10^{-5}\ \text{m}^3\ \text{mol}^{-1}$$

Since 100 cm = 1 m, or 100 cm m^{-1} = 1, b can be expressed in $cm^3\ mol^{-1}$ by multiplying by $(100\ cm\ m^{-1})^3$.

$$b = (2.17 \times 10^{-5}\ \text{m}^3\ \text{mol}^{-1})(100\ \text{cm}\ \text{m}^{-1})^3$$

$$= 21.7\ \text{cm}^3\ \text{mol}^{-1}$$

1.7 What is the molar volume of n-hexane at 660 K and 91 bar according to (a) the ideal gas law and (b) the van der Waals equation?

$T_c = 507.7$ K $\quad P_c = 30.3$ bar

SOLUTION

(a) $$\bar{V} = \frac{RT}{P} = \frac{(0.08314\ \text{L bar K}^{-1}\ \text{mol}^{-1})(660\ \text{K})}{91\ \text{bar}} = 0.603\ \text{L mol}^{-1}$$

(b) $$a = \frac{27 R^2 T_c^2}{64 P_c} = \frac{(27)(0.08314\ \text{L bar K}^{-1}\ \text{mol}^{-1})^2(507.7\ \text{K})^2}{(64)(30.3\ \text{bar})}$$

$$= 24.81\ \text{L}^2\ \text{bar mol}^{-1}$$

$$b = \frac{RT_c}{8\ P_c} = \frac{(0.8314\ \text{L bar K}^{-1}\ \text{mol}^{-1})(507.7\ \text{K})}{8(30.3\ \text{bar})}$$

$$= 0.174\ \text{L mol}^{-1}$$

$$P = \frac{RT}{\bar{V}_c - b} - \frac{a}{\bar{V}^2}$$

$$91\ \text{bar} = \frac{(0.08314)(660\ \text{K})}{\bar{V} - 0.174} - \frac{24.81}{\bar{V}^2}$$

Rather than solving a cubic equation, substituting successive values of V shows that

$$\bar{V} = 0.39\ \text{L mol}^{-1}.$$

1.8 Derive the expressions for van der Waals constants a and b in terms of the critical temperature and pressure; that is, derive equations 1.35 and 1.36 from 1.32 - 1.34.

SOLUTION

Equations 1.35 and 1.36 may be written

$$\frac{RT_c}{(\bar{V}_c - b)^2} = \frac{2a}{\bar{V}_c^3} \qquad (1)$$

$$\frac{2RT_c}{(\bar{V}_c - b)^3} = \frac{6a}{\bar{V}_c^4} \qquad (2)$$

Division of the first equation by the second yields

$\bar{V}_c = 3b$ (3)

Substitution of this expression in equation 1 yields

$T_c \quad = 8a/27Rb$ (4)

Substitution of equations 3 and 4 in equation 1.34 yields

$P_c \quad = a/27b^2$ (5)

Since there are three relations (equations 3 - 5) between the van der Waals constants and the critical constants, a and b may be expressed in terms of T_c and P_c or T_c and V_c. Critical pressures are generally known more accurately than critical volumes, and so a and b are generally calculated using data on the critical temperature and pressure.

1.9 Calculate the second virial coefficient of methane at 300 K and 400 K from its van der Waals constants, and compare these results with Fig. 1.9.

SOLUTION

For methane a = 2.283 L^2 bar mol^{-2} and b = 0.04278 L mol^{-1}. From equation 1.28, the second virial coefficient at 300 K is

$$B \quad = b - a/RT = 0.04278 - \frac{2.283}{(0.08314)(300)}$$

$$= -0.048 \text{ L mol}^{-1}$$

At 400 K, B = -0.026

Fig. 1.9 yields -0.040 L mol^{-1} at 300 K and -0.020 L mol^{-1} at 400 K.

1.10 You want to calculate the molar volume of O_2 at 298.15 K and 50 bar using the van der Waals equation, but you don't want to solve a cubic equation. Use the first two terms of equation 1.29 to obtain an approximate solution. The van der Waals constants of O_2 are a = 0.138 Pa m^6 mol^{-1} and b = 31.8 x 10^{-6} m^3 mol^{-1}. What is the molar volume in L mol^{-1}?.

SOLUTION

$$\frac{p\bar{V}}{RT} = 1 + \frac{1}{RT}\left(b - \frac{a}{RT}\right)P$$

$$\bar{V} = \frac{RT}{P}\left[1 + \frac{1}{RT}\left(b - \frac{a}{RT}\right)P\right]$$

$$= \frac{RT}{P} + b - \frac{a}{RT}$$

$$= \frac{8.31451 \text{ Pa m}^3 \text{ mol}^{-1} \text{ K}^{-1}(298.15 \text{ K})}{50 \times 10^5 \text{ Pa}} + 31.8 \times 10^{-6} \text{ m}^3 \text{ mol}^{-1}$$

$$- \frac{0.138 \text{ Pa m}^6 \text{ m}^{-2}}{(8.314 \text{ Pa m}^3 \text{ K}^{-1} \text{ mol}^{-1})(298 \text{ K})}$$

$$= (495.8 \times 10^{-6} + 31.8 \times 10^{-6} - 55.7 \times 10^{-6})\text{m}^3 \text{ mol}^{-1}$$

$$= 471.9 \times 10^{-6} \text{ m}^3 \text{ mol}^{-1}$$

$$= 0.4719 \text{ L mol}^{-1}$$

1.11 Show that the equation for the dotted line in Fig. 1.13 is

$$P = \frac{a}{\bar{V}^2}\left(1 - \frac{2b}{\bar{V}}\right)$$

This equation gives the boundary of the physically unrealizable region for a van der Waals gas. What is the maximum pressure satisfying this equation?

<u>SOLUTION</u>

The turning points occur at the volumes at which $(\partial P/\partial V)_T = 0$.

$$P = \frac{RT}{\bar{V} - b} - \frac{a}{\bar{V}^2}$$

$$(\partial P/\partial \bar{V})_T = -\frac{RT}{(\bar{V}-b)^2} + \frac{2a}{\bar{V}^3} = 0$$

$$\frac{RT}{\bar{V} - b} = \frac{2a(\bar{V}-b)}{\bar{V}^3}$$

$$P = \frac{2a(\bar{V}-b)}{\bar{V}^3} - \frac{a}{\bar{V}^2} = \frac{a}{\bar{V}^2} - \frac{2ab}{\bar{V}^3} = \frac{a}{\bar{V}^2}\left(1 - \frac{2b}{\bar{V}}\right)$$

The maximum pressure according to this equation is obtained by differentiating with respect to $\bar{V}$.

$\frac{dP}{d\bar{V}} = 0 = \frac{2a}{\bar{V}^3} + \frac{6ab}{\bar{V}^4}$ so that $b = \frac{\bar{V}_c}{3} = \frac{RT_c}{8P_c}$, where equation 1.36 has been used, and $P_c = 8b/RT_c$.

1.12 Calculate the Boyle temperature for hydrogen from its van der Waals constants.

<u>SOLUTION</u>

$$T_B = \frac{a}{bR}$$

$$= \frac{0.2476 \text{ L}^2 \text{ bar mol}^{-2}}{(0.02661 \text{ L mol}^{-1})(0.08314 \text{ L bar K}^{-1} \text{ mol}^{-1})}$$

$$= 112 \text{ K}$$

Table 1.1 gives 110 K.

1.13 What is the equation of state for a liquid for which the coefficient of cubic expansion α and the isothermal compressibility κ are constant?

<u>SOLUTION</u>

$$dV = (\frac{\partial V}{\partial T})_P \, dT + (\frac{\partial V}{\partial P})_T \, dP$$

$$= \beta V dT - \kappa V DP$$

$$\int_{V_1}^{V_2} \frac{dV}{dV} = \beta \int_{T_1}^{T_2} dT - \kappa \int_{P_1}^{P_2} dP$$

$$\ln(V_2/V_1) = \beta(T_2 - T_1) - \kappa(P_2 - P_1)$$

The indefinite integral is

$$\ln V = \beta T - \kappa P + \text{const.}$$

$$V = \kappa^{\beta T} e^{-\kappa P}$$

1.14 The cubic expansion coefficient α is defined by $\alpha = \frac{1}{V}\left(\frac{\partial V}{\partial T}\right)_P$ and the isothermal compressibility is defined by $\kappa = -\frac{1}{V}\left(\frac{\partial V}{\partial P}\right)_T$. Calculate these quantities for an ideal gas.

SOLUTION

$$V = nRT/P$$

$$\left(\frac{\partial V}{\partial T}\right)_P = \frac{nR}{P}$$

$$\alpha = \frac{1}{T}$$

$$\left(\frac{\partial V}{\partial P}\right)_T = -\frac{nRT}{P^2}$$

$$\kappa = \frac{nRT}{VP^2} = \frac{1}{P}$$

1.15 For a liquid cubic expansion coefficient α is nearly constant over a narrow range of temperature. Derive the expression for the volume as a function of temperature and the limiting form for temperatures close to T_0.

SOLUTION

$$\int_{V_o}^{V} \frac{dV}{V} = \alpha \int_{T_o}^{T} dT$$

$$\ln \frac{V}{V_o} = \alpha (T - T_o)$$

$$V = V_o e^{\alpha(T - T_O)}$$

If $\alpha(T - T_o) << 1$,

$$V = V_o[1 + \alpha(T - T_o)]$$

1.16 Assuming that the atmosphere is isothermal at 0 °C and that the average molar mass of air is 29 g mol^{-1}, calculate the atmospheric pressure at 20,000 ft above sea level.

SOLUTION

$h = (2.0 \times 10^4 \text{ ft})(12 \text{ in ft}^{-1})(2.54 \text{ cm in}^{-1})(10^{-2} \text{ m cm}^{-1}) =$ 6096 m

$$P = P_o e^{-gMh/RT}$$

$$P = (1.013 \text{ bar}) \exp[\frac{-(9.8 \text{ m s}^{-2})(29 \times 10^{-3} \text{ kg mol}^{-1})(6096 \text{ m})}{(8.314 \text{ J K}^{-1} \text{ mol}^{-1})(273 \text{ K})}]$$

$= 0.472$ bar

1.17 Calculate the pressure and composition of air on the top of Mt. Everest assuming the atmosphere has a temperature of 0 °C independent of altitude (h = 29,141 ft). Assume that air at sea level is 20% O_2 and 80% N_2.

SOLUTION

$h = (29{,}141 \text{ ft})(12 \text{ in ft}^{-1})(2.54 \text{ cm in}^{-1})(0.01 \text{ m cm}^{-1}) =$ 8,882 m

For O_2, $P = (0.2 \text{ bar}) \exp(-9.8 \times 32 \times 10^{-3} \times 882/8.314 \times 273)$

$= 0.059$ bar

For N_2, $P = (0.8 \text{ bar}) \exp(-9.8 \times 28 \times 10^{-3} \times 8882/8.314 \times 273)$

$= 0.274$ bar

The total pressure is 0.333 bar, and $y_{O_2} = 0.177$ and $y_{N_2} =$ 0.823.

1.18 (a) 0.02479 m^3 (b) 0.02560 m^3 (c) 0.7747, 0.1937, 0.0317

1.19 0.835

1.20 $B = y_1^2 B_{11} + 2y_1y_2 B_{12} + y_2^2 B_{22}$

1.21 276 cm^3 mol^{-1}

1.22 $T_c \approx 435$ K, $V \approx 0.261$ L mol^{-1}, $P_c \approx 39$ bar

1.23 0.324 nm

1.24 -1.96×10^{-3} bar^{-1}, -0.60×10^{-4} bar^{-1}

1.25 (a) 20.1 bar, (b) 1642 bar, (c) 2776 bar

1.26 99.8 bar, 152 bar

1.27 For H_2, P = 0.382 bar, T = 105.2 K

For CO_2, P = 2.17 bar, T = 964.3 K

For H_2O, P = 6.48 bar, T = 2051 K

1.28 (a) 155 bar, (b) 88.6 bar

1.29 For H_2, 6.92×10^{-4} bar^{-1}

For O_2, -1.27×10^{-3} bar^{-1}

1.30 $V_2 = V_1 \exp[-\kappa(P_2 - P_1)]$

1.31 $\alpha = 1/T(1 + bP/RT)$

$\kappa = 1/P(1 + bP/RT)$

1.32 0.843 bar

1.33 3.27×10^{-9} bar, $y_{O_2} = 0.015$, $y_{N_2} = 0.985$

1.34 $M = y_A M_A + y_B M_B$

CHAPTER 2: First Law of Thermodynamics

2.1 How high can a person (assume a weight of 70 kg) climb on one ounce of chocolate, if the heat of combustion (628 kJ) can be converted completely into work of vertical displacement?

SOLUTION

$$w = mgh$$

$$h = \frac{w}{gm} = \frac{628000\ \mathrm{J}}{(9.806\ \mathrm{m\ s^{-2}})(70\ \mathrm{kg})} = 914\ \mathrm{m}$$

2.2 A mole of sodium metal is added to water. How much work is done on the atmosphere by the subsequent reaction if the temperature is 25 °C?

SOLUTION

$Na(s) + H_2O(1) = NaOH(soln) + 1/2\ H_2(g)$

$$\bar{V} = \frac{RT}{P} = \frac{(8.314\ \mathrm{J\ K^{-1}\ mol^{-1}})(298\ \mathrm{K})}{(101325\ \mathrm{Pa})}$$

$$= 0.0245\ \mathrm{m^3\ mol^{-1}}$$

$$w = -P\Delta\bar{V} = -(101325\ \mathrm{Pa})(0.123\ \mathrm{m^3\ mol^{-1}})$$

$$= -1.24\ \mathrm{kJ\ mol^{-1}}$$

2.3 You want to heat 1 kg of water 10 °C, and you have the following four methods under consideration. The heat capacity of water is 4.184 J K^{-1} g^{-1}.

(a) You can heat it with a mechanical eggbeater that is powered by a 1 kg mass on a rope over a pulley. How far does the mass have to descend in the earth's gravitational field to supply enough work?

(b) You can send 1 ampere through a 100 ohm resistor. How long will it take?

(c) You can send the water through a solar collector that has an area of 1 m^2. How long will it take if the sun's intensity on the collector is 4 J cm^{-2} min^{-1}?

(d) You can make a charcoal fire. The heat of combustion of graphite is -393 kJ mol^{-1}. That is, 12 g of graphite will produce 393 kJ of heat when it is burned to $CO_2(g)$ at constant pressure. How much charcoal will you have to burn?

<u>SOLUTION</u>:

(a) $(100\ g)(4.184\ J\ K^{-1}\ g^{-1})(10\ K) = 4.184 \times 10^4\ J =$

$$mg\Delta h = (1\ kg)(9.8\ m\ s^{-2})\Delta h$$

$$\Delta h = \frac{4.184 \times 10^4}{9.8} = 4269\ m$$

(b) $I^2Rt = 4.184 \times 10^4\ J = (1A)^2(100\Omega)t$

$$t = \frac{4.184 \times 10^4}{100} = 418.4\ s = 6.97\ \text{minutes}$$

(c) $4.184 \times 10^4\ J = (4.0\ J\ cm^{-2}\ min^{-1})(100\ cm)^2\ t$

$$t = \frac{4.184 \times 10^4}{4 \times 10^4} = 1.05\ \text{minutes}$$

(d) $4.184 \times 10^4\ J = (393 \times 10^3\ J)\ \frac{m}{12}$

$$m = \frac{12(4.184 \times 10^4)}{393 \times 10^3} = 1.28 \text{ grams}$$

2.4 Show that the differential df is inexact.

$$df = dx - (x/y)dy$$

Thus the integral $\int df$ depends on the path. However, we can define a new function g by

$$dg = (1/y)df$$

which has the property that dg is exact. Show that dg is exact, so that $\oint dg = 0$ In this case y is referred to as an integrating factor.

<u>SOLUTION</u>

To determine whether df is exact take the cross derivatives

$$M(x,y) = 1; (\partial M/\partial y)_x = 0$$

$$N(x,y) = -x/y, (\partial N/\partial x)_y = -1/y$$

Since $(\partial M/\partial y)_x \neq (\partial N/\partial x)_y$, df is inexact.

The new differential is given by

$$dg = (1/y)\ dx - (x/y^2)dy$$

Now take the cross derivatives.

$$M = 1/y,\ (\partial M/\partial y)_x = -1/y^2$$

$$N = -x/y^2,\ (\partial N/\partial x)_y = -1/y^2$$

Since $(\partial M/\partial y)_x = (\partial M/\partial x)_y$, dg is exact.

2.5 Over narrow ranges of temperature and pressure, the differential expression for the volume of a fluid as a function of temperature and pressure can be integrated to obtain

$$V = Ke^{\alpha} e^{-\kappa P}$$

(α and κ are defined in Section 1.10). Show that V is a state function.

SOLUTION

The test of a state function is that it forms exact differentials. For an exact differential the mixed second derivatives are equal.

$$\frac{\partial V}{\partial T} = Ke^{-\kappa P} e^{\alpha T} \alpha$$

$$\frac{\partial}{\partial P}\frac{\partial V}{\partial T} = Ke^{\alpha T} e^{-\kappa P} (-\kappa\alpha)$$

$$\frac{\partial V}{\partial P} = Ke^{\alpha T} e^{-\kappa P} (-\kappa)$$

$$\frac{\partial}{\partial T}\frac{\partial V}{\partial P} = Ke^{\alpha T} e^{-\kappa P} (-\kappa\alpha)$$

Thus V for a substance that can be described in this way is a state function.

2.6 One mole of nitrogen at 25 °C and 1 bar is expanded reversibly and isothermally to a pressure of 0.132 bar. (a) What is the value of w? (b) What is the value of w if the nitrogen is expanded against a constant pressure of 0.132 bar?

SOLUTION

(a) $w = RT \ln(P_2/P_1)$

$= (8.314 \text{ J K}^{-1})(298.15 \text{ K}) \ln 0.132$

$= -5.03 \text{ kJ mol}^{-1}$

(b) $\bar{V}_1 = RT/P_1 = \dfrac{(8.314)(298.15)}{10^5} = 0.248\ m^3\ mol^{-1}$

$\bar{V}_2 = \dfrac{(8.314)(298.15)}{(0.132)(10^5)} = 0.188\ m^3\ mol^{-1}$

$w = -P\Delta\bar{V} = -(0.132 \times 10^5)(0.188 - 0.0248)$

$= -2.15\ kJ\ mol^{-1}$

2.7 (a) Derive the equation for the work of reversible isothermal expansion of a van der Waals gas from V_1 to V_2.

(b) A mole of CH_4 expands reversibly from 1 L to 50 L at 25 °C. Calculate the work in joules assuming

(1) the gas is ideal,

(2) the gas obeys the van der Waals equation.

For CH_4 (g)

$a = 2.283\ L^2\ bar\ mol^{-2}$

$b = 0.04278\ L\ mol^{-1}$

<u>SOLUTION</u>

(a) $w = -\int_{V_1}^{V_2} PdV = -\int_{V_1}^{V_2} \dfrac{nRTdV}{V-nb} + \int_{V_1}^{V_2} \dfrac{an^2dV}{V^2}$

$= -nRT \ln \dfrac{V_2-nb}{V_1-nb} + an^2\left(\dfrac{1}{V_1} - \dfrac{1}{V_2}\right)$

(b)(1) $w = -nRT \ln \dfrac{V_2}{V_1}$

$= -(1\ mol)(8.314\ J\ K^{-1}\ mol^{-1})(298\ K) \ln 50$

$= -9697\ J$

(b)(2) We have to be careful about units in adding these two terms. If we use R = 8.314 J K^{-1} mol^{-1} in the first term it will come out in J, as for the ideal gas. The second will come out in L bar. A convenient way to convert L bar to J is to note that

$$\frac{8.314 \text{ J K}^{-1} \text{ mol}^{-1}}{0.08314 \text{ L bar K}^{-1} \text{ mol}^{-1}} = 100 \text{ J/L bar}$$

$$\begin{aligned} w &= (1 \text{ mol})(8.314 \text{ J K}^{-1} \text{ mol}^{-1})(298 \text{ K}) \ln\frac{50 - 0.04278}{1 - 0.4278} \\ &\quad + (100 \text{ J/L bar})(2.283 \text{ L}^2 \text{ bar mol}^{-2})(1 \text{ mol})^2\left(\frac{1}{1\text{L}} - \frac{1}{50\text{L}}\right) \\ &= -9799 \text{ J} + 224 \text{ J} = -9575 \text{ J} \end{aligned}$$

This is the work done on the gas. The work done on the surroundings is +9575 J. Less work is done by methane than an ideal gas because of intermolecular attractions.

2.8 Liquid water is vaporized at 100 °C and 1.013 bar. The heat of vaporization is 40.69 kJ mol^{-1}. What are the values of (a) w_{rev} per mole, (b) q per mole, (c) $\Delta\bar{U}$, and (d) $\Delta\bar{H}$.

SOLUTION

(a) Assuming that water vapor is an ideal gas and that the volume of liquid water is negligible,

$$\begin{aligned} w &= -P\Delta\bar{V} = -RT \\ &= (8.314 \times 10^{-3} \text{ kJ K}^{-1} \text{ mol}^{-1})(373.15 \text{ K}) \\ &= -3.10 \text{ kJ mol}^{-1} \end{aligned}$$

(b) The heat of vaporization is 40.69 kJ mol^{-1}, and, since heat is absorbed, q has a positive sign.

$$q = 40.69 \text{ kJ mol}^{-1}$$

(c) $\bar{U}$ $= q + w$

$= (40.69 - 3.10)$ kJ mol^{-1}

$= 37.59$ kJ mol^{-1}

(d) $\Delta\bar{H}$ $= \Delta\bar{U} + \Delta(P\bar{V}) = \Delta\bar{U} + P\Delta\bar{V}$

$= \Delta\bar{U} + RT$

$= 37.59$ kJ mol^{-1} + $(8.314 \times 10^{-3}$ kJ K^{-1} mol^{-1})(373.15 K)

$= 40.69$ kJ mol^{-1}

2.9 An ideal gas expands reversibly and isothermally from 10 bar to 1 bar. What are the values of (a) w per mole, (b) q per mole, (c) $\Delta\bar{U}$, and (d) $\Delta\bar{H}$?

SOLUTION

(a) w $= RT \ln(P_2/P_1)$

$= (8.314)(298.15) \ln(1/10)$

$= -5.70$ kJ mol^{-1}

(b) q $= -w = 5.70$ kJ mol^{-1}

(c) $\Delta\bar{U}$ $= 0$ since an ideal gas

(d) $\Delta\bar{H}$ $= \Delta\bar{U} + \Delta P\bar{V} = 0$

2.10 For a gas following $\bar{V} = \frac{RT}{P} + (b - \frac{a}{RT})$,

What is the inversion temperature for the Joule-Thomson effect? What is the sign of ΔT below the inversion temperature?

SOLUTION

$$\left(\frac{\partial \Delta \bar{V}}{\partial T}\right)_P = \frac{R}{P} + \frac{a}{RT^2}$$

$$\mu_{JT} = \frac{\left[T\left(\frac{\partial \Delta \bar{V}}{\partial T}\right)_P - \bar{V}\right]}{C_P}$$

$$= \frac{\left(\frac{2a}{RT} - b\right)}{C_P}$$

At T_{inv}, $\mu_{JT} = 0$ so that

$$T_{inv} = \frac{2a}{bR}$$

Below the inversion temperature, μ_{JT} is positive and ΔT is negative, since ΔP is negative.

2.11 Calculate $\bar{H}^o$ (2000 K) - H^o (0 K) for H(g).

SOLUTION

$$\bar{H}^o(T_2) - \bar{H}^o(T_1) = \int_{T_1}^{T_2} \bar{C}_P^o \, dT$$

$$\bar{H}^o(2000\ K) - \bar{H}^o(0\ K) = \int_0^{2000} \frac{5}{2} R \, dT = \frac{5}{2} R(2000)$$

$$= 41.572 \text{ kJ mol}^{-1}$$

Appendix C.1 yields 6.197 + 35.376 = 41.573 kJ mol^{-1}

(Note that for O(g) a slightly higher value is obtained because there is some absorption of heat by excitation to higher electronic levels.)

2.12 The heat capacity of a gas may be represented by

$$\bar{C}_P = \alpha + \beta T + \gamma T^2$$

For N_2 α = 26.984 J K^{-1} mol^{-1}, β = 5.910 x 10^{-3} J K^{-2} mol^{-1}, and γ = -3.377 x 10^{-7} J K^{-3} mol^{-1}. How much heat is required to heat a mole of N_2 from 300 K to 1000 K?

SOLUTION

$$q = \int_{T_1}^{T_2} (26.984 + 5.910 \times 10^{-3}\ T - 3.377 \times 10^{-7}\ T^2) dT$$

$$= 26.984(1000 - 300) + \frac{1}{2}(5.910 \times 10^{-3})(1000^2 - 300^2)$$

$$- \frac{1}{3}(3.377 \times 10^{-7})(1000^3 - 300^3)$$

$$= 21.468 \text{ kJ mol}^{-1}$$

2.13 Methane (considered to be an ideal gas) initially at 25 °C and 1 bar pressure is heated at constant pressure until the volume has doubled. The variation of the molar heat capacity with absolute temperature is given by

$$\bar{C}_P = 22.34 + 48.1 \times 10^{-3}\ T$$

where $\bar{C}_P$ is in J K^{-1} mol^{-1}. Calculate (a) $\Delta\bar{H}$, and (b) $\Delta\bar{U}$.

SOLUTION

$$\Delta\bar{H} = \int_{T_1}^{T_2} (22.34 + 48.1 \times 10^{-3}\ T)\ dT$$

$$= (22.34)(298.1) + [\frac{(48.1 \times 10^{-3})}{2}](596.2^2 - 298.1^2)$$

$$= 13.07\ \text{kJ mol}^{-1}$$

$$\Delta\bar{U} = \Delta\bar{H} - P\Delta\bar{V} = 13.07 - P(RT/P)$$

$$= 13.07 - (8.314 \times 10^{-3}\ \text{kJ K}^{-1}\ \text{mol}^{-1})(298.1\ \text{K})$$

$$= 10.59\ \text{kJ mol}^{-1}$$

2.14 Calculate the temperature increase and final pressure of helium if a mole is compressed adiabatically and reversibly from 44.8 L at 0 °C to 22.4 L.

SOLUTION

$$\gamma = \frac{C_P}{C_V} = \frac{(\frac{5}{2} R)}{(\frac{3}{2} R)} = \frac{5}{3}$$

$$\frac{T_1}{T_2} = (\frac{V_2}{V_1})^{\gamma - 1}$$

$$T_2 = (273.25\ \text{K})(44.8\ \text{L}/22.4\ \text{L})^{2/3}$$

$$= 433.6\ \text{K or } 160.4\ °\text{C}$$

Thus the temperature increase is 160.4 °C. The final pressure is given by

$$P = \frac{RT}{\bar{V}} = \frac{(0.8314\ \text{L bar K}^{-1}\ \text{mol}^{-1})(433.6\ \text{K})}{22.4\ \text{L mol}^{-1}}$$

2.15 A mole of argon is allowed to expand adiabatically and reversibly from a pressure of 10 bar and 298.15 K to 1 bar. What is the final temperature and how much work is done on the argon?

SOLUTION

$$\gamma = \bar{C}_P/\bar{C}_V = (\frac{5}{2}R)/(\frac{3}{2}R) = \frac{5}{3} \qquad (\gamma - 1)/\gamma = 2/5$$

$$\frac{T_1}{T_2} = (\frac{P_1}{P_2})^{(\gamma - 1)/\gamma}$$

$$T_2 = (298.15\ K)(1/10)^{2/5} = 118.70\ K$$

$$w = \int_{T_1}^{T_2} \bar{C}\,dT = \frac{3}{2}R(T_2 - T_1)$$

$$= \frac{3}{2}(8.314\ J\ K^{-1}\ mol^{-1})(118.70\ K - 298.15\ K) = -2238\ J\ mol^{-1}$$

The maximum work that can be done on the surroundings is 2238 J mol^{-1}.

2.16 A tank contains 20 liters of compressed nitrogen at 10 bar and 25 °C. Calculate w when the gas is allowed to expand reversibly to 1 bar pressure (a) isothermally and (b) adiabatically.

SOLUTION

(a) For the isothermal expansion of 1 mol

$$w_{rev} = RT \ln \frac{P_2}{P_1}$$

$$= (8.314 \text{ J K}^{-1} \text{ mol}^{-1})(298.15 \text{ K}) \ln (1/10)$$

$$= -5708 \text{ J mol}^{-1}$$

There are

(10 bar)(20 L)/(0.08314 L bar K^{-1} mol^{-1})(298 K) = 8.07 mol.

Therefore w = -46.1 kJ.

(b) For the adiabatic expansion we will assume that

$\gamma = \bar{C}_P/\bar{C}_V$ has the value it has at room temperature.

From Table C.2.

$$\gamma = 29.125/(29.125 - 8.314) = 1.399$$

$$\frac{T_1}{T_2} = \left(\frac{P_1}{P_2}\right)^{(\gamma - 1)/\gamma} \qquad T_2 = (298.15 \text{ K})(1/10)^{0.285} = 154.7 \text{ K}$$

$$w = \int_{T_1}^{T_2} \bar{C}_V \, dT = \bar{C}_V(T_2 - T_1)$$

$$= (20.811 \text{ J K}^{-1} \text{ mol}^{-1})(154.7 \text{ K} - 298.15 \text{ K})$$

$$= -2.99 \text{ kJ mol}^{-1}$$

For 8.07 mol, w = -24.1 kJ.

2.17 An ideal monatomic gas at 298.15 K and 1 bar is expanded in reversible adiabatic process to a final pressure of 1/2 bar. Calculate q per mole, w per mole, and $\Delta \bar{U}$.

SOLUTION

$$\frac{T_1}{T_2} = \left(\frac{P_1}{P_2}\right)^{(\gamma - 1)/\gamma}$$

$$\bar{C}_V = \frac{3}{2}R \qquad \bar{C}_P = \frac{5}{2}R \qquad \gamma = \frac{5}{3}$$

$$\frac{298.15}{T_2} = (2)^{0.4}$$

$$T_2 = \frac{298.15}{2^{0.4}} = 226.0 \text{ K}$$

$$q = 0$$

$$W = \bar{C}_V \ (T_2 - T_1) = \frac{3}{2}R(226.0 - 298.15)$$

$$= -899.8 \text{ J mol}^{-1}$$

$$\Delta\bar{U} = q + w = -899.8 \text{ J mol}^{-1}$$

2.18 An ideal monatomic gas at 1 bar and 300 K is expanded adiabatically against a constant pressure of 1/2 bar until the final pressure is 1/2 bar. What are the values of q per mole, w per mole, $\Delta\bar{U}$ and $\Delta\bar{H}$? Given $\bar{C}_V = (3/2)R$.

SOLUTION

For an adiabatic expression at constant pressure

$$\bar{C}_V = -P_{op} \ \Delta\bar{V}$$

$$[(3/2)R][T_2 - 300] = -(\frac{1}{2} \text{ bar})(\bar{V}_2 - 300R/1 \text{ bar})$$

$$= \left(-\frac{1}{2}\ \text{bar}\right)\left(\frac{RT_2}{0.5\ \text{bar}} - \frac{300R}{1\ \text{bar}}\right)$$

$T_2 \quad = 240$ K

the work done on the gas is $w = \bar{C}_V\Delta T = (3/2)(8.314\ \text{J K}^{-1}\ \text{mol}^{-1})(-60\ \text{K}) = -748\ \text{J mol}^{-1}$. The value of q is zero since the process is adiabatic.

$$\Delta\bar{U} \quad = q + w = -748\ \text{J mol}^{-1}$$

$$\Delta\bar{H} \quad = \int_{T_1}^{T_2} (5/2)R\ dT = (5/2)(8.314)(-60) = -1247\ \text{J mol}^{-1}$$

2.19 Calculate $\Delta_r H^0_{298}$ for

$H_2(g) + F_2(g) \quad = 2HF(g)$

$H_2(g) + Cl_2(g) \quad = 2HCl(g)$

$H_2(g) + Br_2(g) \quad = 2HBr(g)$

$H_2(g) + I_2(g) \quad = 2HI(g)$

SOLUTION

(a) $2(-271.1) = -542.2$ kJ mol^{-1}

(b) $2(-92.31) = -184.62$ kJ mol^{-1}

(c) $2(-36.40) - 30.91 = -103.71$ kJ mol^{-1}

(d) $2(26.48) - 62.44 = -9.48$ kJ mol^{-1}

2.20 The following reactions might be used to power rockets.

(1) $H_2(g) + \frac{1}{2} O_2(g) = H_2O(g)$

(2) $CH_3OH(l) + 1\frac{1}{2} O_2(g) = CO_2(g) + 2H_2O(g)$

(3) $H_2(g) + F_2(g) = 2HF(g)$

(a) Calculate the enthalpy changes at 25 °C for each of these reactions per kilogram of reactants.

(b) Since the thrust is greater when the molar mass of the exhaust gas is lower, divide the heat per kilogram by the molar mass of the product (or the average molar mass in the case or reaction (2) and arrange the above reactions in order of effectiveness on the basis of thrust.

SOLUTION

(a) (1) $\Delta_r H = -241.818 \text{ kJ mol}^{-1}$

$$= \frac{(-241.818 \text{ kJ mol}^{-1})(100 \text{ g kg}^{-1})}{(18 \text{ g mol}^{-1})}$$

$$= -13.4 \text{ MJ kg}^{-1}$$

(2) $\Delta_r H = -393.509 + 2(-241.818) + 238.66$

$$= -638.49 \text{ kJ mol}^{-1}$$

$$= \frac{(-638.49 \text{ kJ mol}^{-1})(1000 \text{ g kg}^{-1})}{(80 \text{ g mol}^{-1})}$$

$$= -7.98 \text{ MJ kg}^{-1}$$

(3) $\Delta_r H = 2(-271.1) = -542.2 \text{ kJ mol}^{-1}$

$$= \frac{(-542.2 \text{ kJmol}^{-1})(100 \text{ g kg}^{-1})}{(40 \text{ g mol}^{-1})}$$

(b) (1) -13.4/18 = -0.744

(2) $$\frac{-7.98}{\frac{1}{3}(44 + 2 \times 18)} = -0.299$$

(3) $-13.6/20 = -0.680$

$(1) > (3) > (2)$

2.21 Calculate the enthalpy of formation of $PCl_5(s)$, given the heats of the following reactions at 25 °C.

$$2P(s) + 3Cl_2(g) = 2PCl_3(l) \qquad \Delta_rH^o = -635.13 \text{ kJ mol}^{-1}$$

$$PCl_3(l) + Cl_2(g) = PCl_5(s) \qquad \Delta_rH^o = -137.28 \text{ kJ mol}^{-1}$$

SOLUTION

Multiplying the second reaction by 2 and adding the two reactions yields

$$2P(s) + 5Cl_2(g) = 2PCl_5(s) \qquad \Delta_rH^o = -909.69 \text{ kJ mol}^{-1}$$

$$\Delta_fH^o[PCl_5(s)] = \frac{(-909.69 \text{ kJ mol}^{-1})}{2}$$

$$= -454.85 \text{ kJ mol}^{-1}$$

2.22 Calculate Δ_rH^o for the dissociation $O_2(g) = 2O(g)$ at 0, 298, and 3000 K. In Section 12.2 the enthalpy change for dissociation at 0 K will be found to be equal to the spectroscopic dissociation energy D_0.

SOLUTION

$\Delta_r H^o(0\ K)$ = 2(246.790) = 493.580 kJ mol^{-1}

$\Delta_r H^o$ (298 K) = 2(249.173) = 498.346 kJ mol^{-1}

$\Delta_r H^o$ (3000 K) = 2(256.741) = 513.482 kJ mol^{-1}

The spectroscopic dissociation energy of O_2 is given as 5.115 eV in Table 14.4. this can be converted to kJ mol^{-1} by multiplying by 96.485 kJ V^{-1} mol^{-1} to obtain 493.521 kJ mol^{-1}.

2.23 Methane may be produced from coal in a process represented by the following steps, where coal is approximated by graphite:

$2C(s) + 2H_2O(g) = 2CO(g) + 2H_2(g)$

$CO(g) + H_2O(g) = CO_2(g) + H_2(g)$

$CO(g) + 3H_2(g) = CH_4(g) + H_2O(g)$

the sum of these three reactions is

$2C(s) + 2H_2O(g) = CH_4(g) + CO_2(g)$

What is $\Delta_r H^o$ at 500 K for each of these reactions? Check that the sum of the $\Delta_r H^o$'s of the first three reactions is equal to $\Delta_r H^o$ for the fourth reaction. From the standpoint of heat balance would it be better to develop a process to carry out the overall reactions in three separate reactors, or in a single reactor?

SOLUTION

$\Delta_r H^o_{500}$ = 2(-110.00) - 2(-243.83) = 267.66 kJ mol^{-1}

$\Delta_r H^o_{500}$ = -393.67 - (-110.00) - (243.83) = -39.84 kJ mol^{-1}

$\Delta_r H^o_{500}$ = -80.82 - 243.83 - (-110.00) = -214.65 kJ mol^{-1}

$\Delta_r H^o_{500} \quad = \quad -80.82 - 393.67 - 2(-243.83) = 13.17 \text{ kJ mol}^{-1}$

Since the first reaction is very endothermic, there is an advantage in carrying the subsequent reactions out in the same reactor so that they can provide heat.

2.24 What is the heat of freezing water at -10 °C given that

$H_2O(1) = H_2O(s) \qquad \Delta H^o\,(273\text{ K}) = -6004\text{ J mol}^{-1}$

$\bar{C}_P(H_2O,1) = 75.3\text{ J K}^{-1}\text{ mol}^{-1}$ and $\bar{C}_P(H_2O,s) = 36.8\text{ J K}^{-1}\text{ mol}^{-1}$

SOLUTION

$\Delta H^o \quad = \Delta H^o(273\text{ K}) + [\bar{C}_{P,H_20(cr)} - \bar{C}_{P,H_20(1)}] \times (263\text{ K} - 273\text{ K})$

$= -6004\text{ J mol}^{-1} + (-38.5\text{ J K}^{-1}\text{ mol}^{-1})(-10\text{ K})$

$= -5619\text{ J mol}^{-1}$

2.25 What is the enthalpy change for the vaporization of water at 0 °C? This value may be estimated from Appendix C.1 by assuming that the heat capacities of $H_2O(1)$ and $H_2O(g)$ are independent of temperature from 0 to 25 °C.

SOLUTION

$H_2O(1) \quad = H_2O(g)$

$\Delta_{vap}H^o(298\text{ K}) \quad = -241.818 - (-385.830) = 44.011\text{ kJ mol}^{-1}$

$\Delta_{vap}H^o_2 \quad = \Delta_{vap}H^o_1 + \Delta_{vap}C_P(T_2 - T_1))$

$$\Delta_{vap}H^o \text{ (273 K)} = \Delta_{vap}H^o\text{(298 K)} + [\bar{C}_P(H_2O,g) - \bar{C}_P(H_2O,l)](273-298)$$

$$= 44{,}011 + (33.577 - 75.291)(-25)$$

$$= 45{,}054 \text{ J mol}^{-1}$$

2.26 Calculate the standard enthalpy of formation of methane at 1000 K from the value at 298.15 K using the $\bar{H}^0 - \bar{H}^0_{298}$ data in Appendix C.2.

SOLUTION

1000 K: C(graphite) + $2H_2(g)$ = $CH_4(g)$

$(\bar{H}^o_{1000} - \bar{H}^o_{298})C$ $\quad 2(\bar{H}^o_{1000} - \bar{H}^o_{298})H_2$ $\quad (\bar{H}^o_{1000} - \bar{H}^o_{298})CH_4$

298 K : C(graphite) + $2H_2(g)$ = $CH_4(g)$

The change in state represented by the first reaction can be accomplished by cooling one mole of graphite and two moles of hydrogen to 298 K, converting them to methane, and then heating the methane to 1000 K.

$$\Delta_f H^o_{1000} = -(\bar{H}^o_{1000} - \bar{H}^o_{298})C - 2(\bar{H}^o_{1000} - \bar{H}^o_{298})H_2 + \Delta_f\bar{H}^o(CH_4, 298\text{ K})$$

$$+ (\bar{H}^o_{1000} - \bar{H}^o_{298})CH_4$$

$$= -11.795 - 2(20.680) - 74.873 + 38.179$$

$$= -89.849 \text{ kJ mol}^{-1}$$

2.27 For a diatomic molecule the bond energy is equal to the change in internal energy for the reaction

$$X_2(g) = 2X(g)$$

at 0 K. Of course, the change in internal energy and the change in enthalpy are the same at 0 K. Calculate the enthalpy of dissociation of $O_2(g)$ at 0 K. The enthalpy of formation of O(g) at 298.15 K is 249.173 kJ mol^{-1}. In the range 0-298 K the average value of the heat capacity of $O_2(g)$ is 29.1 J K^{-1} mol^{-1} and the average value of the heat capacity of O(g) is 22.7 J K^{-1} mol^{-1}. What is the value of the bond energy in electron volts? (When the changes in heat capacities in the range 0-298 K are taken into account, the enthalpy of dissociation at 0 K is 493.58 kJ mol^{-1}.)

<u>SOLUTION</u>

$$\begin{array}{ccc} O_2(g) & \xrightarrow[298\ K]{2(249.173\ kJ\ mol^{-1})} & 2O(g) \\ \| \ \bar{C}_P = 29.1\ J\ K^{-1}\ mol^{-1} & & \| \ \bar{C}_P = 22.7 J\ K^{-1} mol^{-1} \\ O_2(g) & \xlongequal{\qquad\qquad} & 2O(g) \end{array}$$

$$\Delta_r H^o_o = \Delta_r H^o_{298} - \Delta_r C_P(-298.15)$$

$$= 2(249.173) - (2 \times 22.7 - 29.1)(-298.15) \times 10^{-3}$$
$$= \Delta U^o_0 = \text{bond energy for } O_2(g)$$

$$\frac{493486\ J\ mol^{-1}}{96485\ C\ mol^{-1}} = 5.115\ eV$$

2.28 One gram of liquid benzene is burned in a bomb calorimeter. The temperature before ignition was 20.826 °C and the temperature after the combustion was 25.000 °C. This was an adiabatic calorimeter. The heat capacity of the bomb, the water around it, and the contents of the bomb after the combustion was 10,000 J K^{-1}. Calculate $\Delta_f H^o$ for $C_6H_6(1)$ at 298.15 K from these data. Assume that the water produced in the combusion is in the liquid state and the carbon dioxide produced in the combustion is in the gas state.

SOLUTION

$$C_6H_6(l) + 7\frac{1}{2} O_2(g) = 6CO_2(g) + 3H_2O(l)$$

$$\Delta U_{T_1} = -\int_{T_1}^{T_2} [C_V(K) + C_V(P)]dT$$

$$= -(10,000 \text{ J K}^{-1})(4.174 \text{ K})$$

$$= -41.74 \text{ kJ for 1 gram } C_6H_6$$

$$= -41.74 \text{ kJ/g} \times 78 \text{ g/mol} = -3255.7 \text{ kJ mol}^{-1}$$

$$q_P = q_V + (n_P - n_R)RT = q_v - 1.5\ RT$$

$$\Delta_c H(298) = -3255.7 - 1.5\ (8.314 \times 10^{-3})(298)$$

$$= -\ 3259.4 \text{ kJ mol}^{-1}$$

$$\Delta_c H^0 = 6\Delta_f H^o(CO_2) + 3\Delta_f H^o(H_2O,\ell) - \Delta_f H^o(C_6H_6,\ell)$$

$$-3259.4 = 6(-393.51) + 3(-285.83) - \Delta_f H^o(C_6H_6,\ell)$$

$$\Delta_f H^o = 41.4 \text{ kJ mol}^{-1} \text{ for } C_6H_6(\ell)$$

2.29 An aqueous solution of unoxygenated hemoglobin containing 5 g of protein (M = 64,000 g mol^{-1}) in 100 cm^3 of solution is placed in an insulated vessel. When enough molecular oxygen is added to the solution to completely saturate the hemoglobin, the temperature rises 0.031 °C. Each mole of hemoglobin binds 4 moles of oxygen. What is the enthalpy of reaction per mole of oxygen bound? The heat capacity of the solution may be assumed to be 4.18 J K^{-1} cm^{-3}.

SOLUTION

Oxygenated $Hb(T_2)$

$\Delta H = 0$ $\qquad\qquad$ $\Delta H = C_P\Delta T$

Unoxygenated $Hb(T_1) + O_2(T_1) \rightarrow$ Oxygenated $Hb(T_1)$

$\Delta H(T_1)$

$$\Delta H(T_1) + C_P\Delta T = 0$$

$$\Delta H(T_1) = -(100\ cm^3)(4.18\ J\ K^{-1}\ cm^{-3})(0.031\ K)$$

$$= -13.0\ J$$

$\Delta_r H$ per mole of oxygen bound

$$= \frac{-13.0\ J}{(4\ x\ 5\ g)/(64000\ g\ mol^{-1})}$$

$$= -\ 41.6\ kJ\ mol^{-1}$$

2.30 Calculate the heat of hydration of $Na_2SO_4(s)$ from the integral heats of solution of $Na_2SO_4(s)$ and $Na_2SO_4 \cdot 10\ H_2O(s)$ in infinite amounts of H_2O, which are -2.34 kJ mol^{-1} and 78.87 kJ mol^{-1}, respectively. Enthalpies of hydration cannot be measured directly because of the slowness of the phase transition.

SOLUTION

$Na_2SO_4(s) \quad = Na_2SO_4(ai) \qquad \Delta_rH^o = -2.43 \text{ kJ mol}^{-1}$

$Na_2SO_4(ai) \quad = Na_2SO_4 \cdot 10H_2O(s) \qquad \Delta_rH^o = -78.87 \text{ kJ mol}^{-1}$

$Na_2SO_4(s) + 10\ H_2O(1) = Na_2SO_4 \cdot 10\ H_2O(s) \quad \Delta_rH^o = -81.21 \text{ kJ mol}^{-1}$

2.31 We want to determine the enthalpy of hydration of $CaCl_2$ to form $CaCl_2 \cdot 6H_2O$.

$$CaCl_2(s) + 6H_2O(1) = CaCl_2 \cdot 6H_2O(s)$$

We cannot do this directly for a couple of reasons: (1) reactions in the solid state are slow, and (2) there is a series of hydrates and so a mixture of different hydrates would probably be obtained. We can, however, determine the heats of solution of $CaCl_2(s)$ and $CaCl_2 \cdot 6H_2O(s)$ in water at 298 K and take the difference. The experimental heats of solution are as follows:

$CaCl_2(s) + Aq = CaCl_2(ai) \qquad \Delta_rH = -81.33 \text{ kJ mol}^{-1}$

$CaCl_2 \cdot 6H_2O(s) + Aq = CaCl_2(ai) + 6H_2O(1) \quad \Delta_rH = 15.79 \text{ kJ mol}^{-1}$

What is the enthalpy of hydration of $CaCl_2$ to form $CaCl_2 \cdot 6H_2O$?

SOLUTION

Reverse the second reaction and add

$$CaCl_2(s) + 6H_2O(l) = CaCl_2 \cdot 6H_2O(s)$$

$$\Delta_rH = -81.33 - 15.79$$

$$= -97.12 \text{ kJ mol}^{-1}$$

2.32 Calculate the integral heat of solution of one mole of HCl(g) in 200 $H_2O(1)$.

$$HCl(g) + 200\ H_2O(1) = HCl \text{ in } 200\ H_2O$$

SOLUTION

$\Delta_r H^{o}_{298}$ = $\Delta_f H^o$ [HCl in 200 H_2)] - $\Delta_f H^o$[HCl(g)]

= -166.272 - (-92.307) = -73.965 kJ mol^{-1}

2.33 Calculate the enthalpies of reaction at 25 °C for the following reactions in dilute aqueous solutions:

(a) HCl(ai) + NaBr(ai) = HBr(ai) + NaCl(ai)

(b) $CaCl_2$(ai) + Na_2CO_3(ai) = $CaCO_3$(s) + 2 NaCl(ai)

SOLUTION

(a) $\Delta_r H^o$ = 0 because all of the reactants and products are completely ionized.

$H^+ + Cl^- + Na^+ + Br^- = H^+ + Br^- + Na^+ + Cl^-$

(b) Ca^{2+}(ai) + CO_3^{2-} (ai) = $CaCO_3$ (s)

$\Delta_r H^o$ = -1206.92 - (-542.83) - (-677.14) = 13.05 kJ mol^{-1}

2.34 Estimate the molar heat capacity at constant pressure at 298 K for ethane using the Benson method. Compare this result with the value calculated using values in Table 2.2.

SOLUTION

$\bar{C}_P$ = 2(25.90) = 51.80 J K^{-1} mol^{-1}

$\bar{C}_P$ = 9.404 + (159.836 x 10^{-3})(298.15)-(462.28 x 10^{-7})$(298.15)^2$

= 52.95 J K^{-1} mol^{-1}

2.35 Estimate the enthalpies of formation of n-butane and iso-butane at 298.15 K using the Benson method.

SOLUTION

For n-butane, $\Delta_f H^o$ = 2(-42.68) + 2(-20.63) = -126.62 kJ mol^{-1}

For iso-butane, $\Delta_f H^o$ = 3(-42.68) - 7.95 = -135.99 kJ mol^{-1}

The values in R. A. Alberty and C. Gehrig, *J. Phys. Chem Ref. Data* **13**, 1173 (1984) are -125.7 and -134.6 kJ mol^{-1}, respectively.

2.36 124.4, 497.6 kJ

2.37 2.91 kW-hr, 0.172, 4.4 kg

2.38 3.89 J

2.39 (a) No, (b) Yes

2.40 $\partial C_V/\partial \ln V \neq \partial(RT)/\partial T = R$

$\partial C_V/\partial \ln V = \partial R/\partial \ln T = 0$

2.41 (a) - 5.71, (b) -5.67 kJ mol^{-1}

2.42 (a) -7.42 kJ mol^{-1} , (b) 7.42 kJ mol^{-1}, (c) 0, (d) 0

2.43 (a) 1.99, (b) =-23.30, (c) -21.31, (d) -23.30 kJ mol^{-1}

2.44 (a) -1.72, (b) -1.16 kJ mol^{-1}

2.45 $\mu_{JT} = -b/C_P$, $\Delta T = +$

2.46 98.098 kJ mol^{-1}

2.47 99.371 kJ mol^{-1}

2.48 (a) 570.2 K, (b) 9.48 bar, (c) 5.580 kJ mol^{-1}

2.49 (a) 0.496 bar -1739 J mol^{-1}

(b) 0.310 bar -1390 J mol^{-1}

2.50 (a) -46.1 kJ, (b) -24.1 kJ

2.51 -802.303, -807.513 kJ mol^{-1}

2.52 The sum of the two enthalpies of reaction is more negative than the enthalpy change for the first reaction by the enthalpy of formation of methane.

2.53 12.2 kJ mol^{-1}

2.54 (a) -119.9 kJ g^{-1}, (b) -50.0 kJ g^{-1}, (c) -19.9 kJ g^{-1}, (d) -45.2 kJ g^{-1}

2.55 206.11 kJ is liberated.

Burning CH_4 liberates 802.34 kJ.

Burning CO + $3H_2$ liberates 1008.45 kJ.

2.56 432.070, 435.998, 459.580 kJ mol^{-1}

2.57 (I) 214.63, -97.94; (II) 116.69 kJ mol^{-1}

The disadvantage of the first process is the large demand for heat in the first step.

2.58 43.8 kJ mol^{-1}

2.59 459.579 kJ mol^{-1} (4.74551 eV)

2.60 1663.54 kJ mol^{-1}

2.61 -89.849 kJ mol^{-1}

2.62 About 2600 K

2.63 $\Delta_r U$ = -5163 kJ mol^{-1}, $\Delta_r H$ = -5168 kJ mol^{-1}, $\Delta_f H(C_{10}H_8)$ = 78 kJ mol^{-1}

2.64 (a) -253.4, (b) -249.7 kJ mol^{-1}

2.65 17.9 kJ mol^{-1}

2.66 303 g

2.67 (a) 0.038 kJ mol^{-1}, (b) 10.03 kJ mol^{-1}

2.68 -85.36, -105.99 kJ mol^{-1}

2.69 (a) -1412.6 kJ mol^{-1}, -1430.6 kJ mol^{-1}

(b) 11.5 kbar

CHAPTER 3: Second and Third Laws of Thermodynamics

3.1 Theoretically, how high could a gallon of gasoline lift an automobile weighing 2800 lb against the force of gravity, if it is assumed that the cylinder temperature is 2200 K and the exit temperature 1200 K? (Density of gasoline = 0.80 g cm^{-3}; 1 lb = 453.6 g; 1 ft = 30.48 cm; 1 L = 0.2642 gal. Heat combustion of gasoline = 46.9 kJ g^{-1}.)

SOLUTION

$$q = \frac{(46.9 \times 10^3 \text{ J g}^{-1})(1 \text{ gal})(10^3 \text{ cm}^3 \text{ L}^{-1})(0.80 \text{ g cm}^{-3})}{0.2642 \text{ gal L}^{-1}}$$

$$= 14.2 \times 10^7 \text{ J}$$

$$w = q\frac{T_2 - T_1}{T_2} = \frac{(14.2 \times 10^7 \text{ J})(2200 \text{ K} - 1200 \text{ K})}{(2200 \text{ K})} = 6.45$$

$\times 10^7$ J

$= mgh = (2800 \text{ lb})(0.4536 \text{ kg lb}^{-1})(9.8 \text{ m s}^{-2})(0.3048 \text{ m ft}^{-1})h$

$h = 17{,}000$ ft

3.2 (a) What is the maximum work that can be obtained from 100 J of heat supplied to a steam engine with a high-temperature reservoir at 100 °C if the condenser is at 20 °C? (b) If the boiler temperature is raised to 150 °C by the use of superheated steam under pressure, how much more work can be obtained?

SOLUTION

(a) $$w = q\frac{T_2 - T_1}{T_2} = (1000 \text{ J})\frac{80 \text{ K}}{373.1 \text{ K}} = 214 \text{ J}$$

(b) w $= (1000\ \text{J}) \dfrac{130\ \text{K}}{423.1\ \text{K}}$ = 307 J or 93 J more than (a)

3.3 A heat pump is to be used to maintain the temperature of a building at 18 °C when the outside temperature is -5 °C. For a frictionless heat pump how much work must be expended to obtain a joule of heat?

SOLUTION

$|w| = |q_1|(T_1 - T_2)/T_1 = (1\ \text{J})(23\ \text{K})/(291\ \text{K})$

$= 0.079\ \text{J}$

3.4 Water is vaporized reversibly at 100 °C and 1.01325 bar. The heat of vaporization is 40.69 kJ mol^{-1}. (a) What is the value of ΔS for the water? (b) What is the value of ΔS for the water plus the heat reservoir at 100 °C?

SOLUTION

(a) $\Delta \bar{S}_{H_2O} = \dfrac{40690\ \text{J mol}^{-1}}{373.15\ \text{K}} = 109.04\ \text{J K}^{-1}\ \text{mol}^{-1}$

(b) $\Delta \bar{S}_{res} = \dfrac{-40690\ \text{J mol}^{-1}}{373.15\ \text{K}} = -109.04\ \text{J K}^{-1}\ \text{mol}^{-1}$

$\Delta \bar{S}_{syst} = 0$

This is necessarily true for a reversible process in an isolated system.

3.5 Assuming that CO_2 is an ideal gas, calculate ΔH^o and ΔS^o for the following process:

$$1\ CO_2(g,\ 298.15\ K,\ 1\ bar) \rightarrow 1\ CO_2(g,\ 1000\ K,\ 1\ bar)$$

Given: $\bar{C}_p^o = 26.648 + 42.262 \times 10^{-3}\ T - 142.40 \times 10^{-7}\ T^2$ in J K^{-1} mol^{-1}

SOLUTION

$$\Delta\bar{H} = \int_{298.15}^{1000} \bar{C}_p dT = 26.648\ (1000 - 298.15)$$

$$+ \left(\frac{42.264 \times 10^{-3}}{2}\right)(1000^2 - 298.15^2)$$

$$- \left(\frac{142.4 \times 10^{-7}}{3}\right)(1000^3 - 298.15^3)$$

$$= 33.34\ \text{kJ mol}^{-1}$$

$$\Delta\bar{S} = \int_{298.15}^{1000} \frac{\bar{C}}{T}\ dT = 26.648\ \ln \frac{1000}{298.15}$$

$$+ 42.262 \times 10^{-3}(1000 - 298.15)$$

$$- \left(\frac{142.4 \times 10^{-7}}{2}\right)(1000^2 - 298.15^2)$$

$$= 55.42\ \text{J K}^{-1}\ \text{mol}^{-1}$$

3.6 Calculate the increase in molar entropy of silver that is heated at constant pressure from 0 to 30 °C if the value of $\bar{C}_P$ in this temperature range is considered to be constant at 25.48 J K^{-1} mol^{-1}.

SOLUTION

$$\Delta\bar{S} = \bar{C}_P \ln\frac{T_2}{T_1} = (25.48 \text{ J K}^{-1} \text{ mol}^{-1}) \ln\frac{303}{273}$$

$$= 2.657 \text{ J K}^{-1} \text{ mol}^{-1}$$

3.7 The temperature of an ideal monatomic gas is increased from 300 K to 500 K. What is the change in molar entropy of the gas (a) if the volume is held constant and (b) if the pressure is held constant?

SOLUTION

(a) $$\Delta\bar{S} = \bar{C}_V \ln\frac{T_2}{T_1} = \frac{3}{2}(8.314 \text{ J K}^{-1} \text{ mol}^{-1}) \ln\frac{500 \text{ K}}{300 \text{ K}}$$

$$= 6.371 \text{ J K}^{-1} \text{ mol}^{-1}$$

(b) $$\Delta\bar{S} = \bar{C}_P \ln\frac{T_2}{T_1} = \frac{5}{2}(8.314 \text{ J K}^{-1} \text{ mol}^{-1}) \ln\frac{500 \text{ K}}{300 \text{ K}}$$

$$= 10.618 \text{ J K}^{-1} \text{ mol}^{-1}$$

3.8 Ammonia (considered to be an ideal gas) initially at 25 °C and 1 bar pressure is heated at constant pressure until the volume has trebled. Calculate (a) q per mole, (b) w per mole, (c) $\Delta\bar{H}$, (d)$\Delta\bar{U}$, and (e) $\Delta\bar{S}$. Given: $\bar{C}_P$ = 25.895 + 32.999 x 10^{-3} T - 30.46 x 10^{-7} T^2 in J K^{-1} mol^{-1}.

SOLUTION

(a) $q = \int_{T_1}^{T_2} \bar{C}_P \, dT$

$= \int_{298}^{894} [25.895 + 32.999 \times 10^{-3} T - 30.46 \times 10^{-7} T^2] dT$

$= (25.895)(596) + \frac{32.999 \times 10^{-3}}{2} (894^2 - 298^2) - \frac{30.46 \times 10^{-3}}{3} (894^3 - 298^3) = 26.4 \text{ kJ mol}^{-1}$

(b) $w = -P\Delta\bar{V} = -R(T_2 - T_1) = -(8.314 \text{ J K}^{-1} \text{ mol}^{-1})(596 \text{ K})$

$= -4.96 \text{ kJ mol}^{-1}$

(c) $\Delta\bar{H} = q_p = 26.4 \text{ kJ mol}^{-1}$

(d) $\Delta\bar{U} = q + w = 26.4 - 5.0 = 21.4 \text{ kJ mol}^{-1}$

(e) $\bar{S} = \int_{298}^{894} \frac{\bar{C}_P}{T} dT$

$= \int_{298}^{894} [\frac{25.895}{T} + 32.999 \times 10^{-3} - 30.46 \times 10^{-7} T] dT$

$= 25.895 \ln \frac{894}{298} + 32.999 \times 10^{-3} (894 - 298) - \frac{30.46 \times 10^{-7}}{2} (894^2 - 298^2)$

$= 46.99 \text{ J K}^{-1} \text{ mol}^{-1}$

3.9 Two blocks of the same metal are of the same size but are at different temperatures, T_1 and T_2. These blocks of metal are brought together and allowed to come to the same temperature. Show that the entropy change is given by

$$\Delta S = C_P \ln \left[\frac{(T_1 + T_2)^2}{4T_1T_2}\right]$$

if C_P is constant. How does this equation show that the change is spontaneous?

<u>SOLUTION</u>

If we designate the blocks as A and B, the total entropy change is given by

$$\Delta S = \int_{T_1}^{(T_1+T_2)/2} \frac{C_P dT_A}{T_A} + \int_{T_1}^{(T_1+T_2)/2} \frac{C_P dT_B}{T_B}$$

$$= C_P \ln\left(\frac{T_1+T_2}{2T_1}\right) + C_P \ln\left(\frac{T_1+T_2}{2T_2}\right)$$

$$= C_P \ln \left[\frac{(T_1+T_2\,)}{2T_1}\right]$$

Since this quantity is always positive, the change is spontaneous.

3.10 In the reversible isothermal expansion of an ideal gas at 300 K from 1 to 10 liters, where the gas has an initial pressure of 20.27 bar, calculate (a) ΔS for the gas and (b) ΔS for all systems involved in the expansion.

<u>SOLUTION</u>

(a) $n = \frac{PV}{RT} = \frac{(20.27 \text{ bar})(1 \text{ L})}{(0.08314 \text{ L bar K}^{-1} \text{ mol}^{-1})(300 \text{ K})} = 0.812$ mol

$$\Delta S = nR \ln \frac{V_2}{V_1} = (0.812 \text{ mol})(8.314 \text{ J K}^{-1} \text{ mol}^{-1}) \ln \frac{10}{1}$$

$$= 15.56 \text{ J K}^{-1}$$

(b) $\Delta S = 0$ since the process is carried out reversibly. The heat gained by the gas is equal to the heat lost by the heat reservoir, and both bodies are at the same temperature.

3.11 One mole of an ideal gas is expanded isothermally and reversibly from 30 L to 100 L at 300 K. (a) What are the values of $\Delta\bar{U}$, $\Delta\bar{S}$, w per mole and q per mole? (b) If the expansion is carried out irreversibly by allowing the gas to expand into an evacuated container, what are the values of $\Delta\bar{U}$, $\Delta\bar{S}$, w per mole and q per mole?

SOLUTION

(a) $\Delta\bar{U} = -\ 0 \text{ Jmol}^{-1}$

$$\Delta\bar{S} = R \ln \frac{\bar{V}_2}{\bar{V}_1} = (8.314 \text{ J K}^{-1} \text{ mol}^{-1}) \ln \frac{100 \text{ L}}{30 \text{ L}}$$

$$= 10.01 \text{ J K}^{-1} \text{ mol}^{-1}$$

$$w = -RT \ln \frac{\bar{V}_2}{\bar{V}_1} = -(8.314 \text{ J K}^{-1} \text{ mol}^{-1})(300 \text{ K}) \ln \frac{100 \text{ K}}{30 \text{ L}}$$

$$= 3.003 \text{ kJ mol}^{-1}$$

$$q = 3.003 \text{ kJ mol}^{-1}$$

(b) $\Delta\bar{U} = 0$ $\quad \Delta\bar{S} = 10.01 \text{ J K}^{-1} \text{ mol}^{-1}$

$w = 0$ $\quad q = 0$

3.12 (a) A system consists of a mole of ideal gas that undergoes the following change in state

$$1\ X(g,\ 298\ K,\ 10\ bar) = 1\ X(g,\ 298\ K, 1\ bar)$$

What is the value of $\Delta\bar{S}$ if the expansion is reversible? What is the value of $\Delta\bar{S}$ if the gas expands into a larger container so that the final pressure is 1 bar?

<u>SOLUTION</u>

(a) $w = RT \ln(10\ bar/1\ bar) = (8.314)(298) \ln 10$

$= 5705\ J\ mol^{-1} = q$

$$\Delta\bar{S} = \frac{5705\ J\ mol^{-1}}{298\ K} = 19.14\ J\ K^{-1}\ mol^{-1}$$

This is $\Delta\bar{S}$, whether the expansion is reversible or irreversible.

(b) If the expansion is reversible, $\Delta\bar{S} = 0$, as it must be for any reversible process in an isolated system. The gas gains q and the surroundings lose q. If the expansion is irreversible, $\Delta\bar{S} = 19.14\ J\ K^{-1}\ mol^{-1}$, which is positive, as it must be for any irreversible process in an isolated system.

3.13 An ideal gas at 298 K expands isothermally from a pressure of 10 bar to 1 bar. What are the value of w per mole, q per mole, $\Delta\bar{U}$, $\Delta\bar{H}$, and $\Delta\bar{S}$ in the following cases?:

(a) The expansion is reversible.

(b) The expansion is free.

(c) The gas and its surroundings form an isolated system, and the expansion is reversible.

(d) The gas and its surroundings for an isolated system, and the expansion is free.

SOLUTION

	Reversible	Irreversible	Isolated Rev.	Isol. Irr.
w	-5.71 kJ mol^{-1}	0	0	0
q	5.71 kJ mol^{-1}	0	0	0
$\Delta\bar{U}$	0	0	0	0
$\Delta\bar{H}$	0	0	0	0
$\Delta\bar{S}$	19.1 JK^{-1}mol^{-1}	19.1 JK^{-1}mol^{-1}	0	19.1 J K^{-1} mol^{-1}

$$w = -\int_{\bar{V}_2}^{\bar{V}_1} P d\bar{V} = RT \ln \frac{\bar{V}_2}{\bar{V}_1} = -RT \ln \frac{P_1}{P_2} = (8.315)(298) \ln 10 = -4.71 \text{ kJ mol}^{-1}$$

$$\Delta\bar{S} = \frac{5710 \text{ J mol}^{-1}}{298 \text{ K}} = 19.1 \text{ J K}^{-1} \text{ mol}^{-1}$$

3.14 An ideal monatomic gas is heated from 300 to 1000 K and the pressure is allowed to rise to 1 to 2 bar. What is the change in molar entropy?

SOLUTION

$$\Delta\bar{S} = \bar{C}_P \ln(\frac{T_2}{T_1}) - R \ln (\frac{P_2}{P_1})$$

$$= \frac{5}{2} (8.314 \text{ J K}^{-1} \text{ mol}^{-1}) \ln \frac{1000 \text{ K}}{300 \text{ K}} - (8.314 \text{ J K}^{-1} \text{ mol}^{-1}) \ln \frac{2 \text{ bar}}{1 \text{ bar}}$$

$$= 19.26 \text{ J K}^{-1} \text{ mol}^{-1}$$

3.15 The purest acetic acid is often called glacial acetic acid because it is purified by fractional freezing at its melting point of 16.6 °C. A flask containing several moles of liquid acetic acid at 16.6 °C is lowered into an ice-water bath briefly. When it is removed it is found that exactly 1 mol of acetic acid has frozen. Given: $\Delta_{fus}H$ (CH_3CO_2H) = 11.45 kJ mol^{-1} and $\Delta_{fus}H(H_2O)$ = 5.98 kJ mol^{-1}.

(a) What is the change in entropy of the acetic acid?

(b) What is the change in entropy of the water bath?

(c) Now consider the water bath and acetic acid are in the same system. What is the entropy change for the combined system? Is the process reversible or irreversible? Why?

SOLUTION

(a) $$\Delta S_{aa} = \frac{\Delta H}{T_{fp}} = \frac{-11450\ J}{298.75\ K} = -39.52\ J\ K^{-1}$$

(b) $$\Delta S_{H_20} = \frac{11450\ J}{273.15\ K}$$

$$= +\ 41.92\ J\ K^{-1}$$

(c) $$\Delta S_{syst} = 41.92 - 39.52 = 2.40\ J\ K^{-1}$$

The process is irreversible because S increases in our isolated system. Once the freezing of acetic acid starts, it cannot be stopped by infinitesimal change.

3.16 In Problem 2.18 an ideal monatomic gas at 1 bar and 300 K was expanded adiabatically against a constant pressure of 1/2 bar until the final pressure was 1/2 bar; a temperature of 240 K was reached. What is the value of $\Delta\bar{S}$ for this process?

SOLUTION

In order to calculate $\Delta\bar{S}$ we have to find a reversible path for the change in state which is

$$1\ X(300\ K,\ 1\ bar) \rightarrow 1\ X(240\ K,\ 1/2\ bar)$$

This same change in state can be accomplished in two reversible steps.

Reversible isothermal expansion: 1 X(300 K, 1 bar) $\rightarrow$ 1 X(300 K, 1/2 bar)

$$\Delta\bar{S}_1 = R\ln 2 = 5.76\ J\ K^{-1}\ mol^{-1}$$

Cooling process: 1 X(300 K, 1/2 bar) = 1 X(240 K, 1/2 bar)

$$\Delta\bar{S}_2 = \bar{C}_P \ln\ (T_2/T_1)$$

$$= (5/2)(8.314)\ln(240/300)$$

$$= -464\ J\ K^{-1}\ mol^{-1}$$

The entropy change for the reversible two-step process is

$$\Delta\bar{S} = \Delta\bar{S}_1 + \Delta\bar{S}_2 = 5.76 - 4.64 = 1.12\ J\ K^{-1}\ mol^{-1}$$

Note that the entropy of the system increases in this irreversible expansion, even though q = 0.

3.17 Ten moles of H_2 and two moles of D_2 are mixed at 25 °C and 1 bar. What is the value of $\Delta\bar{S}^o$?

SOLUTION

$\Delta S^o = -R(n_1 \ln y_1 + n_2 \ln y_2)$

$= -(8.314 \text{ J K}^{-1} \text{ mol}^{-1})[(10 \text{ mol}) \ln \frac{10}{12} + (2 \text{ mol}) \ln \frac{2}{12}]$

$= 44.95 \text{ J K}^{-1}$

3.18 Use the microscopic point of view of Section 3.12 to show that for the expansion of amount n of an ideal gas by a factor of two, $\Delta S = nR\ln 2$. In this expression S is an extensive property.

SOLUTION

The probability that the molecules will all be found in one half of the container is

$$\Omega'/\Omega' = (1/2)^{N_A n}$$

$$\Delta S = k\ln(\Omega'/\Omega)$$

$$= kN_A n \ \ln(1/2)$$

$$= -nR\ln 2$$

where $R = N_A k$. For expansion from half the container to the full container,

$$\Delta S = nR\ln 2$$

ΔS has units of J K^{-1}, n has units of mol, and R has units of J K^{-1} mol^{-1}.

3.19 Calculate the change in molar entropy of aluminum which is heated from 600 °C to 700 °C. The melting point of aluminum is 660 °C, the heat of fusion is 393 J g^{-1}, and the heat capacities of the solid and liquid may be taken as 31.8 and 34.3 J K^{-1} mol^{-1}, respectively.

SOLUTION

$$\Delta\bar{S} = \int_{T_1}^{T_f} \bar{C}_{P,s}/T\ dT + \frac{\Delta_{fus}H}{T_{fus}} + \int_{T_f}^{T_2} \bar{C}_{P,\ell}/T\ dT$$

$$= \bar{C}_{P,s} \ln \frac{T_f}{T_1} + \frac{\Delta_{fus}H}{T_{fus}} + \bar{C}_{P,\ell} \ln \frac{T_2}{T_f}$$

$$= (31.8 \text{ J K}^{-1} \text{ mol}^{-1}) \ln \frac{933 \text{ K}}{873 \text{ K}} + \frac{(27 \text{ g mol}^{-1})(393 \text{ J g}^{-1})}{933 \text{ K}}$$

$$+ (34.3 \text{ J K}^{-1} \text{ mol}^{-1}) \ln \frac{973 \text{ K}}{933 \text{ K}} = 14.92 \text{ J K}^{-1} \text{ mol}^{-1}$$

3.20 Steam is condensed at 100 °C and the water is cooled to 0 °C and frozen to ice. What is the molar entropy change of the water? Consider that the average specific heat of liquid water is 4.2 J K^{-1} g^{-1}. The heat of vaporization at the boiling point and the heat of fusion at the freezing point are 2258.1 and 33.5 J g^{-1}, respectively.

SOLUTION

$$\Delta S = \frac{\Delta_{vap}H}{T_{vap}} + \int_{373 \text{ K}}^{273 \text{ K}} \frac{\bar{C}_P}{T} dT - \frac{\Delta_{fus}H}{T_{fus}}$$

$$= -\frac{(2258.1 \text{ J g}^{-1})(18.016 \text{ g mol}^{-1})}{373.15 \text{ K}}$$

$$+ (75.379 \text{ J K}^{-1} \text{ mol}^{-1}) \int_{373 \text{ K}}^{273 \text{ K}} d \ln T - \frac{(333.5 \text{ J g}^{-1})(18.016 \text{ g mol}^{-1})}{273.15 \text{ K}}$$

$$= -154.4 \text{ J K}^{-1} \text{ mol}^{-1}$$

3.21 Calculate the molar entropy of carbon disulfide at 25 °C from the following heat-capacity data and the heat of fusion, 4389 J mol^{-1}, at the melting point (161.11 K).

T/K	15.05	20.15	29.76	42.22	57.52	75.54	89.37
$\bar{C}_P$/J K^{-1} mol^{-1}	6.90	12.01	20.75	29.16	35.56	40.04	43.14

T.K	99.00	108.93	119.91	131.54	156.83	161-298
$\bar{C}_P$/J K^{-1} mol^{-1}	45.94	48.49	50.50	52.63	56.62	75.48

SOLUTION

$$\bar{S}° (298.15 \text{ K}) = \frac{\bar{C}_P(15.05 \text{ K})}{3} + \int_{15.05 \text{ K}}^{161.11 \text{ K}} \frac{\bar{C}_P}{T} dT + \frac{\Delta_{fus}H}{161.11 \text{ K}} + \int_{161.11 \text{ K}}^{298.15 \text{ K}} \frac{\bar{C}_P}{T} dT$$

The first integral may be approximated by multiplying the average value of $\frac{\bar{C}_P}{T}$ for each temperature interval by the width of the interval. Thus the first contribution is

$$\frac{1}{2} \left(\frac{6.90}{15.05} + \frac{12.01}{20.15}\right)(20.15 - 15.05) = 2.69 \text{ J K}^{-1} \text{ mol}^{-1}$$

The first integral has the value 74.69 J K^{-1} mol^{-1}. Thus

$$\bar{S}° (298.15 \text{ K}) = \frac{6.90}{3} + 74.69 + \frac{4389}{161.11} + 75.45 \ln \frac{298.15}{161.11}$$

$= 150.67 \text{ J K}^{-1} \text{ mol}^{-1}$

3.22 What is the standard change in entropy for the dissociation of molecular oxygen at 298.15 K and 1000 K? Use Appendix C.2. How do the results compare with those for molecular hydrogen in Example 3.9?

<u>SOLUTION</u>

$O_2(g) = 2\ O(g)$

$\Delta_r S(298.15)\ K = 2(161.058) - 205.147$

$= 116.969 \text{ J K}^{-1} \text{ mol}^{-1}$

$\Delta_r S(1000\ K) = 2(186.790) - 243.578$

$= 130.002 \text{ J K}^{-1} \text{ mol}^{-1}$

The values are quite similar to those for $H_2(g) = 2H(g)$ since the main effect on the entropy change comes from the increase in the number of particles.

3.23 Using molar entropies from Appendix C.1, calculate $\Delta_r S^o$ for the following reactions at 25 °C.

(a) $H_2(g) + \frac{1}{2} O_2(g) = H_2O(1)$

(b) $H_2(g) + Cl_2(g) = 2HCl(g)$

(c) Methane (g) + $\frac{1}{2} O_2(g)$ = Methanol(1)

<u>SOLUTION</u>

(a) $\Delta_r S^o = 69.91 - 130.68 - \frac{1}{2}(205.13) = -163.34 \text{ J K}^{-1} \text{ mol}^{-1}$

(b) $\Delta_r S^o = 2(186.908) - 130.684 - 223.066 = 20.066 \text{ J K}^{-1} \text{ mol}^{-1}$

(c) $\Delta_r S^o = 126.8 - 186.264 - \frac{1}{2}(205.138) = -162.0\ J\ K^{-1}$ mol^{-1}

3.24 What is $\Delta_r S^o$ (298 K) for

$$H_2O(l) = H^+(ao) + OH^-(ao)$$

Why is this change negative and not positive?

SOLUTION

$\Delta_r S^o = -10.75 - 69.92 = -80.67\ J\ K^{-1}\ mol^{-1}$

The ions polarize neighboring water molecules and attract them. For this reason the product state is more ordered than the reactant state.

3.25 Estimate the standard entropy of n-butane and iso-butane at 298.15 K using the Benson method.

SOLUTION

For n-butane, $\bar{S}^o = 2(127.24) + 2(39.41) - 8.314 \ln (3^2 \times 2)$

$= 309.3\ J\ K^{-1}\ mol^{-1}$

For iso-butane, $\bar{S}^o = 3(127.24) - 50.50 - 8.314 \ln 3^4$

$= 303.8\ J\ K^{-1}\ mol^{-1}$

3.26 21.0×10^6 J

3.27 0.114 kWh

2.38 A single expansion is not a basis for an engine.

3.29 -22.00 J K^{-1} mol^{-1}

3.30 126 J K^{-1} mol^{-1}

3.31 (a) 45.25, (b) 33.18 J K^{-1} mol^{-1}

3.32 1.82 J K^{-1} mol^{-1}

3.33 13.38 J K^{-1} mol^{-1}

3.36 16.48 J K^{-1} mol^{-1}

3.37 (a) -35.4, (b) 36.1, (c) 0.7 J K^{-1} mol^{-1}, irreversible

3.38 9.13 J K^{-1} mol^{-1}

3.40 107.75 J K^{-1} mol^{-1}

3.41 -5.082, 0.070 J K^{-1} mol^{-1}

3.42 98.752, 113.526, 122.521 J K^{-1} mol^{-1}

3.43 (a) -33.0, (b) -77.4 J K^{-1} mol^{-1}

3.44 56.6 J K^{-1} mol^{-1}

3.45 393.8, 365.0 J K^{-1} mol^{-1}

CHAPTER 4: Gibbs Energy and Helmholtz Energy

4.1 One mole of nitrogen is allowed to expand from 0.5 to 10 L. Calculate the change in entropy using (a) the ideal gas law and (b) the van der Waals equation.

SOLUTION

(a) $$\Delta\bar{S} = R \ln\left(\frac{\bar{V}_2}{\bar{V}_1}\right)$$

$$= (8.314 \text{ J K}^{-1} \text{ mol}^{-1}) \ln(10/0.5)$$

$$= 24.91 \text{ J K}^{-1} \text{ mol}^{-1}$$

(b) $$\Delta\bar{S} = R \ln\left[\frac{(\bar{V}_2 - b)}{(\bar{V}_1 - b)}\right]$$

$$= (8.314 \text{ J K}^{-1} \text{ mol}^{-1}) \ln\left[\frac{(10 - 0.039)}{(0.5 - 0.039)}\right]$$

$$= 25.55 \text{ J K}^{-1} \text{ mol}^{-1}$$

4.2 Derive the relation for $\bar{C}_P - \bar{C}_V$ for a gas that follows van der Waals' equation.

SOLUTION

$$\bar{C}_P - \bar{C}_V = T\bar{V}\alpha^2/\kappa = -T\left(\frac{\partial \bar{V}}{\partial P}\right)_T\left(\frac{\partial P}{\partial T}\right)_{\bar{V}}$$

The van der Waals equation can be written in the following form:

$$T = \frac{P\bar{V}}{R} - \frac{bP}{R} + \frac{a}{\bar{V}R} - \frac{ab}{R\bar{V}^2}$$

$$(\partial T/\partial \bar{V})_P = \frac{P}{R} - \frac{a}{R\bar{V}^2} + \frac{2ab}{R\bar{V}^3}$$

$$\alpha^{-1} = \bar{V}(\partial T/\partial \bar{V})_P = \frac{P\bar{V}}{R} - \frac{a}{R\bar{V}} + \frac{2ab}{R\bar{V}^2}$$

Eliminating P with the van der Waals equation yields

$$\alpha^{-1} = \frac{T\bar{V}}{\bar{V}-b} - \frac{2a}{R\bar{V}^2}(\bar{V} - b)$$

κ is calculated from $\quad (\partial P/\partial \bar{V})_T = -\frac{RT}{(\bar{V}-b)^2} + \frac{2a}{\bar{V}^3}$

$$\kappa^{-1} = \frac{R\alpha^{-1}}{(\bar{V} - b)}$$

$$\bar{C}_P - \bar{C}_V = \frac{R}{1 - \frac{2a}{RT}\frac{(\bar{V} - b)^2}{\bar{V}^3}}$$

4.3 Earlier we derived the expression for the entropy of an ideal gas as a function of T and P. Now that we have the Maxwell relations, derive the expression for dS for any fluid.

<u>SOLUTION</u>

$$dS = (\partial S/\partial T)_P\, dT + (\partial S/\partial P)_T\, dP$$

$$= (C_P/T)\, dT - \left(\frac{\partial V}{\partial T}\right)_P dP$$

$$= (C_P/T)\, dT - \alpha V dP$$

4.4 What is the change in molar entropy of liquid benzene at 25 °C when the pressure is raised to 1000 bar? The coefficient of thermal expansion α is 1.237×10^{-3} K^{-1}, the density is 0.879 g cm^{-3}, and the molar mass is 78.11 g mol^{-1}.

SOLUTION

Equation 4.45 can be written as

$$\left(\frac{\Delta\bar{S}}{\Delta P}\right)_T = -\left(\frac{\partial\bar{V}}{\partial T}\right)_P = -\bar{V}\alpha$$

where α is defined by equation 1.48.

$$\Delta\bar{S} = -\bar{V}\alpha\Delta P$$

$$= -\left(\frac{78.11 \text{ g mol}^{-1}}{0.879 \text{ g cm}^{-3}}\right)(10^{-2} \text{ m cm}^{-1})^3 \text{-} (1.237 \times 10^{-3} \text{ K}^{-1})(1000 \text{ bar})$$

$$= -10.99 \text{ J K}^{-1} \text{ mol}^{-1}$$

4.5 Derive the expression for $\bar{C}_P - \bar{C}_V$ for a gas with the following equation of state.

$$(P + \frac{a}{\bar{V}^2}) = RT$$

SOLUTION

$$\bar{C}_P - \bar{C}_V = T\left(\frac{\partial P}{\partial T}\right)_V \left(\frac{\partial V}{\partial T}\right)_P$$

$$P = \frac{RT}{\bar{V}} - \frac{a}{\bar{V}^2}$$

$$\left(\frac{\partial P}{\partial T}\right)_{\bar{V}} = \frac{R}{\bar{V}}$$

$$P\bar{V}^2 - RT\bar{V} + a = 0$$

$$2\bar{V}P\left(\frac{\partial \bar{V}}{\partial T}\right)_P - R\bar{V} - RT\left(\frac{\partial \bar{V}}{\partial T}\right)_P = 0$$

$$\left(\frac{\partial \bar{V}}{\partial T}\right)_P = \frac{R\bar{V}}{2P\bar{V} - RT}$$

$$\bar{C}_P - \bar{C}_V = R\left(\frac{2P\bar{V}}{RT} - 1\right)^{-1} = R\left(1 - \frac{2a}{\bar{V}RT}\right)^{-1}$$

4.6 What is the difference between the molar heat capacity of iron at constant pressure and constant volume at 25 °C? Given: $\alpha = 35.1 \times 10^{-6}\ K^{-1}$, $\kappa = 0.52 \times 10^{-6}\ bar^{-1}$, and the density is 7.86 g cm^{-3}.

SOLUTION

$$\bar{V} = \frac{55.847\ g\ mol^{-1}}{7.86\ g\ cm^{-3}} = \frac{7.11\ cm^3\ mol^{-1}}{(10^2\ cm\ m^{-1})^3} = 7.11 \times 10^{-6}\ m^3\ mol^{-1}$$

$$\bar{C}_P - \bar{C}_V = \alpha^2 T\bar{V}/\kappa$$

$$= \frac{(35.1 \times 10^{-6}\ K^{-1})^2(298\ K)(7.11 \times 10^{-6}\ m^3\ mol^{-1})(10^5\ Pa\ bar^{-1})}{(0.52 \times 10^{-6}\ bar^{-1})}$$

$= 0.51 \text{ J K}^{-1} \text{ mol}^{-1}$

4.7 In equation 1.29 we saw that the compressibility factor of a van der Waals gas can be written as

$$Z = 1 + \frac{1}{RT}(b - \frac{1}{RT}) P + \cdots$$

(a) To this degree of approximation, derive the expression for $(\partial \bar{H}/\partial P)_T$ for a van der Waals gas . (b) Calculate $(\partial \bar{H}/\partial P)_T$ for $CO_2(g)$ in J bar^{-1} mol^{-1} at 298 K. Given: $a = 3.640$ L^2 bar mol^{-2} and $b = 0.04267$ L mol^{-1}.

SOLUTION

(a) $\partial \bar{H}/\partial P)_T = \bar{V} - T(\partial \bar{V}/\partial T)_P$

$$P\bar{V} \cong RT/P + (b - a/RT)P$$

$$\bar{V} \cong RT/P + b - a/RT$$

$$(\frac{\partial \bar{V}}{\partial T})_P \cong R/P + a/RT^2$$

$$(\partial \bar{H}/\partial P)_T \cong \bar{V} - RT/P - a/RT$$

$$\cong b - 2a/RT$$

(b) $$(\frac{\partial \bar{H}}{\partial P})_T \cong 0.04267 \text{ L mol}^{-1} - \frac{2(3.64 \text{ L}^2 \text{ bar mol}^{-2})}{(0.08314 \text{ L bar K}^{-1} \text{ mol}^{-1})(298 \text{ K})}$$

$$\cong -0.251 \text{ L mol}^{-1}$$

$$\cong \frac{-0.251 \text{ L mol}^{-1}}{0.01 \text{ L bar J}^{-1}} = -25 \text{ J bar}^{-1} \text{ mol}^{-1}$$

Since $R = 8.314\ J\ K^{-1}\ mol^{-1} = 0.08314\ L\ bar\ K^{-1}\ mol^{-1}$.

Therefore, $1\ J = 0.01\ L\ bar$ or $0.01\ L\ bar\ J^{-1} = 1$.

4.8 (a) Integrate the Gibbs-Helmholtz equation to obtain an expression for ΔG_2 at temperature T_2 in terms of ΔG_1 at T_1, assuming ΔH is independent of temperature. (b) Obtain an expression for ΔG_2 using the more accurate approximation that $\Delta H = \Delta H_1 + (T - T_1)\Delta C_P$ where T_1 is an arbitrary reference temperature.

SOLUTION

(a) Using equation 4.57 $\int_{\Delta G_1/T}^{\Delta G_2/T} d(\Delta G/T) = - \int_{T_1}^{T_2} (\Delta H/T^2)dT$

$$\frac{\Delta G_2}{T_2} - \frac{\Delta G_1}{T_1} = -\Delta H(\frac{1}{T_1} - \frac{1}{T_2})$$

$$\Delta G_2 = \Delta G_1 T_2/T_1 + \Delta H[1 - (T_2/T_1)]$$

$$\text{(b)} \int d(\Delta G/T) = - \int \frac{\Delta H_1}{T^2} dT + \Delta C_P \int \frac{(T - T_1)}{T^2} dT$$

$$\frac{\Delta G_2}{T_2} - \frac{\Delta G_1}{T_1} = -\Delta H(\frac{1}{T_1} - \frac{1}{T_2}) + \Delta C_P \ln (\frac{T_2}{T_1}) - T_1 \Delta C_P(\frac{1}{T_1} - \frac{1}{T_2})$$

$$\Delta G_2 = \Delta G_1 T_2/T_1 + [\Delta H_1 + T_1 \Delta C_P] (1 - \frac{T_2}{T_1}) + T_2 \Delta C_P \ln \frac{T_2}{T_1}$$

4.9 When a liquid is compressed its Gibbs energy is increased. To a first approximation the increase in molar Gibbs energy can be calculated using $(\partial\bar{G}/\partial P)_T = \bar{V}$, assuming a constant molar volume. What is the change in the molar Gibbs energy for liquid water when it is compressed to 1000 bar?

SOLUTION

$$\int_{\bar{G}_1}^{\bar{G}_2} d\bar{G} = \int_{P_1}^{P_2} \bar{V} dP$$

$$\Delta\bar{G} = \bar{V}\,\Delta P = (18 \text{ x } 10^{-6} \text{ m}^{-3} \text{ mol}^{-1})(999 \text{ bar})(10^5 \text{ Pa bar}^{-1})$$

$$= 1.8 \text{ kJ mol}^{-1}$$

4.10 Show that when a liquid is compressed, the change in Gibbs energy is given by

$\Delta G = V\Delta P - \frac{1}{2}\kappa V(\Delta P)^2$ if $\kappa\Delta P << 1$. The isothermal compressibility is represented by κ

SOLUTION

$$\int_{G_1}^{G_2} dG = \int_{P_1}^{P_2} VdP = \int_{P_1}^{P_2} V\, e^{-\kappa(P_2 - P_1)}\, dP$$

$$\Delta G = \frac{V}{\kappa}\,[e^{-\kappa(P_2 - P_1)} - 1]$$

If $\kappa\,\Delta P << 1$, $e^{-\kappa\Delta P} = 1 - \kappa\Delta P + \frac{1}{2}(\kappa\Delta P)^2 - \cdots$

$$\Delta G \quad = \frac{V}{\kappa}[\kappa\Delta P - \frac{1}{2}\kappa^2(\Delta P)^2 + \cdots] = V\Delta P - \frac{1}{2}\kappa V(\Delta P)^2$$

4.11 Given the isothermal compressibility κ of water is 45.0×10^{-6} bar^{-1}, calculate a more accurate value for $\Delta \bar{G}$ for the compression of 1000 bar than that calculated in Problem 4.9 (See Problem 4.10).

SOLUTION

$$\Delta\bar{G} \quad = \bar{V}\Delta P - \frac{1}{2}\kappa\bar{V}(\Delta P)^2$$

$$= (18 \times 10^{-6}\ m^3\ ml^{-1})(999\ bar)(10^5\ Pa\ bar^{-1})$$

$$- \frac{(45.0 \times 10^{-6})(18 \times 10^{-6}\ m^3\ mol^{-1})(999\ bar)^2(10^5\ Pa\ bar^{-1})^2}{2(10^5\ Pa\ bar^{-1})}$$

$$= 1798\ J\ mol^{-1} - 40\ J\ mol^{-1}$$

$$= 1758\ J\ mol^{-1}$$

4.12 An ideal gas is allowed to expand reversibly and isothermally (25 °C) from a pressure of 1 bar to a pressure of 0.1 bar: (a) What is the change in molar Gibbs energy? (b) What would be the change in molar Gibbs energy if the process occurred irreversibly?

SOLUTION

(a) $$\left(\frac{\partial\bar{G}}{\partial P}\right)_T \quad = \bar{V} = \frac{RT}{P}$$

$$\Delta\bar{G} = RT \ln \frac{P_2}{P_1} = (8.314 \text{ J K}^{-1} \text{ mol}^{-1})(298.15 \text{ K}) \ln 0.1$$

$$= -5708 \text{ J mol}^{-1}$$

(b) $\Delta\bar{G} = -5708 \text{ J mol}^{-1}$ because G is a state funciton and depends only on the initial state and the final state.

4.13 The standard entropy of $O_2(g)$ at 1 bar is listed in Appendix C.2 as 205.137 J K^{-1} mol^{-1}, and the standard Gibbs energy of formation is listed as 0 kJ mol^{-1}. Assuming O_2 is an ideal gas, what will be the molar entropy and molar Gibbs energy of formation at 100 bar?

SOLUTION

$$\bar{S} = \bar{S}^o - R \ln \frac{P}{P^o} = 205.137 - 8.314 \ln 100 = 166.848 \text{ J K}^{-1} \text{ mol}^{-1}$$

$$\Delta_f G^o = 0 + RT \ln (P/P^o) = (8.314 \times 10^{-3})(298) \ln 100$$

$$= 11.41 \text{ kJ mol}^{-1}$$

4.14 Helium is compressed isothermally and reversibly at 100 °C from a pressure of 2 to 10 bar. Calculate (a) q per mole, (b) w per mole, (c) $\Delta\bar{G}$, (d) $\Delta\bar{A}$, (e) $\Delta\bar{H}$, (f) $\Delta\bar{U}$, and (g) $\Delta\bar{S}$, assuming helium is an ideal gas.

SOLUTION

(a) $\Delta\bar{U} = q + w = 0$

$$q = -w = -RT \ln \frac{P_2}{P_1}$$

$$= -(8.314 \text{ J K}^{-1} \text{ mol}^{-1})(373.15 \text{ K}) \ln \frac{10 \text{ bar}}{2 \text{ bar}} = -4993 \text{ J mol}^{-1}$$

(b) w $= -q = 4993$ J mol^{-1}

(c) $\Delta\bar{G} = \int_{P_1}^{P_2} V dP = RT \ln \frac{P_2}{P_1} = 4993 \text{ J mol}^{-1}$

(d) $\Delta\bar{A} = w_{max} = 4993$ J mol^{-1}

(e) $\Delta\bar{H} - \Delta\bar{U} + \Delta (P\bar{V}) = 0$

(f) $\Delta\bar{U} = 0$

(g) $\Delta\bar{S} = \Delta\bar{H} - \frac{\Delta\bar{G}}{T} = 0 - \frac{4933 \text{ J mol}^{-1}}{373.15 \text{ K}} = -13.38 \text{ J K}^{-1} \text{ mol}^{-1}$

4.15 Toluene is vaporized at its boiling point, 111 °C. The heat of vaporization at this temperature is 361.9 J g^{-1}. For the vaporization of toluene, calculate (a) w per mole, (b) q per mole, (c) $\Delta\bar{H}$, (d) $\Delta\bar{U}$, (e) $\Delta\bar{G}$, and (f) $\Delta\bar{S}$.

SOLUTION

(a) Assuming that toluene vapor is an ideal gas and that the volume of the liquid is negligible, the work on the toluene is

$w = -P\Delta\bar{V} = -RT = -(8.314 \text{ J K}^{-1} \text{ mol}^{-1})(384 \text{ K})$

$= -3193$ J mol^{-1}

(b) $q_p = \Delta\bar{H} = (361.9 \text{ J g}^{-1})(92.13 \text{ g mol}^{-1}) = 33{,}340 \text{ J mol}^{-1}$.

(c) $\Delta\bar{H} = 3{,}340$ J mol^{-1}

(d) $\Delta\bar{U} = q + w = 33{,}340 - 3193 = 30{,}147$ J mol^{-1}

(e) $\Delta\bar{G}$ = 0 because the evaporation is reversible at the boiling point and 1 atm.

(f) $\Delta\bar{S} = \frac{q_{rev}}{T} = \frac{33340 \text{ J mol}^{-1}}{384 \text{ K}} = 86.8 \text{ J K}^{-1} \text{ mol}^{-1}$

4.16 If the Gibbs energy varies with temperature according to

$$G/T = a + b/T + c/T^2$$

How will the enthalpy and entropy vary with temperature? Check that these three equations are consistent.

SOLUTION

$G = aT + b + c/T$

$\left(\frac{\partial G}{\partial T}\right)_P = -S = a - c/T^2$

$S = -a + c/T^2$

$\left[\frac{\partial (G/T)}{\partial T}\right]_P = -H/T^2 = -b/T^2 - 2c/T^3$

$H = b + 2c/T$

$G = H - TS = aT + b + c/T$

4.17 Calculate the change in molar Gibbs energy $\bar{G}$ when supercooled water at -3 °C freezes at constant T and P. The density of ice at -3 °C is 0.917×10^3 kg m^{-3}, and its vapor pressure is 475 Pa. The density of supercooled water at -3 °C is 0.9996 kg m^{-3} and its vapor pressure is 489 Pa.

SOLUTION

$$\bar{V}_s = \frac{(18.015 \times 10^{-3}\ \text{kg mol}^{-1})}{(0.917 \times 10^{3}\ \text{kg m}^{-3})}$$

$$= 1.965 \times 10^{-5}\ \text{m}^3\ \text{mol}^{-1}$$

$$\bar{V}_l = \frac{(18.015 \times 10^{-3}\ \text{kg mol}^{-1})}{(0.9996 \times 10^{3}\ \text{kg m}^{-3})}$$

$$= 1.802 \times 10^{-5}\ \text{m}^3\ \text{mol}^{-1}$$

Since the actual process is irreversible, the calculation uses the following reversible isothermal path

$H_2O(l, 270.15\ \text{K}, 10^5\ \text{Pa})$ $\qquad$ $H_2O(s, 270.15\ \text{K}, 10^5\ \text{Pa})$

$\downarrow \Delta\bar{G}_1 = \int_{10^5}^{489} \bar{V}_l\, dP = -1.7\ \text{J mol}^{-1}$ $\qquad$ $\uparrow \Delta\bar{G}_5 = \int_{475}^{10^5} \bar{V}_s\, dP = -1.9\ \text{J mol}^{-1}$

$H_2O(l, 270.15\ \text{K}, 489\ \text{Pa})$ $\qquad$ $H_2O(s, 270.15\ \text{K}, 475\ \text{Pa})$

$\downarrow \Delta\bar{G}_2 = 0$ $\qquad$ $\uparrow \Delta\bar{G}_4 = 0$

$H_2O(g, 270.15\ \text{K}, 489\ \text{Pa})$ $\quad = \quad$ $H_2O(g, 270.15\ \text{K}, 475\ \text{Pa})$

$$\Delta\bar{G}_3 = RT \ln (475/489) = -65.2\ \text{J mol}^{-1}$$

$$\Delta\bar{G} = \Delta\bar{G}_1 + \Delta\bar{G}_2 + \Delta\bar{G}_3 + \Delta\bar{G}_4 + \Delta\bar{G}_5$$

$$= -1.7 + 40 - 65.2 + 1.9$$

$$= -65.1\ \text{J mol}^{-1}$$

4.18 Calculate the molar Gibbs energy of fusion when supercooled water at -3 °C freezes at constant T and P. The molar enthalpy of fusion of ice is 6000 J mol^{-1} at 0 °C. The heat capacities of water and ice in the vicinity of the freezing point are 75.3 and 38 J K^{-1} mol^{-1}, respectively.

SOLUTION

Since the actual process in irreversible, the calculation uses the following reversible isobaric path.

$$\begin{array}{ccc} H_2O(l,\ 273.15\ K,\ 1\ bar) & \rightarrow & H_2O(s,\ 273.15,\ 1\ bar) \\ \uparrow & & \uparrow \\ H_2O(l,\ 270.15\ K,\ 1\ bar) & \rightarrow & H_2O(s,\ 270.15,\ 1\ bar) \end{array}$$

To obtain $\Delta_{fus}G$ (270.15 K), we have to calculate $\Delta_{fus}H$ (270.15 K) and $\Delta_{fus}S$ (270.15 K) separately and use $\Delta_{fus}G = \Delta_{fus}H - 270.15\ \Delta_{fus}S$.

$$\Delta_{fus}H(270.15\ K) = \Delta_{fus}H(273.15\ K) + \int_{273.15}^{270.15} \Delta_{fus}C_P\ dT$$

$$= -6000 + (38 - 75.3)(-3)$$

$$= -5888\ J\ mol^{-1}$$

$$\Delta_{fus}S(270.15\ K) = -6000/273.15 + \Delta_{fus}C_P\ \ln(270.15/273.15)$$

$$= -21.56\ J\ mol^{-1}$$

$$\Delta_{fus}G = \Delta_{fus}H - T\Delta_{fus}S = -5888 - (270.15)(-21.56)$$

$$= -63.6\ J\ mol^{-1}$$

4.19 As shown in Example 4.9, the fugacity of a van der Waals gas is given by a fairly simple expression if only the second virial coefficient is used. To this degree of approximation derive the expressions for $\bar{G}$, $\bar{S}$, $\bar{A}$, $\bar{U}$, $\bar{H}$, and $\bar{V}$.

SOLUTION

$$\bar{G} = \bar{G}^o + RT\ \ln\left(\frac{f}{P^o}\right) = \bar{G}^o + RT\ \ln\left(\frac{P}{P^o}\right) + \left(b - \frac{a}{RT}\right)P$$

$$\bar{S} = -\left(\frac{\partial \bar{G}}{\partial T}\right)_P = \bar{S}^o - R \ln\left(\frac{P}{P^o}\right) + \frac{aP}{RT^2}$$

$$\bar{A} = \bar{G} - P\left(\frac{\partial \bar{G}}{\partial P}\right)_T = \bar{G}^o + RT \ln\left(\frac{P}{P^o}\right) - RT$$

$$\bar{U} = \bar{G} - T\left(\frac{\partial \bar{G}}{\partial T}\right)_P - P\left(\frac{\partial \bar{G}}{\partial P}\right)_T = \bar{G}^o + T\bar{S}^o - RT - \frac{aP}{RT}$$

$$= \bar{H}^o - RT - \frac{aP}{RT} = \bar{U}^o - \frac{aP}{RT}$$

Since the second term is small, we can use the ideal gas law to obtain

$$\bar{U} = \bar{U}^o - \frac{a}{\bar{V}}$$

Note that this agrees with the earlier result (Section 4.4) that

$$\left(\frac{\partial \bar{U}}{\partial \bar{V}}\right)_T = \frac{a}{\bar{V}^2}$$

$$\bar{H} = \bar{U} + P\bar{V} = \bar{U}^o + \frac{aP}{RT} + RT + bP - \frac{aP}{RT}$$

$$= \bar{H}^o + \left(b - \frac{2a}{RT}\right)P$$

$$\bar{V} = \left(\frac{\partial \bar{G}}{\partial P}\right)_T = \frac{RT}{P} + b - \frac{a}{RT}$$

4.20 Using the relation derived in Example 4.9, calculate the fugacity of $H_2(g)$ at 100 bar at 298 K.

SOLUTION

$$f = P \exp\left(b - \frac{a}{RT}\right)\frac{P}{RT}$$

$= (100 \text{ bar}) \exp [0.02661 -$

$0.2476/(0.08314)(298)]100/(0.08314)(298)=106.9 \text{ bar}$

4.21 Show that if the compressibility factor is given by $Z = 1 + BP/RT$ the fugacity is given by $f = Pe^{Z-1}$. If Z is not very different from unity, $e^{Z-1} = 1 + (Z - 1) + \cdots \cong Z$ so that $f = PZ$. Using this approximation, what is the fugacity of $H_2(g)$ at 50 bar and 298 K using its van der Waals constants?

SOLUTION

Substituting this expression for the compressibility factor in equation 4.74,

$$\frac{f}{P} = \exp[\int_0^P \frac{B}{RT} dP]$$

$$= \exp[\int_0^P (B/RT)dP]$$

$$= \exp(BP/RT)$$

$$= \exp(Z-1)$$

If Z is not very different from unity, $f = PZ$. For $H_2(g)$ at 50 bar and 298 K,

$$Z = 1 + (b - a/RT) P/RT$$

$$= 1 + (0.02661 \text{ L mol}^{-1} - \frac{0.02476 \text{ L}^2 \text{ bar mol}^{-2}}{0.08314 \text{ L bar K}^{-1} \text{ mol}^{-1}} \frac{50 \text{ bar}}{(0.08314)(298)}$$

$$= 1.0335$$

$$f = (50 \text{ bar})(1.0335) = 5.17 \text{ bar}$$

4.22 Calculate the partial molar volume of zinc chloride in 1-molal $ZnCl_2$ solution using the following data.

% by weight of $ZnCl_2$	2	6	10	14	18
Density/g cm^{-3}	1.0167	1.0532	1.0891	1.1275	1.1665

SOLUTION

Taking the first solution as an example, 1 g of solution contains 0.02 g of $ZnCl_2$ (M = 136.28 g mol^{-1}) and 0.98 g of H_2O. The mass of solution containing 1000 g of water is

$$\frac{1000}{0.98} = 1020 \text{ g}$$

$$\text{molality } ZnCl_2 = \frac{(0.02)(1020 \text{ g})}{136.28 \text{ g mol}^{-1}} = 0.1497 \text{ m}$$

Volume of solution containing 1000 g of H_2O

$$\frac{1020 \text{ g}}{1.0167 \text{ g cm}^{-3}} = 1003.2 \text{ cm}^3$$

Wt %	Molality	Volume Containing 1000 g of H_2O
2	0.1497 m	1003.2
6	0.4683	1010.1
10	0.8152	1020.2
14	1.194	1031.3
18	1.610	1045.5
20	1.834	1053.4

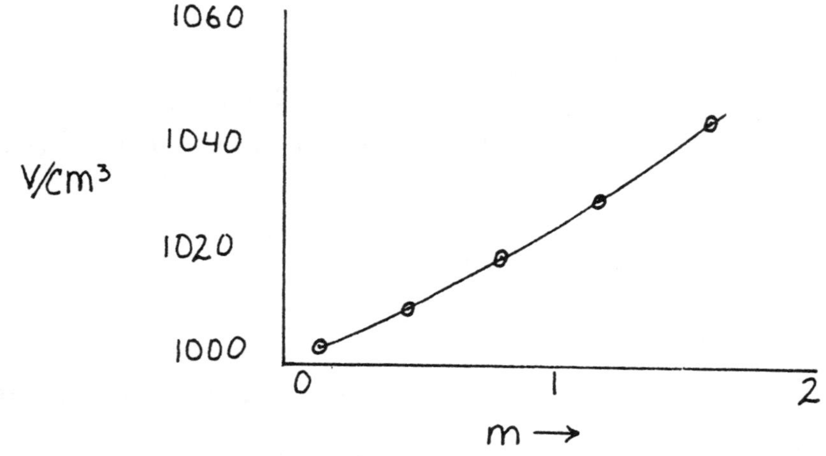

The slope of this plot at m = 1 molar is 29.3 $cm^3\ mol^{-1}$ and so this is the partial molar volume of $ZnCl_2$.

4.23 Calculate $\Delta_{mix}G$ and $\Delta_{mix}S$ for the formation of a quantity of air containing 1 mol of gas by mixing nitrogen and oxygen at 298.15 K. Air may be taken to be 80% nitrogen and 20% oxygen by volume.

SOLUTION

$$\Delta_{mix}G = RT(y_1 \ln y_1 + y_2 \ln y_2)$$

$$= (8.314\ J\ K^{-1}\ mol^{-1})(298.15\ K)(0.8 \ln 0.8 + 0.2 \ln 0.2)$$

$$= -1239\ J\ mol^{-1}$$

$$\Delta_{mix}S = -R(y_1 \ln y_1 + y_2 \ln y_2) = 4.159\ J\ K^{-1}\ mol^{-1}$$

4.24 A mole of gas A is mixed with a mole of gas B at 1 bar and 298 K. How much work is required to separate these gases to produce a container of each at 1 bar and 298 K?

SOLUTION

The mixture can be separated by diffusion through a perfect semipermeable membrane. The highest partial pressure of each that can be reached is 1/2 bar. These gases then have to be compressed to 1 bar.

$$w = (2\ mol) \int_{1/2}^{1} \bar{V} dP = (2\ mol) \int_{1/2}^{1} (RT/P) dP = (2\ mol)\ RT \ln 2$$

$$= 2(8.314)(298) \ln 2$$

$$= 3.4\ kJ$$

The actual process will require more work.

4.25 The surface tension of toluene at 20 °C is 0.0284 N m^{-1}, and its density at this temperature is 0.866 g cm^{-3}. What is the radius of the largest capillary that will permit the liquid to rise 2 cm?

<u>SOLUTION</u>

$$\gamma = \frac{1}{2} h \rho g r$$

$$r = \frac{2\gamma}{\rho g h} = \frac{2(0.284 \text{ N m}^{-1})}{(0.866 \times 10^3 \text{ kg m}^{-3})(9.80 \text{ m s}^{-2})(0.02 \text{ m})}$$

$$= 3.35 \times 10^{-4} \text{ m} = 3.35 \times 10^{-2} \text{ cm}$$

4.26 Mercury does not wet a glass surface. Calculate the capillary depression if the diameter of the capillary is (a) 0.1 mm, and (b) 2 mm. The density of mercury is 213.5 g cm^{-3}. The surface tension of mercury at 25 °C is 0.520 N m^{-1}.

<u>SOLUTION</u>

$$\gamma = hg\rho r/2 \qquad h = 2\gamma/g\rho r$$

(a) $$h = \frac{(2)(0.520 \text{ N m}^{-1})}{(9.80 \text{ m s}^{-2}) \times (13.5 \times 20^3 \text{ kg m}^{-3})(0.05 \times 10^{-3} \text{ m})}$$

$$= 15.7 \text{ cm}$$

(b) $$h = \frac{(2)(0.520)}{(9.80)(13.5 \times 10^3 \text{ kg m}^{-3} \text{ m})} = 7.86 \text{ mm}$$

4.27 If the surface tension of a soap solution is 0.050 N m^{-1}, what is the difference in pressure across the film for (a) a soap bubble of 2 mm in diameter and (b) a bubble 2 cm in diameter.

SOLUTION

(a) $$\Delta P = \frac{2\gamma}{R} = \frac{4(0.050\ N\ m^{-1})}{0.001\ m} = 200\ Pa$$

(b) $$\Delta P = \frac{4(0.050\ N\ m^{-1})}{0.01\ m} = 20\ Pa$$

4.28 Using experimental data from Fig. 4.9 calculate $(\partial S/\partial l)_T$ and $(\partial U/\partial l)_T$ using only the data at 10 °C and 780 °C, assuming these quantities are independent of temperature. These thermodynamic quantities do depend on the elongation, so calculate them for 200% and 400% elongation. (For f and T simply use averages for this temperature range.)

SOLUTION

At 200% elongation

$$\left(\frac{\partial S}{\partial l}\right)_T = -\left(\frac{\partial f}{\partial T}\right)_l = -\frac{(10.8 - 9.0)(10^5\ N\ m^{-2})}{(70 - 10)K}$$

$$= -3.0 \times 10^3\ N\ K^{-1}\ m^{-2}$$

$$\left(\frac{\partial U}{\partial l}\right)_T = f - T\left(\left(\frac{\partial f}{\partial T}\right)_l = 9.5 \times 10^5\ N\ m^{-2} - (313\ K)(3.0 \times 10^3\ N\ K^{-1}\ m^{-2})\right.$$

$$= 0.1\ N\ m^{-2}$$

At 400% elongation

$$\left(\frac{\partial S}{\partial l}\right)_T = -\frac{(26 - 21.3) \times 10^5 \text{ N m}^{-2}}{60 \text{ K}} = -7.8 \times 10^3 \text{ N K}^{-1} \text{ m}^{-2}$$

$$\left(\frac{\partial U}{\partial l}\right)_T = 23 \times 10^5 \text{ N m}^{-2} - (313 \text{ K})(7.8 \times 10^3 \text{ N K}^{-1} \text{ m}^{-2})$$

$$= -1.4 \text{ N m}^{-2}$$

4.29 Show that $\left(\frac{\partial U}{\partial S}\right)_V = \left(\frac{\partial H}{\partial S}\right)_P$ and $\left(\frac{\partial H}{\partial P}\right)_S = \left(\frac{\partial G}{\partial P}\right)_T$

SOLUTION:

$dU = TdS - PdV = \left(\frac{\partial U}{\partial S}\right)_V dS + \left(\frac{\partial U}{\partial V}\right)_S dV$, therefore $\left(\frac{\partial U}{\partial S}\right)_V = T$

$dH = TdS + VdP = \left(\frac{\partial H}{\partial S}\right)_P dS + \left(\frac{\partial H}{\partial P}\right)_S dP$, therefore $\left(\frac{\partial H}{\partial P}\right)_S = V$

and $\left(\frac{\partial H}{\partial S}\right)_P = T$

$dG = -SdT + VdP = \left(\frac{\partial G}{\partial T}\right)_P + \left(\frac{\partial G}{\partial P}\right)_T dP$, therefore $\left(\frac{\partial G}{\partial P}\right)_T = V$

$$\left(\frac{\partial H}{\partial P}\right)_S = \left(\frac{\partial G}{\partial T}\right)_P$$

4.30 $dS = (C_V/T)dT + (\alpha/\kappa)dV$

4.31 -0.0252 J K^{-1} mol^{-1}

4.32 $(\partial S/\partial P)_T = 0$, $(\partial H/P)_T = V$, $(\partial U/\partial P)_T = 0$

4.36 88.2 J mol^{-1}

4.37 (a) 4993 (b) 3655 J mol^{-1}

4.38 168.97 J K^{-1} mol^{-1}, 0, -11.42 kJ mol^{-1}

4.39 (a) 6820 J mol^{-1} (b) 6072 J mol^{-1} (c) 75.9 J K^{-1} mol^{-1} (d) 0

4.40 (a) -5229 J mol^{-1} (b) 5229 J mol^{-1} (c) 0 (d) -5229 J mol^{-1}
(e) 19.14 J K^{-1} mol^{-1} (f) 0 (g) 0 (h) 0 (i) -5229 J mol^{-1}
(j) 19.14 J K^{-1} mol^{-1} (k) 0 (l) 19.14 J K^{-1} mol^{-1}

4.41 3100 J mol^{-1}, -40,690 J mol^{-1}, -37.6 kJ mol^{-1}, -34.5 kJ mol^{-1}, 3.1 kJ mol^{-1}, -109.0 J K^{-1} mol^{-1}, 0 kJ mol^{-1}

4.42 (a) -6.754 kJ mol^{-1}, (b) 0, (c) 0, (d) -6.754 kJ mol^{-1}, (e) 22.51 J K^{-1} mol^{-1}

4.43 0, 19.16 J K^{-1} mol^{-1}

4.44 0, -13.38 J K^{-1} mol^{-1}, 4991 J mol^{-1}

4.45 28.23 kJ mol^{-1}, 73.52 J K^{-1} mol^{-1}, 0

4.46 -02121 J mol^{-1}

4.47 $\Delta\bar{G} = RT \ln (P_2/P_1) + (b - a/RT)(P_2 - P_1)$

4.49 0.79×10^{-3} m^3 kg^{-1}

4.50 (a) 0.2033×10^{-3} m^{-3} kg^{-1} (b) 19.36×10^{-6} m^3 mol^{-1}

4.51 -4731 J, 15.88 J K^{-1}

4.52 0.0233 N m^{-1}

4.54 (a) 1.002878, (b) 1.02878 bar

4.55 6.0×10^{25}, 5.8×10^{25}

4.56 $\Delta_{vap}S = 109.3 \text{ J K}^{-1} \text{ mol}^{-1}$

$\Delta_{vap}H = 39.9 \text{ kJ mol}^{-1}$

$\Delta_{vap}G = -3.6 \text{ kJ mol}^{-1}$

CHAPTER 5 Chemical Equilibrium

5.1 For the reaction $N_2(g) + 3H_2(g) = 2NH_3(g)$, $K = 1.60 \times 10^{-4}$ at 400 °C. Calculate (a) $\Delta_r G^o$ and (b) $\Delta_r G$ when the pressures of N_2 and H_2 are maintained at 10 and 30 bar, respectively, and NH_3 is removed at a partial pressure of 3 bar. (c) Is the reaction spontaneous under the latter conditions?

SOLUTION

(a) $\Delta_r G^o = -RT \ln K$

$= -(8.314 \text{ J K}^{-1} \text{ mol}^{-1})(673 \text{ K}) \ln 1.60 \times 10^{-4}$

$= 48.9 \text{ kJ mol}^{-1}$

(b) $$\Delta_r G = \Delta_r G^o + RT \ln \frac{\left(\frac{P_{NH_3}}{P^o}\right)^2}{(P_{N_2}/P^o)(P_{H_2}/P^o)^3}$$

$$= 48.9 + (8.314 \times 10^{-3})(673) \ln \frac{3^3}{10(30)^3}$$

$= -2.63 \text{ mol}^{-1}$

(c) Yes

5.2 At 1:3 mixture of nitrogen and hydrogen was passed over a catalyst at 450 °C. It was found that 2.04% by volume of ammonia was formed when the total pressure was maintained at 10.13 bar [A. T. Larson and R. L. Dodge, *J. Am. Chem. Soc.* **45**, 2918 (1923)]. Calculate the value of K for $\frac{3}{2} H_2(g) + \frac{1}{2} N_2(g) = NH_3(g)$ at this temperature.

SOLUTION

At equilibrium $P_{H_2} + P_{N_2} + P_{NH_3} = 10.13$ bar

P_{N_2} = (10.13 bar)(0.0204) = 0.207 bar

$P_{H_2} + P_{N_2}$ = 10.13 bar - 0.207 bar = 9.923 bar

$P_{H_2} = 3P_{N_2}$ because this initial ratio is not changed by reaction

$$P_{N_2} = \frac{9.923 \text{ bar}}{4} = 2.481 \text{ bar}$$

$$P_{H_2} = \frac{3}{4}(9.923 \text{ bar}) = 7.442 \text{ bar}$$

$$K = \frac{(P_{NH_3}/P^o)}{(P_{H_2}/P^o)^{3/2}(P_{N_2}/P^o)^{1/2}}$$

$$= \frac{0.207}{7.442^{3/2}\ 2.482^{1/2}} = 6.47 \times 10^{-3}$$

5.3 At 55 °C and 1 bar the average molar mass of partially dissociated N_2O_4 is 61.2 g mol^{-1}. Calculate (a) ξ and (b) K for the reaction $N_2O_4(g) = 2NO_2(g)$. (c) Calculate ξ at 55 °C if the total pressure is reduced to 0.1 bar.

SOLUTION

(a) $$\xi = \frac{M_1 - M_2}{M_2} = \frac{92.0 - 61.2}{61.2} = 0.503$$

(b) $$K = \frac{4\xi^2(P/P^o)}{1 - \xi^2} = \frac{4(0.503)^2}{1 - 0.503^2} = 1.36$$

(c) $$\frac{\xi^2}{1 - \xi^2} = \frac{K}{4(P/P^o)} = \frac{1.36}{4(0.1)}$$

$$\xi = 0.879$$

(Note that ξ is the dimensionless extent of reaciton.)

5.4 A liter reaction vessel containing 0.233 mol of N_2 and 0.341 of PCl_5 is heated to 250 °C. The total pressure at equilibrium is 29.33 bar. Assuming that all the gases are ideal, calculate K for the only reaction that occurs.

$$PCl_5(g) = PCl_3(g) + Cl_2(g)$$

SOLUTION

	PCl_5 =	PCl_3 +	Cl_2	
initial	0.341	0	0	total = 0.341 + ξ
eq.	0.341 - ξ	ξ	ξ	

$$K = \frac{\xi^2 (P/P^o)^2}{(0.341 + \xi)(0.341 - \xi)}$$

$$= \frac{\xi^2 (P/P^o)}{0.341^2 - \xi^2} \text{ where } P = P_{PCl_5} + P_{Cl_3}$$

$$P = 29.33 \text{ bar} - \frac{(0.233 \text{ mol})(0.08314 \text{ L bar K}^{-1}\text{mol}^{-1})(523 \text{ K})}{1 \text{ L}}$$

$$= 19.20 \text{ bar} = \frac{(0.341 + \xi)(0.08314)(523)}{1}$$

$$\xi = 0.1005$$

$$K = \frac{(0.1005)^2 (19.2)}{0.341^2 - 0.1005^2} = 1.83$$

5.5 An evacuated tube containing 5.96×10^{-3} mol L^{-1} of solid iodine is heated to 973 K. The experimentally determined pressure is 0.496 [M. L. Perlman and G. K. Rollefson, *J. Chem.*

Phys. **9**, 362 (1941)]. Assuming ideal gas behavior, calculate K for $I_2(g) = 2I(g)$.

SOLUTION

$$P = \frac{n}{V} RT$$

$$0.496 \text{ bar} = (1 + \xi)(5.96 \times 10^{-3} \text{ L}^{-1})(0.08314 \text{ L bar K}^{-1} \text{ mol}^{-1})(973 \text{ K})$$

$$\xi = \frac{0.496 \text{ bar}}{(5.96 \times 10^{-3} \text{ mol L}^{-1})(0.08314 \text{ L bar}^{-1} \text{ K}^{-1} \text{ mol}^{-1})(973 \text{ K})} - 1$$

$$= 0.02867$$

$$K = \frac{4\xi^2(P/P^o)}{1 - \xi^2} = \frac{4(0.0287)^2(0.496)}{1 - 0.0287^2} = 1.64 \times 10^{-3}$$

5.6 Nitrogen trioxide dissociates according to the reaction

$$N_2O_3(g) = NO_2(g) + NO(g)$$

When one mole of $N_2O_3(g)$ is held at 25 °C and 1 bar total pressure until equilibrium is reached, the extent of reaction is 0.30. What is Δ_rG^o for this reaction at 25 °C?

SOLUTION

	$N_2O_3(g)$	$= NO_2(g)$	$+ NO(g)$	
int.	1	0	0	
eq.	$1 - \xi$	ξ	ξ	total $= 1 + \xi$
mole fr.	$\frac{1-\xi}{1+\xi}$	$\frac{\xi}{1+\xi}$	$\frac{\xi}{1+\xi}$	

$$K = \frac{\xi^2P^2(1 + \xi)}{(1 + \xi)^2(1 - \xi)P} = \frac{\xi^2 P}{1 - \xi^2}$$

$$= \frac{0.30^2\ 1}{0.91} = 0.099$$

$\Delta_r G^o$ $= -RT \ln K = -8.314\ (298) \ln 0.099$

$= 5.73\ kJ\ mol^{-1}$

5.7 For the reaction

$$2HI(g) = H_2(g) + I_2(g)$$

at 698.6 K, K = 1.83 x 10^{-2}. (a) How many grams of hydrogen iodide will be formed when 10 g of iodine and 0.2 g of hydrogen are heated to this temperature in a 3-L vessel? (b) What will be the partial pressures of H_2, I_2, and HI?

SOLUTION

(a) Pressures due to reactants prior to reaction:

$$P_{I_2} = \frac{(10\ g)(0.08314\ L\ bar\ K^{-1}\ mol^{-1})(698.6\ K)}{(254\ g\ mol^{-1})(3\ L)}$$

$= 0.762$ bar

$$P_{H_2} = \frac{(0.2)(0.08314\ L\ bar\ K^{-1}\ mol^{-1})(698.6\ K)}{2 \times 3}$$

$= 1.936$ bar

$$K = \frac{(0.751 - x)(1.936 - x)}{(2x)^2} = 1.83 \times 10^{-2}$$

$0.0732\ x^2 = 1.454 - 2.687\ x + x^2$

$0.9268\ x^2 - 2.687x + 1.454 = 0$

$x = 0.720$ bar

$P_{HI} = 2x = 1.440$ bar

$$m_{HI} = \frac{P_{HI} V M}{RT}$$

$$= \frac{(1.440)(3)(128)}{(0.08314)(698.6)} = 9.520 \text{ g}$$

(b) $P_{H_2} = 1.936 - 0.720 = 1.216$ bar

$P_{I_2} = 0.762 - 0.720 = 0.042$ bar

$P_{HI} = 1.440$ bar

5.8 Express K for the reaction

$$CO(g) + 3H_2(g) = CH_4(g) + H_2O(g)$$

in terms of the equilibrium extent of reaction ξ when one mole of CO is mixed with one mole of hydrogen.

SOLUTION

	6)	CO	+	$3H_2$	=	CH_4	+	H_2O	
int.		1		1		0		0	
eq.		$1-\xi$		$1-3\xi$		ξ		ξ	total = $2-2\xi$

$$K = \frac{\left(\frac{\xi}{2-2\xi}\right)^2 \left(\frac{P}{P^o}\right)^2}{\left(\frac{1-\xi}{2-2\xi}\right)\left(\frac{1-3\xi}{2-2\xi}\right)^3 \left(\frac{P}{P^o}\right)^4}$$

$$K = \frac{(\xi)^2(2-2\xi)^2}{(1-\xi)(1-3\xi)^3\left(\frac{P}{P^o}\right)^2}$$

5.9 What are the percentage dissociations of $H_2(g)$, $O_2(g)$, and $I_2(g)$ at 2000 K and a total pressure of 1 bar?

SOLUTION

$H_2(g) \quad = 2H(g)$

$\Delta_r G^o = 2(106{,}653 \text{ J mol}^{-1})$

$= -RT \ln K$

$= -(8.314 \text{ J K}^{-1} \text{ mol}^{-1})(2000 \text{ K}) \ln K$

$$K = 2.69 \times 10^{-6} = \frac{4\xi^2}{1-\xi^2}$$

$$\xi = \left(\frac{K_p}{K_p + 4}\right)^{1/2} = \left(\frac{2.69 \times 10^{-6}}{4}\right)^{1/2} = 0.000819 \text{ or } 0.0819\%$$

$O_2(g) = 2O(g)$

$\Delta_r G^o = 2(121{,}448 \text{ J mol}^{-1})$

$K = 4.53 \times 10^{-7} \qquad \xi = 0.0337\%$

$I_2(g) = 2I(g)$

$\Delta_r G^o = 2(-29{,}637 \text{ J mol}^{-1})$

$K = 35.32 \qquad \xi = 94.8\%$

5.10 In order to produce more hydrogen from "synthesis gas" ($CO + H_2$) the water gas shift reaction is used.

$$CO(g) + H_2O(g) = CO_2(g) + H_2(g)$$

Calculate K at 1000 K and the equilibrium extent of reaction starting with an equimolar mixture of CO and H_2O.

SOLUTION

ΔG^o = -395,886 - (-200,275) - (-192,590) = -3021 J mol^{-1}

= -(8.314 J K^{-1} mol^{-1})(1000 K) ln K

$$K = 1.44 = \frac{\left(\frac{P_{H_2}}{P^o}\right)\left(\frac{P_{CO_2}}{P^o}\right)}{\left(\frac{P_{CO}}{P^o}\right)\left(\frac{P_{H_2O}}{P^o}\right)} = \frac{\xi^2}{(1 - \xi)^2}$$

ξ = 0.545, fractional conversion of reactants to products

(Note that this reaction is exothermic so that there will be a larger extent of reaction at lower temperatures. In practice this reaction is usually carried out at about 700 K.)

5.11 Calculate the extent of reaction ξ of 1 mol of $H_2O(g)$ to form $H_2(g)$ and $O_2(g)$ at 2000 K and 1 bar. (Since the extent of reaction is small, the calculation may be simp.ified by assuming that P_{H_2O} = 1 bar.)

SOLUTION

$$H_2O(g) = H_2(g) + \frac{1}{2}O_2(g)$$

init.	1	0	0
eq.	$1 - \xi$	ξ	$\frac{1}{2}\xi$

$\Delta_r G^o$ = 135,528 J mol^{-1}

= -(8.1315 J K^{-1} mol^{-1})(2000 K) ln K

K = 2.887 x 10^{-4}

$$= \frac{(\frac{P_{H_2}}{P^o})(\frac{P_{O_2}}{P^o})^{1/2}}{(\frac{P_{H_2O}}{P^o})}$$

$$(\frac{P_{H_2}}{P^o})(\frac{P_{H_2}}{2P^o})^{1/2} = \frac{1}{\sqrt{2}}(\frac{P_{H_2}}{P^o})^{3/2}$$

$$\xi = (\frac{P_{H_2}}{P^o}) = (\sqrt{2}\ K)^{2/3} = 0.0055$$

5.12 At 500 K CH_3OH, CH_4 and other hydrocarbons can be formed from CO and H_2. Until recently the main source of the CO mixture for the synthesis of CH_3OH was methane.

$$CH_4(g) + H_2O(g) = CO(g) + 3H_2(g)$$

When coal is used as the source, the "synthesis gas" has a different composition.

$$C(graphite) + H_2O(g) = CO(g) + H_2(g)$$

Suppose we have a catalyst that catalyzes only the formation of CH_3OH. (a) What pressure is required to convert 25% of the CO to CH_3OH at 500 K if the "synthesis gas" comes from CH_4? (b) If the synthesis gas comes from coal?

SOLUTION

(a) $CO + 2H_2 = CH_3OH$

	CO	$2H_2$	CH_3OH	
Initial	1	3	0	
Equil.	$1-\xi$	$3-2\xi$	ξ	Total = $4-2\xi$

$$K = \frac{y_{CH_3OH}}{y_{CO_3} y_{H_2}^2 P^2} = \frac{\xi(4-2\xi)^2}{(1-\xi)(3-2\xi)^2 P^2}$$

$\Delta_r G$ = -134.27 - (-155.41) = 22.14 kJ mol^{-1}

$K = 6.188 \times 10^{-3}$

$$P = \left[\frac{\xi(4-2\xi)^2}{K(1-\xi)(3-2\xi)^2}\right]^{1/2}$$

$$= \left[\frac{(0.25)(3.5)^2}{6.188 \times 10^{-3}(0.75)(2.5)^2}\right]^{1/2}$$

(b)	$CO + 2H_2 = CH_3OH$			
Initial	1	1	0	
Equil.	$1-\xi$	$1-2\xi$	ξ	Total = $2-2\xi$

$$K = \frac{\xi(2-2\xi)^2}{(1-\xi)(1-2\xi)^2 P^2} \qquad P = \left[\frac{(0.25)(1.5)^2}{K(0.75)(0.5)^2}\right]^{1/2} = 22.0 \text{ bar}$$

5.13 Many equilibrium constants in the literature were calculated with a standard state pressure of 1 atm (1.01325 bar). Show that the corresponding equlibrium constant with a standard pressure of 1 bar can be calculated using

$$K(\text{bar}) = K(\text{atm})(1.01325)^{\Sigma \nu_i}$$

where the ν_i are the stoichiometric numbers of the gaseous reactants.

SOLUTION

$$K(bar) = \prod_i [P_i/(1 \ bar)]^{\nu_i}$$

$$K(atm) = \prod_i [P_i/(1 \ atm)]^{\nu_i}$$

$$\frac{K(bar)}{K(atm)} = \prod_i [(\frac{1 \ atm}{1 \ bar})^{\nu_i} = (\frac{1 \ atm}{1 \ bar})^{\Sigma\nu_i}$$

$$= 1.01325^{\Sigma\nu_i}$$

5.14 Older tables of chemical thermodynamic properties are based on a standard state pressure of 1 atm. Show that the corresponding $\Delta_f G_i^o$ with a standard state pressure of 1 bar can be calculated using

$$\Delta_f G_j^o(bar) = \Delta_f G_j^o(atm) - (0.109 \times 10^{-3} \ kJ \ K^{-1} \ mol^{-1}) \ T \ \Sigma\nu_i$$

where the ν_i are the stoichiometric numbers of the gaseous reactants and products in the formation reaction.

SOLUTION

$$\Delta_f G_j^o(bar) = -RT \ln\{\prod_i [P_i/(1 \ atm)]^{\nu_i}\}$$

$$= -RT \ln\{\prod_i [P_i/(1 \ bar)]^{\nu_i}[(1 \ atm)/(1 \ bar)]^{\Sigma\nu_i}\}$$

$$= -RT \ln\{\prod_i [P_i/(1 \ atm)]^{\nu_i}\} - RT \ln[(1 \ atm)/(1 \ bar)]^{\Sigma\nu_i}$$

$$= \Delta_f G_j^o(atm) - RT\ln \ (1.01325)^{\Sigma\nu_i}$$

5.15 Show that the equilibrium mole fractions of n-butane and iso-butane are given by

$$y_n = \frac{e^{-\Delta_f G_n^o/RT}}{(e^{-\Delta_f G_n^o/RT} + e^{-\Delta_f G_{iso}^o/RT})}$$

$$y_{iso} = \frac{e^{-\Delta_f G_{iso}^o/RT}}{(e^{-\Delta_f G_n^o/RT} + e^{-\Delta_f G_{iso}^o/RT})}$$

SOLUTION

$4C(\text{graphite}) + 5H_2(g) = C_4H_{10}(g,n)$

$$K_n = P_n/P_{H_2}^5 = e^{-\Delta_f G_n^o/RT}$$

$4C(\text{graphite}) + 5\ H_2(g) = C_4H_{10}(g,iso)$

$$K_{iso} = P_{iso}/P_{H_2}^5 = e^{-\Delta_f G_{iso}^o/RT}$$

$$y_n = \frac{P_n}{P_n + P_{iso}} = \frac{P_{H_2}^5\ e^{-\Delta_f G_n^o/RT}}{P_{H_2}^5 e^{-\Delta_f G_n^o/RT} + P_{H_2}^5\ e^{-\Delta_f G_{iso}^o/RT}}$$

$$= \frac{e^{-\Delta_f G_n^o/RT}}{e^{-\Delta_f G_n^o/RT} + e^{-\Delta_f G_{iso}^o/RT}}$$

5.16 Calculate the molar Gibbs energy of butane isomers for extents of reaction of 0.2, 04, 0.6, and 0.8 for the reaction

n-butane = iso-butane

at 1000 K and 1 bar. At 1000 K,

$\Delta_f G^o$ (n-butane) = 279 kJ mol^{-1}

$\Delta_f G^o$ (iso-butane) = 276.6 kJ mol^{-1}

Make a plot and show that the minimum corresponds with the equilibrium extent of reaction.

SOLUTION

n-butane = iso-butane

1	0	moles at k = 0
1-ξ	ξ	moles at equilibrium; n_T= 1 mole

$$G = n_n\mu_n + n_i\mu_i = n_n\mu_n^o + n_nRT\ln(x_n) + n_i\mu_i^o + n_iRT\ln(x_i)$$

$$= (1 - \xi)\,\Delta_f G_n^o + (1 - \xi)RT\ \ln(1-\xi) + \xi\Delta_f G_i^o + \xi RT\ \ln(\xi)$$

$$= 270 + (276.6 - 270)\ \xi + RT[(1-\xi)\ln(1-\xi) + \xi\ln(\xi)]$$

$$G(\xi) = 270 + 6.6\xi + 8.314[(1-\xi)\ln(1-\xi) + \xi\ln(\xi)]\ \text{kJ mol}^{-1}$$

G(0.2) = 267.2 G(0.6) = 268.4

G(0.4) m= 267.0 G(0.8) = 271.1 in kJ/mol

$$\Delta G^o = \Delta_f G_i^o - \Delta_f G_n^o = 6.6\ \text{kJ mol}^{-1} = -RT\ln K$$

$$K = \frac{\xi}{1-\xi} = \exp\left[\frac{-\Delta_r G^o}{RT})\right] = 0.4521 \qquad (T = 10^3K)$$

$$\xi = 0.311$$

G(0.311) = 266.9 kJ/mol which corresponds to the minimum of the following graph of G(ξ)

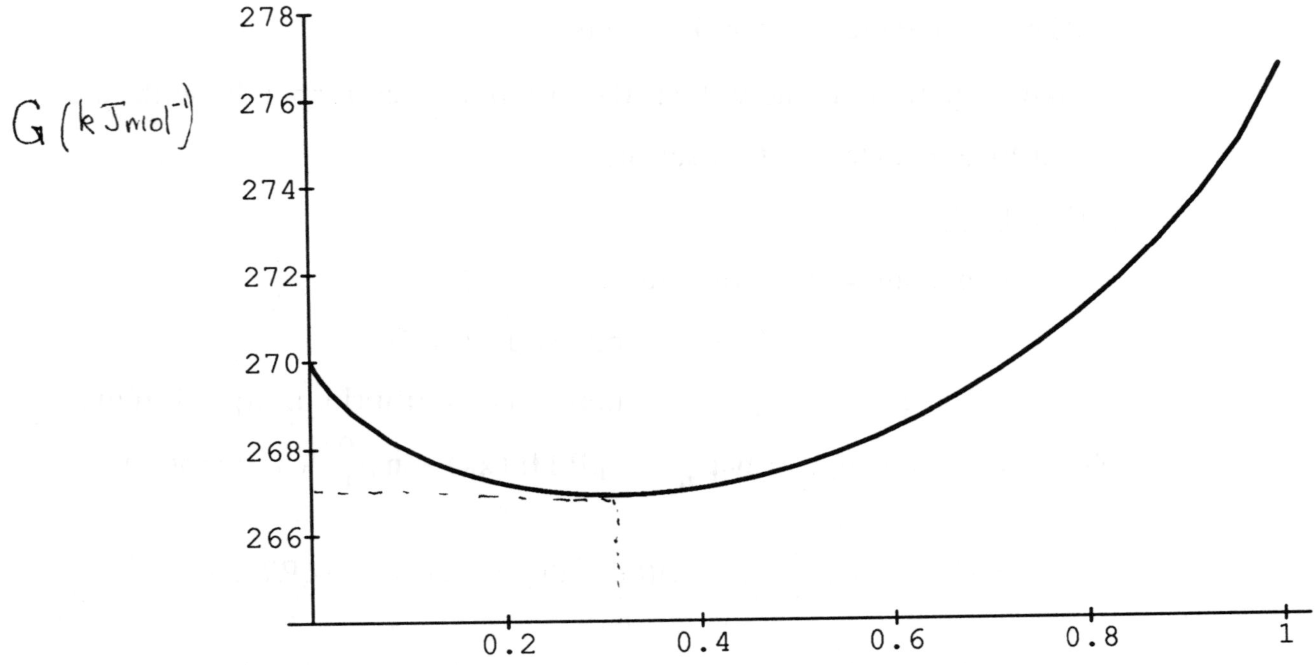

5.17 In the synthesis of methanol by $CO(g) + 2H_2(g) = CH_3OH(g)$ at 500 K, calculate the total pressure required for a 90% conversion to methanol if CO and H_2 are initially in a 1:2 ratio. Given: $K = 6.09 \times 10^{-3}$.

<u>SOLUTION</u>

	$CO(g)$	$+ 2H_2(g) =$	$CH_3OH(g)$	
initial moles	1	2	0	
equil. moles	0.1	0.2	0.9	Total 1.2

$$K = \frac{(P_{CH_3OH}/P^o)}{(P_{CO}/P^o)(P_{H_2}/P^o)^2} = 6.09 \times 10^{-3}$$

$$= \frac{\frac{0.9}{1.2}\frac{P}{P^o}}{\frac{0.1}{1.2}\frac{P}{P^o}(\frac{0.2}{1.2}\frac{P}{P^o})^2}$$

$$\frac{P}{P^o} = \sqrt{\frac{(0.9)(1.2)^2}{(0.1)(0.04)(6.09 \times 10^{-3})}} = 231$$

P = 231 bar = total pressure for 90% conversion to CH_3OH

5.18 At 1273 K and at a total pressure of 30.4 bar the equilibrium in the reaction $CO_2(g) + C(s) = 2CO(g)$ is such that 17 mole % of the gas is CO_2.

(a) What percentage would be CO_2 if the total pressure were 20.3 bar? (b) What would be the effect on the equilibrium of adding N_2 to the reaction mixture in a closed vessel until the partial pressure of N_2 is 10 bar? (c) At what pressure of the reactants will 25% of the gas be CO_2?

SOLUTION

(a) $P_{CO_2} = (30.4\ \text{bar})(0.17) = 5.2\ \text{bar}$

$P_{CO} = (30.4\ \text{bar})(0.83) = 25.2\ \text{bar}$

$$K = \frac{(24.2)^2}{5.2} = 122$$

Let x = mole fraction CO_2

$$K = \frac{[20.3(1 - x)]^2}{20.3x}) = 122 \qquad x = 0.127$$

Percentage CO_2 at equilibrium = 12.7%

(b) No effect for ideal gases because the partial pressures of the reactants are not affected.

(c) $$K = \frac{[0.75(P/P^o)]^2}{0.25(P/P^o)}$$

$$P = 54 \text{ bar}$$

5.19 When alkanes are heated up, they lose hydrogen and alkenes are produced. For example,

$$C_2H_6(g) = C_2H_4(g) + H_2(g) \quad K = 0.36 \text{ at } 1000 \text{ K}$$

If this is the only reaction that occurs when ethane is heated to 1000 K, at what total pressure will ethane be (a) 10% dissociated and (b) 90% dissociated to ethylene and hydrogen?

<u>SOLUTION</u>

C_2H_6	$= C_2H_4$	$+ H_2$	
1	0	0	
$1-\xi$	ξ	ξ	$tot = 1 + \xi$

$$K = \frac{\xi^2(P/P^o)^2}{(1+\xi)(1-\xi)} = \frac{\xi^2(P/P^o)^2}{1-\xi^2}$$

(a) $$0.36 = \frac{0.1^2(P/P^o)^2}{1-0.1^2} = \frac{0.01(P/P^o)^2}{0.99}$$

$$\frac{P}{P^o} = \left[\frac{(0.99)(0.36)}{0.01}\right]^{1/2} = 5.97$$

(b) $$0.36 = \frac{0.9^2(P/P^o)^2}{1-0.9^2} = \frac{0.81(P/P^o)}{0.19}$$

$$P = [\frac{(0.19)(0.36)}{0.81}]^{1/2} = 0.29 \text{ bar}$$

5.20 At 2000 °C water is 2% dissociated into oxygen and hydrogen at a total pressure of 1 bar.

(a) Calculate K for $H_2O(g) = H_2(g) + \frac{1}{2} O_2(g)$

(b) Will the extent of reaction increase or decrease if the pressure is reduced? (c) Will the extent of reaction increase or decrease if argon gas is added, holding the total pressure equal to 1 bar? (d) Will the extent of reaction change if the pressure is raised by addition of argon at constant volume to the closed system containing partially dissociated water vapor? (e) Will the extent of reaction increase or decrease if oxygen is added while holding the total pressure constant at 1 bar?

SOLUTION

(a)	$H_2O(g)$	=	$H_2(g)$	$+\frac{1}{2} O_2(g)$	
Initially	1		0	0	
Equilibrium	$1-\xi$		ξ	$\xi/2$	Total = $1 + \xi/2$

$$P_{H_2O} = \frac{1-\xi}{1+\frac{\xi}{2}} P \qquad P_{H_2} = \frac{\xi}{1+\frac{\xi}{2}} P \qquad P_{O_2} = \frac{\xi/2}{1+\frac{\xi}{2}} P$$

$$K = \frac{[\frac{\xi/2}{1+\xi/2}\frac{P}{P^o}]^{1/2}[\frac{\xi}{1+\xi/2}\frac{P}{P^o}]}{\frac{1-\xi}{1+\xi/2}\frac{P}{P_o}} = \frac{\xi^{3/2}(P/P^o)^{1/2}}{\sqrt{2}(1+\xi/2)^{1/2}(1-\xi)}$$

$$= \frac{0.02^{3/2}\ 1^{1/2}}{\sqrt{2}\ (1.01)^{1/2}\ (0.98)} = 2.03 \times 10^{-3}$$

(b) If the total pressure is reduced, the extent of reaction will increase because the reaction will produce more molecules to fill the volume.

(c) If argon is added at constant pressure, the extent of reaction will increase because the partial pressure due to the reactants will decrease.

(d) If argon is added at constant volume, the extent of reaction will not be changed because the partial pressure due to the reactants will not change.

(e) If oxygen is added at constant total pressure, the extent of reaction of H_2O will decrease because the reaction will be pushed to the left.

5.21 At 250 C PCl_5 is 80% dissociated at a pressure of 1.013 bar, and so K = 1.80. What is the extent of reaction at equilibrium after sufficient nitrogen has been added at constant pressure to produce a nitrogen partial pressure of 0.9 bar? The total pressure is maintained at 1 bar.

<u>SOLUTION</u>

$$K = \frac{0.8^2\ (1.013)}{1 - 0.8^2} = 1.80$$

$$K = \frac{\xi^2 (0.1)}{1 - \xi^2}$$

ξ = 0.973 or 97.3% dissociated.

5.22 The following exothermic reaction is at equilibrium at 500 K and 10 bar.

$$CO(g) + 2H_2(g) = CH_3OH(g)$$

Assuming the gases are ideal, what will happen to the amount of methanol at equilibrium when (a) the temperature is raised, (b) the pressure is increased, (c) an inert gas is pumped in at constant volume, (d) an inert gas is pumped in at constant pressure, and (e) hydrogen gas is added at constant pressure?

SOLUTION

(a) $n(CH_3OH)$ decreases because heat is involved.

(b) $n(CH_3OH)$ increases because there are fewer molecules of gas in the product.

(c) No effect

(d) $n(CH_3OH)$ decreases because the volume increases

(e) $n(CH_3OH)$ increases because there is more of a reactant. Note: the effect of the addition of CO is more complicated.)

5.24 The following reaction is nonspontaneous at room temperature and endothermic.

$$3C(graphite) + 2H_2O(g) = CH_4(g) + 2CO(g)$$

As the temperature is raised, the equilibrium constant will become equal to unity at some point. Estimate this temperature using data from Appendix C.2.

SOLUTION

At 1000 K

$\Delta_r G^o = 19.492 + 2(-200.275) - 2(-192.590) = 4.122$ kJ mol^{-1}

$= -(8.3145 \times 10^{-3}\ \text{kJ K}^{-1}\ \text{mol}^{-1})(1000\ \text{K})\ \ln K$

$K = 0.609$

$\Delta_r H^o = -89.849 + 2(-111.983) - 2(-247.857) = 181.899\ \text{kJ mol}^{-1})$

$$\ln \frac{K_2}{K_1} = \frac{\Delta H^o}{R}\left(\frac{1}{T_1} - \frac{1}{T_2}\right)$$

$$\ln \frac{1}{0.609} = \frac{181899\ \text{J mol}^{-1}}{8.3145\ \text{J K}^{-1}\ \text{mol}^{-1}}\left(\frac{1}{1000\ \text{K}} - \frac{1}{T_2}\right)$$

$$T_2 = \frac{1}{\frac{1}{1000\ \text{K}} - \frac{8.3145}{181899}\ln\frac{1}{0.609}} = 1023\ \text{K}$$

5.24 The measured density of an equilibrium mixture of N_2O_4 and NO_2 at 15 °C and 1.013 bar is 3.62 g L^{-1}, and the density at 75 °C and 1.013 bar is 1.84 g L^{-1}. What is the enthalpy change of the reaction $N_2O_4(g) = 2NO_2(g)$?

SOLUTION

At 15 °C

$$M = \frac{RT}{P}\frac{g}{V} = \frac{(0.08314)(288)(3.62)}{1.013} = 85.57\ \text{g mol}^{-1}$$

$$\xi = \frac{M_1 - M_2}{M_2} = \frac{92.01 - 85.57}{85.57} = 0.0753$$

$$K = \frac{4\xi^2 P}{1 - \xi^2} = \frac{4(0.0753)^2(1.013)}{1 - 0.0753^2} = 0.0231$$

At 75 °C

$$M = \frac{(0.08314)(348)(1.84)}{1.013} = 52.55 \text{ g mol}^{-1}$$

$$\xi = \frac{92.01 - 52.55}{52.55} = 0.751 \qquad K = 5.24$$

$$\Delta_r H = \frac{RT_1T_2}{(T_2 - T_1)} \ln \frac{K_2}{K_1} = \frac{(8.314)(288)(348)}{60} \ln \frac{5.24}{0.0231}$$

$$= 75 \text{ kJ mol}^{-1}$$

5.25 Calculate K_c for the reaction problem 5.19 at 1000 K and describe what it is equal to.

SOLUTION

$$K_c = K_P\left(\frac{P^o}{c^oRT}\right)^{\Sigma\nu_i}$$

$$= 0.36 \frac{1 \text{ bar}}{(1 \text{ L mol}^{-1})(0.08314 \text{ L bar K}^{-1} \text{ mol}^{-1})(238 \text{ K})}$$

$$= 0.0145$$

$$= \frac{\left(\frac{[C_2H_4]}{c^o}\right)\left(\frac{[H_2]}{c^o}\right)}{\frac{[C_2H_6]}{c^o}}$$

where [ ] represents concentrations in moles per liter and $c° = 1$ mol L^{-1}.

5.26 The equilibrium constant for the reaction

$$N_2(g) + 3H_2(g) = 2NH_3(g)$$

is 35.0 at 400 K when partial pressures are expressed in bars. Assume the gases are ideal. (a) What is the equilibrium volume when 0.25 mol N_2 is mixed with 0.75 mol H_2 at a temperature of 400 K and a pressure of 1 bar? (b) What is the equilibrium composition and equilibrium pressure if this mixture is held at a constant volume of 33.26 L at 400 K?

SOLUTION

(a)	N_2	+	$3H_2$	$= 2NH_3$	
initial	0.25		0.75	0	
equil.	$0.25 - \xi$		$0.75 - 3\xi$	2ξ	total= $1-2\xi$

$K_p = (2\xi)^2(1 - 2\xi)^2/(0.25 - \xi)(0.75 - 3\xi)^3 = 35.0$

The method of successive approximations or the use of an equation solver yields $\xi = 0.1652$.

$P(N_2)/P^o = (0.25 - \xi)/(1 - 2\xi) = 0.1266$

$P(H_2)/P^o = 3(0.1266) = 0.3798$

$P(NH_3)/P^o = 2\xi/(1 - 2\xi) = 0.493$

Since the total pressure is 1 bar, these numbers also give the equilibrium mole fractions.

$$V = (1 - 2\xi)RT/P$$

$$= (0.6696 \text{ mol})(0.8314 \text{ L bar K}^{-1} \text{ mol}^{-1})(400 \text{ K})/(1 \text{ bar})$$

$$= 22.27 \text{ L}$$

(b) In order to calculate the equilibrium composition at constant volume, it is convenient to use K_c.

$$K_c = (P^o/c^oRT)^{\Sigma\nu_i} K_P$$

$$= \left[\frac{1 \text{ bar}}{(1 \text{ mol L}^{-1})(0.08314 \text{ L bar K}^{-1} \text{ mol}^{-1})(400 \text{ K})}\right]^{-2} 35.0$$

$$= 3.871 \times 10^4$$

$$K_c = [c(NH_3)/c^o]^3/[c(N_2)/c^o][c(H_2)/c^o]^3$$

$$= (2\xi)^2\ 33.16^2/(0.25 - \xi)(0.75 - 3\xi)^3$$

$$= 3.871 \times 10^4$$

The method of successive approximation or the use of an equation solver yields $\xi = 0.151$.

$n(N_2) = 0.25 - \xi = 0.0990$

$n(H_2) = 0.75 - 3\xi = 0.2970$

$n(NH_3) = 2\xi = 0.302$

$n(\text{total}) = 0.698$

$$P = \frac{(0.698 \text{ mol})RT}{33.26 \text{ L}} = 0.698 \text{ bar}$$

5.27 Show that to a first approximation the equation of state of a gas that dimerizes to a small extent is given by

$$\frac{P\bar{V}}{RT} = 1 - \frac{K_c}{\bar{V}}$$

SOLUTION

$2A(g) = A_2(g)$

$$K_c = \frac{[A_2]}{[A]^2} = \frac{n_{A_2}V}{n_A^2}$$

$$n_o = \text{Total number of moles of A} = n_A + 2n_{A_2} = n_A + \frac{2K_c n_A^2}{\bar{V}}$$

$$= n_A[1 + 2K_c/\bar{V}]$$

$$\frac{PV}{RT} = n_A + n_{A_2} = n_A + \frac{K_c n_A^2}{\bar{V}} = n_A[1 + \frac{2K_c}{\bar{V}}]$$

$$= \frac{n_o}{1 + \frac{2K_c}{\bar{V}}}[1 + \frac{K_c}{\bar{V}}] = n_o[1 - \frac{K_c}{\bar{V}}]$$

5.28 Water vapor is passed over coal (assumed to be pure graphite in this problem) at 1000 K. Assuming that the only reaction occurring is the water gas reaction

$$C(\text{graphite}) + H_2O(g) = CO(g) + H_2(g) \qquad K = 2.52$$

calculate the equilibrium pressures of H_2O, CO, and H_2 at a total pressure of 1 bar. (Actually the water gas shift reaction

$$CO(g) + H_2O(g) = CO_2(g) + H_2(g)$$

occurs in addition, but it is considerably more complicated to take this additional reaction into account.)

SOLUTION

$$K = \frac{(P_{CO}/P^o)(P_{H_2}/P^o)}{(P_{H_2O}/P^o)} = \frac{x^2}{y} = \frac{x^2}{1-2x} = 2.52$$

$2x + y = 1 \qquad x^2 = 2.52 - 5.04x$

$x^2 + 5.04x - 2.52 = 0$

$$x = \frac{-5.04 \pm \sqrt{5.04^2 + 4(2.52)}}{2} = 0.458 = \frac{P_{CO}}{P^o} = \frac{P_{H_2}}{P^o}$$

$$1 - 2x = 0.084 = \frac{P_{H_2O}}{P^o}$$

$P_{H_2O} = 0.084$ bar $\qquad P_{CO} = 0.458$ bar $\qquad P_{H_2} = 0.458$ bar

5.29 Mercuric oxide dissociates according to the reaction $2HgO(s) = 2Hg(g) + O_2(g)$. At 420 °C the dissociation pressure is 5.16×10^4 Pa, and at 450 °C it is 10.8×10^4 Pa. Calculate (a) the equilibrium constants, and (b) the enthalpy of dissociation per mole of HgO.

<u>SOLUTION</u>

(a) $P_{Hg} = 2P_{O_2}$ $\qquad P_{Hg} = \frac{2}{3}P \qquad P_{O_2} = \frac{1}{3}P$

$$K_{420} = P_{Hg}^2 P_{O_2} = (\frac{2}{3})^2 (\frac{1}{3})P^3$$

$$= (\frac{4}{27}) (\frac{5.16 \times 10^4 \text{ Pa}}{1.013 \times 10^5 \text{ Pa}})^3 = 0.0196$$

$$K_{450} = (\frac{4}{27}) (\frac{10.8 \times 10^4 \text{ Pa}}{1.013 \times 10^5 \text{ Pa}})^3 = 0.1794$$

(b) $$\Delta_r H^o = \frac{RT_1T_2}{T_2 - R_1} \ln \frac{K_2}{K_1}$$

$$= \frac{(8.314 \text{ J K}^{-1} \text{ mol}^{-1})(693 \text{ K})(723 \text{ K})}{30\text{K}} \ln \frac{0.1794}{0.0196}$$

$= 308$ kJ mol^{-1} for the reaction as written

$= 154$ kJ mol^{-1} of HgO(s)

5.30 The decomposition of silver oxide is represented by

$$2Ag_2O(s) = 4Ag(s) + O_2(g)$$

Using data from Appendix C.1 and assuming $\Delta_r C_P^o = 0$ calculate the temperature at which the equilibrium pressure of O_2 is 0.2 bar. This temperature is of interest because Ag_2O will decompose to yield Ag at temperatures above this value if it is in contact with air.

SOLUTION

$\Delta_r H^o = -2(-31.05) = 62.10 \text{ kJ mol}^{-1}$

$\Delta_r S^o = 4(42.55) + 205.138 - 2(121.3) = 132.7 \text{ J K}^{-1} \text{ mol}^{-1}$

$\Delta_r G = \Delta G^o + RT \ln P_{O2}$

$O = \Delta H^o - T\Delta S^o + RT \ln 0.2$

$\Delta_r H^o = T(\Delta S^o - R \ln 0.2)$

$$T = \frac{\Delta_r H^o}{\Delta_r S^o - R \ln 0.2} = \frac{62100 \text{ J mol}^{-1}}{(132.7 - 8.314 \ln 0.2) \text{ J K}^{-1} \text{ mol}^{-1}}$$

$= 425 \text{ K}$

5.31 The dissociation of ammonium carbamate takes place according to the reaction

$$(NH_2)CO(ONH_4)(s) = 2NH_3(g) + CO_2(g)$$

When an excess of ammonium carbamate is placed in a previously evacuated vessel, the partial pressure generated by NH_3 is twice the partial pressure of the CO_2, and the partial pressure of $(NH_2)CO(ONH_4)$ is negligible in comparison. Show that

$$K = \left(\frac{P_{NH_3}}{P^o}\right)^2 \left(\frac{P_{CO_2}}{P^o}\right) = \frac{4}{27}\left(\frac{P}{P^o}\right)^3$$ where P is the total pressure.

<u>SOLUTION</u>

$P \quad = P_{NH_3} + P_{CO_2} = 3P_{CO_2}$ since $P_{NH_3} = 2P_{CO_2}$

$P_{CO_2} = \frac{P}{3} \qquad P_{NH_3} = \frac{2}{3}P$

$$K = \left(\frac{P_{NH_3}}{P^o}\right)^2 \left(\frac{P_{CO_2}}{P^o}\right) = \left(\frac{2}{3}\frac{P}{P^o}\right)^2 \left(\frac{1}{3}\frac{P}{P^o}\right) = \frac{4}{27}\left(\frac{P}{P^o}\right)^3$$

5.32 At 1000 K methane at 1 bar is in the presence of hydrogen. In the presence of a sufficiently high partial pressure of hydrogen, methane does not decompose to form graphite and hydrogen. What is this partial pressure?

<u>SOLUTION</u>

$CH_4(g) \quad = C(\text{graphite}) + 2H_2(g)$

$\Delta G^o \quad = -RT \ln K = -19.46 \text{ kJ mol}^{-1}$

$$K \quad = 10.39 = \frac{\left(\frac{P_{H_2}}{P^o}\right)^2}{\frac{P_{CH_4}}{P^o}}$$

$P_{H_2} \quad = P^o[(10.39)(1)]^{1/2} = 3.2 \text{ bar}$

5.33 For the reaction

$$Fe_2O_3(s) + 3CO(g) = 2Fe(s) + 3CO_2(g)$$

the following values of K are known.

t/°C	250	1000
K	100	0.0721

At 1120 °C for the reaction $2CO_2(g) = 2CO(g) + O_2(g)$, K = 1.4 x 10^{-12}. What equilibrium partial pressure of O_2 would have to be supplied to a vessel at 1120 °C containing solid Fe_2O_3 just to prevent the formation of Fe?

<u>SOLUTION</u>

$$\ln\left(\frac{K_2}{K_1}\right) = \frac{\Delta_r H^o(T_2 - T_1)}{RT_1T_2}$$

$$\Delta_r H^o = \frac{(8.314)(523)(1273)\ \ln\ (0.0721/100)}{(750)}$$

$$= -53.4\ \text{k J mol}^{-1}$$

$$\ln\left(\frac{K_{1393}}{K_{1273}}\right) = \frac{(53400)(120)}{8.314\ (1393)(1273)}$$

$$= -0.4346$$

$$K_{1393} = (0.0721)\ \exp(-0.4346)$$

$$= 0.0467 = \frac{P^3_{CO_2}}{P^3_{CO}}$$

For $2CO_2(g) = 2CO(g) + O_2$

$$K_{II,1393} = \frac{(P_{CO}/P^o)^2(P_{O_2}/P^o)}{(P_{CO}/P^o)^2}) = 1.4 \times 10^{-12}$$

$$\frac{P_{O_2}}{P^o} = \frac{(1.4 \times 10^{-12})(P_{CO_2}/P^o)^2}{(P_{CO}/P^o)}$$

$$= (1.4 \times 10^{-12})(0.0467)^{2/3}$$

$$= 1.82 \times 10^{-13}$$

$$P_{O_2} = 1.82 \times 10^{-13} \text{ bar}$$

5.34 Calculate the partial pressure of $CO_2(g)$ over $CaCO_3$(calcite) - CaO(s) at 500 °C using equation 5.45 and data in Appendix C.1.

<u>SOLUTION</u>

$$CaCO_3(\text{calcite}) = CaO(s) + CO_2(g)$$

$\Delta_r H^o = -635.09 + (-393.51) - (-1206.92) = 178.32 \text{ kJ mol}^{-1}$

$\Delta_r C_P^o = 42.80 + 37.11 - 81.88 = -1.97 \text{ J K}^{-1} \text{ mol}^{-1}$

$\Delta_r S^o = 39.75 + 213.74 - 92.9 = 160.6 \text{ J K}^{-1} \text{ mol}^{-1}$

Substituting in equation 5.45,

$$\ln K_{773} = -\frac{178320}{(8.314)(773.15)} + \frac{160.6}{8.314} + \frac{1.97}{8.314}\left(1 - \frac{298.15}{773.15} - \ln \frac{773.19}{298.15}\right)$$

$$K_{773} = \frac{P_{CO_2}}{P^o} = 20 \times 10^{-5}$$

Table 5.1 gives $\frac{P_{CO_2}}{P^o} = 9.2 \times 10^{-5}$ bar

5.35 The NBS Tables contain the following data at 298 K:

	$\Delta_f H^o$/kJ mol^{-1}	$\Delta_f G^o$/kJ mol^{-1}
$CuSO_4(s)$	-771.36	-661.8
$CuSO_4{\cdot}H_2O(s)$	-1085.83	-918.11
$CuSO_4{\cdot}3H_2O(s)$	-1684.31	-1399.96
$H_2O(g)$	-241.818	-228.572

(a) What is the equilibrium partial pressure of H_2O over a mixture of $CuSO_4(s)$ and $CuSO_4{\cdot}H_2O(s)$ at 25 °C?

(b) What is the equilibrium partial pressure of H_2O over a mixture of $CuSO_4{\cdot}H_2O(s)$ and $CuSO_4{\cdot}3H_2O(s)$ at 25 °C?

(c) What are the answers to (a) and (b) if the temperature is 100 °C and ΔC_P^o is assumed to be zero?

SOLUTION

(a) $CuSO_4{\cdot}H_2O(s) = CuSO_4(s) + H_2O(g)$

$$\Delta G^o = -228.572 - 661.8 + 918.11$$

$$= 27.7 \text{ kJ mol}^{-1}$$

$$K = \exp\left(\frac{-\Delta G^o}{RT}\right) = 1.4 \times 10^{-5} = \frac{P_{H_2O}}{P^o}$$

$$P_{H_2O} = 1.4 \times 10^{-5} \text{ bar}$$

(b) $CuSO_4{\cdot}3H_2O(s) = CuSO_4{\cdot}H_2O(s) + 2H_2O(g)$

$$\Delta G^o = 2(-228.572) - 918.11 + 1399.46$$

$$= 24.71 \text{ kJ mol}^{-1}$$

$$P_{H_2O} = \exp\left[\frac{-24710}{(2)(8.314)(298)}\right]$$

$$= 6.83 \times 10^{-3} \text{ bar}$$

(c) for the first reaction

$$\ln \frac{K_{100}}{1.4 \times 10^{-5}} = \frac{72650(75)}{8.314(298)(373)}$$

$$\Delta H^\circ = -241.818 - 771.36 + 1085.83 = 72.65 \text{ kJ mol}^{-1}$$

$$\ln \frac{K_{100}}{1.4 \times 10^{-5}} = 5.896$$

$$K_{100} = 363.6(1.4 \times 10^{-5}) = 0.0051 \text{ bar}$$

for the second reaction

$$\Delta H^\circ = 2(-241.818) - 1085.83 + 1684.31$$

$$= 114.84 \text{ kJ mol}^{-1}$$

$$\ln \frac{K_{100}}{4.7 \times 10^{-5}} = \frac{114840(75)}{8.314(298)(373)} = 9.32$$

$$K_{100} = 0.52 \text{ bar}^2 = P_{H_2O}^2$$

$$P_{H_2O} = 0.72 \text{ bar}$$

5.36 One micromole of CuO(s) and 0.1 μmole of Cu(s) are placed in a 1-L container at 1000 K. Determine the identity and quantity of each phase present at equilibrium if $\Delta_f G^\circ$ of CuO is -66.66 kJ mol^{-1} and of Cu_2O is -77.94 kJ mol^{-1} at 1000 K. (From H. F. Franzen, *J. Chem. Ed.* **65**, 146 (1988).)

SOLUTION

$$Cu(s) + CuO(s) = Cu_2O(s)$$

$$\Delta_r G^\circ = -77.94 - (-66.66) = -11.28 \text{ kJ mol}^{-1}$$

Therefore, this reaction goes to completion to the right. The two solids are in equilibrium with $O_2(g)$.

$$2CuO(s) \quad = Cu_2O(s) + \frac{1}{2} O_2(g)$$

$$\Delta_r G^o = -77.94 - 2(-66.66) = 55.38 \text{ kJ mol}^{-1}$$

$$= -RT\ln P_{O_2}^{1/2}$$

$$P_{O_2} = e^{-(2 \times 55.38/(8.314 \times 10^{-3})(1000))}$$

$$= 1.64 \times 10^{-6}$$

The amount of $O_2(g)$ at equilibrium is

$$n_{O_2} = \frac{PV}{RT} = \frac{(1.64 \times 10^{-6})(1)}{(0.08314)(1000)}$$

$$= 1.97 \times 10^{-8} \text{ mol}$$

Thus the amounts at equilibrium are essentially

$$n_{CuO} = 0.9 - 4(0.02) = 0.82\ \mu\text{mol}$$

$$n_{Cu_2} = 0.1 + 2(0.02) = 0.14\ \mu\text{mol}$$

$$n_{O_2} = 0.02\ \mu\text{mol}$$

5.37 Acetic acid and ethanol react to form ethyl acetate and water according to

$$CH_3CO_2H(1) + C_2H_5OH(1) = CH_3CO_2H_5(1) + H_2O(1)$$

Analytical data obtained by titrating the acetic acid remaining at equilibrium at 298 K indicates that K = 40 when mole fractions are used, and the solution is assumed to be ideal. The standard Gibbs energies of formation of ethanol, acetic acid, and water are given in Appendix C.1. Calculate the standard Gibbs energy of formation of ethyl acetate.

SOLUTION

$\Delta_r G^o \quad = -RT \ln K = -(8.314)(298) \ln 4.0 = -3.44 \text{ kJ mol}^{-1}$

$\quad\quad = \Delta_f G^o_{EtAc} - 228.57 + 389.9 + 174.78$

$\Delta_f G^o_{EtAc} \quad = -339.6 \text{ kJ mol}^{-1}$

5.38 Amylene, C_5H_{10}, and acetic acid react to give the ester according to the reaction

$$C_5H_{10}(l) + CH_3COOH(l) + CH_3COOC_5H_{11}(l)$$

What is the value of K_c if 0.006 45 mol of amylene and 0.001 mol of acetic acid, dissolved in a certain inert solvent to give a total volume of 0.845 L, react to give 0.000 784 mol of ester? Use the molar concentration scale.

SOLUTION

$$K = \frac{\frac{(7.84 \times 10^{-4})}{0.845}}{[\frac{(64.5 \times 10^{-4} - 7.84 \times 10^{-4})}{0.845}][\frac{(10 \times 10^{-4} - 7.84 \times 10^{-4})}{0.845}]}$$

$= 541$

5.39 (a) A system contains $CO(g)$, $CO_2(g)$, $H_2(g)$, and $H_2O(g)$. How many chemical reactions are required to describe chemical changes in this system? Give an example. (b) If solid carbon is present in the system in addition, how many independent chemical reactions are there. Give a suitable set.

SOLUTION

(a)

	CO	CO_2	H_2	H_2O
C	1	1	0	0
O	1	2	0	2
H	0	0	2	2

To perform a Gaussian elimination, subtract the first row from the second, and divide the third row by 2.

1	1	0	0
0	1	0	1
0	0	1	1

Subtract the second row from the first.

1	0	0	- 1
0	1	0	1
0	0	1	1

The rank of the **A** matrix is 3, and so the number of independent reactions is

$$R = S - \text{rank } \mathbf{A} = 4 - 3 = 1$$

where S is the number of species.

The stoichiometric numbers for a suitable reaction is obtained by changing the sign of the numbers in the last column, and extending the vector with a 1; that is, 1, -1, -1, 1. These are the stoichiometric numbers for the species across the top of **A** .

$$CO - CO_2 - H_2 + H_2O = O$$

$$CO_2 + H_2 = CO + H_2O$$

(b)

	CO	CO_2	H_2	H_2O	C
C	1	1	0	0	1
O	1	2	0	1	0
H	0	0	2	2	0

Subtract the first row from the second and divide the third row by 2.

1	1	0	0	1
0	1	0	1	-1
0	0	1	1	0

Subtract the second row from the first

1	0	0	-1	2
0	1	0	1	-1
0	0	1	1	0

Rank **A** = 3 and R = S - rank **A** = 5 - 3 = 2. To obtain a suitable set of reactions, change the signs in the last two columns and put an identity matrix at the bottom.

1	-2
-1	1
-1	0
1	0
0	1

These two independent reactions are

$$CO - CO_2 - H_2 + H_2O = 0$$
$$-2CO + CO_2 + C = 0$$

or

$$CO_2 + H_2 = CO + H_2O$$
$$2CO = CO_2 + C$$

5.40 For a closed system containing C_2H_2, H_2, C_6H_6, and $C_{10}H_8$, use a Gaussian elimination to obtain a set of independent chemical reactions. Perform the matrix multiplication to verify $\mathbf{A}\nu = \mathbf{0}$.

SOLUTION

	C_2H_2	H_2	C_6H_6	$C_{10}H_8$
C	2	0	6	10
H	2	2	6	8
	1	0	3	5
	1	1	3	4
	1	0	3	5
	0	1	0	-1

$$\nu = \begin{matrix} C_2H_2 \\ H_2 \\ C_6H_6 \\ C_{10}H_8 \end{matrix} \begin{bmatrix} -3 & -5 \\ 0 & 1 \\ 1 & 0 \\ 0 & 1 \end{bmatrix}$$

$3C_2H_2 = C_6H_6$

$5C_2H_2 = C_{10}H_8 + H_2$

$$\begin{bmatrix} 2 & 0 & 6 & 10 \\ 2 & 2 & 6 & 8 \end{bmatrix} \begin{bmatrix} -3 & -5 \\ 0 & 1 \\ 1 & 0 \\ 0 & 1 \end{bmatrix} = \begin{bmatrix} 0 & 0 \\ 0 & 0 \end{bmatrix}$$

5.41 17.1 kJ mol^{-1}

5.42 0.696

5.43 0.0166

5.44 0.351 bar

5.45 3.74 bar

5.46 0.803, 1.84

5.47 $K = (2\xi)^2(4 - 2\xi)^2/(1 - \xi)(3 - 2\xi)^3(P/P^o)^2$

5.48 26.3

5.49 (a) 3.81 x 10^{-2}, (b) 0.348

5.50 (a) 0.0788 (b) 0.0565

5.51 0.0273, 0.0861

5.52 0.465, 0.494, 0.041. The pressure has no effect.

5.54 (a) 16.69 kJ mol^{-1} (b) 0.787 bar

5.55 1.84 x 10^6

5.56 30.1 bar

5.57 (a) 0.5000, 0.4363, 0.0637

(b) 0.5481, 0.3946, 0.0574

(c) When additional N_2 is added, the equilibrium shifts so that the mole fraction of N_2 is reduced below what it otherwise would have been.

5.58 (a) Yield of CH_4 will increse. (b) Yield of CH_4 willdecrease.

(c) Mole fraction CH_4 computed without including N_2 will increase.

5.59 (a) NO (b) 0.286

5.60 (a) 0.0050 bar (b) 0.0220 bar

5.62 199.2 kJ mol^{-1}

5.63 (a) 0.1852, 0.3775, 0.6284 (b) 60.7 kJ mol^{-1} (c) 0.320 (d) 0.371

5.64 K_P = 0.0024, K_c = 4.15

5.65 Hydrogen dissolves as atoms.

5.66 Fe_2O_3

5.67 7.56 bar

5.68 16.06 kJ mol^{-1}

5.69 (a) 6.66 x 10^{-3} (b) 12.1 kJ mol^{-1}

5.70 (a) 57.3 kJ mol^{-1} (b) 5.82 kJ mol^{-1} (c) 154.7 J K^{-1} mol^{-1}

5.71 -14.8 kJ mol^{-1}

5.72 $K_1 = 2.5$, $K_2 = 0.61$, $K_3 = 0.35$, $y_{CH_4} = 0.27$, $y_{CO} = 0.27$, $y_{H_2O} = 0.46$

5.73 Two, for example

$2C_2H_6 = 2CH_4 + C_2H_4$

$3C_2H_6 = 3CH_4 + C_3H_6$

CHAPTER 6 Phase Equilibrium: One Component Systems

6.1 Derive

$$\bar{C}_P = -T\left(\frac{\partial^2 \mu}{\partial T^2}\right)_P$$

from equation 4.40,

$$H = G - T\left(\frac{\partial G}{\partial T}\right)_P$$

<u>SOLUTION</u>

For one mole of substance

$$\bar{H} = \mu - T\left(\frac{\partial \mu}{\partial T}\right)_P$$

$$\bar{C}_P = \left(\frac{\partial \bar{H}}{\partial T}\right)_P = \left(\frac{\partial \mu}{\partial T}\right)_P - T\left(\frac{\partial^2 \mu}{\partial T^2}\right)_P - \left(\frac{\partial \mu}{\partial T}\right)_P$$

$$= -T\left(\frac{\partial^2 \mu}{\partial T^2}\right)_P$$

6.2 The boiling point of hexane at 1 atm is 68.7 °C. What is the boiling point at 1 bar? Given: The vapor pressure of hexane at 49.6 °C is 53.32 k Pa.

<u>SOLUTION</u>

$$\ln \frac{P_2}{P_1} = \frac{\Delta_{vap}H(T_2 - T_1)}{RT_1T_2}$$

$$\Delta_{vap}H = \frac{RT_1T_2}{(T_2 - T_1)} \ln\left(\frac{P_2}{P_1}\right)$$

$$= \frac{(8.3145)(322.8)(341.9)}{19.1} \ln\left(\frac{101.325 \text{ k Pa}}{53.32 \text{ k Pa}}\right)$$

$$= 30{,}850 \text{ J mol}^{-1}$$

$$\ln\left(\frac{101.325}{100}\right) = \frac{30850}{8.3145}\left(\frac{1}{T_1} - \frac{1}{341.9}\right)$$

$$T_1 = \frac{1}{\frac{8.3145}{30850}\ln\left(\frac{101.325}{100}\right) + \frac{1}{341.9}} = 344.5\ \text{K}$$

Thus the boiling point is reduced 0.4 °C to 68.3 °C.

6.3 What is the boiling point of water 2 miles above sea level? Assume the atmosphere follows the barometric formula (equation 1.65) with $M = 0.0289\ \text{kg mol}^{-1}$ and $T = 300$ K. Assume the enthalpy of vaporization of water is 44.0 kJ mol^{-1} independent of temperature.

SOLUTION

h = (2 miles)(5280 ft/mile)(12 in/ft)(2.54 cm/in)(0.01 m/cm) = 3219 m

$$P = P_o \exp(-gMh/RT)$$

$$= (1.01325\ \text{bar})\exp\left[-\frac{(9.8\ \text{ms}^{-2})(0.0289\ \text{kg mol}^{-1})(3219\ \text{m})}{(8.314\ \text{J K}^{-1}\ \text{mol}^{-1})(300\ \text{K})}\right]$$

$$= 0.703\ \text{bar}$$

$$\ln\left(\frac{P_2}{P_1}\right) = \Delta_{vap}H(T_2 - T_1)/RT_2T_2$$

$$\ln\left(\frac{1.01325}{0.703}\right) = \frac{44000(373 - T_1)}{(8.314)(373)T_1}$$

$$\frac{373 - T_1}{T_1} = 2.576 \times 10^{-2}$$

$$T_1 = 364\ \text{K} = 90.6\ °\text{C}$$

6.4 Liquid mercury has a density of 13.690 g cm^{-3}, and solid mercury has a density of 14.193 g cm^{-3}, both being measured at the melting point, -38.87 °C, at 1 bar pressure. The heat of fusion is 9.75 J g^{-1}. Calculate the melting points of mercury under a pressure of (a) 10 bar, and (b) 3540 bar. The observed melting point under 3540 bar is -19.9 °C.

SOLUTION

$$\frac{\Delta T}{\Delta P} = \frac{T(V_l - V_s)}{\Delta_{fus}H} = \left[\frac{(234.3\ K)(\frac{1}{13.690} - \frac{1}{14.193})\ cm^3\ g^{-1}}{9.75\ J\ g^{-1}}\right](10^{-2}\ m\ cm^{-1})^3$$

$= 6.22 \times 10^{-8}$ K Pa^{-1}

(a) $\Delta T = (6.22 \times 10^{-8}$ K $Pa^{-1})(9 \times 101{,}325$ Pa).

$= 0.06$ K

t $= -38.87 + 0.06 = -38.81$ °C

(b) $\Delta T = (6.22 \times 10^{-8}$ K $Pa^{-1})(3539 \times 101{,}325$ Pa)

$= 22.3$ K

t $= -38.87 + 22.3 = -16.6$ °C

6.5 From the $\Delta_f G^o$ of $Br_2(g)$ at 25 °C, calculate the vapor pressure of $Br_2(l)$. The pure liquid at 1 bar and 25 °C is taken as the standard state.

SOLUTION

$Br_2(l) = Br_2(g) \qquad K = \frac{P_{Br_2}}{P^o}$

$\Delta G^o = -RT \ln \left(\frac{P_{Br_2}}{P^o}\right)$

$$\frac{P_{Br_2}}{P^o} = \exp\left[\frac{-3110\ J\ mol^{-1}}{(8.314\ J\ K^{-1}\ mol^{-1})(298.15\ K)}\right]$$

$P_{Br_2} = 0.258$ bar

6.6 Calculate ΔG^o for the vaporization of water at 0 °C using data in Appendix C.1 and assuming that ΔH^o for the vaporization is independent of temperature. Use ΔG^o to calculate the vapor pressure of water at 0 °C.

SOLUTION

$H_2O(\ell) = H_2O(g)$

$\Delta G^o(287.15\ K) = -227.572 + 237.129 = 8.577\ kJ\ mol^{-1}$

$\Delta H^o(298.15\ K) = -241.818 + 285.830 = 44.012\ kJ\ mol^{-1}$

Using the equation in problem 4.8(a)

$$\Delta G_2 = \frac{\Delta G_1 T_2}{T_1} + \Delta H[(1 - \frac{T_2}{T_1})]$$

$$\Delta G(0\ °C) = 8.577(\frac{273.15}{298.15}) + 44.012[(1 - (\frac{273.15}{298.15})]$$

$$= 7.858 + 3.690$$

$$= 11.548\ kJ\ mol^{-1}$$

$$= -RT \ln(\frac{P}{P_o})$$

$$P = P_o \exp(\frac{-11.548}{8.3145 \times 10^{-3} \times 273.15})$$

$$= 6.190 \times 10^{-3}\ bar$$

6.7 The change in Gibbs energy for the conversion of aragonite to calcite at 25 °C is -1046 J mol^{-1}. The density of aragonite is 2.93 g cm^{-3} at 15 °C and the density of calcite is 2.71 g cm^{-3}. At what pressure at 25 °C would these two forms of $CaCO_3$ be in equlibrium?

SOLUTION

aragonite = calcite $\qquad \Delta G^o = -1046\ J\ mol^{-1}$

$$\Delta\bar{V} = \frac{100.09\ g\ mol^{-1}}{2.71\ g\ cm^{-3}} - \frac{100.09\ g\ mol^{-1}}{2.73\ g\ cm^{-3}}$$

$$= 2.77\ cm^3\ mol^{-1} = 2.77 \times 10^{-6}\ m^3\ mol^{-1}$$

$$\left(\frac{\partial \Delta G}{\Delta P}\right)_T = \Delta\bar{V} \quad \int_1^2 d\Delta G = \int_1^P \Delta\bar{V} dP$$

$\Delta G_2 - \Delta G_1 = \Delta\bar{V}(P - 1)$

$0 + 1046\ J\ mol^{-1} = (2.77 \times 10^{-6}\ m^3\ mol^{-1})(P - 1)$

$P = 3780$ bar

6.8 n-Propyl alcohol has the following vapor pressures.

t/°C	40	60	80	100
P/kPa	6.69	19.6	50.1	112.3

Plot these data so as to obtain a nearly straight line, and calculate (a) the enthalpy of vaporization, (b) the boiling point at 1 bar, and (c) the boiling point at 1 atm.

SOLUTION: log P is plotted versus 1/T

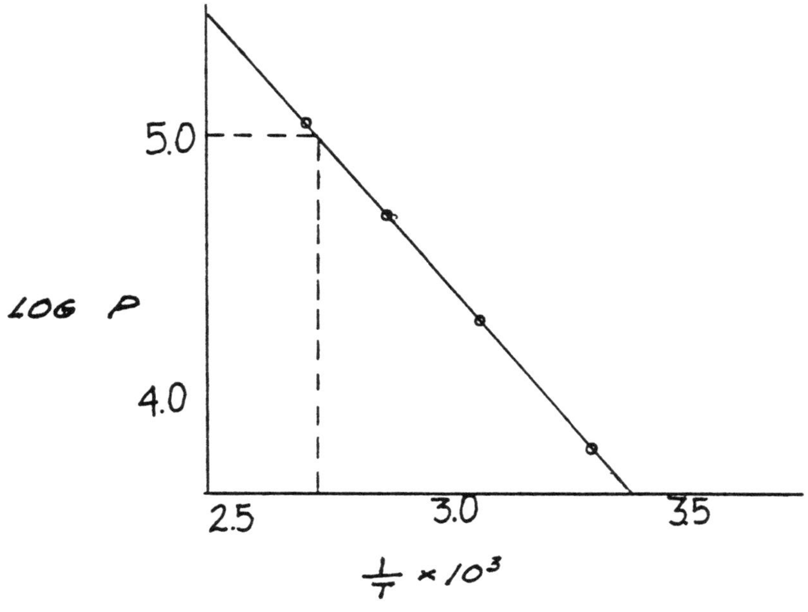

(a) slope $= -2.34 \times 10^3\ \text{K} = \dfrac{\Delta_{vap}H}{2.303\ (8.314\ \text{J K}^{-1}\ \text{mol}^{-1})}$

$\Delta_{vap}H = 44.8\ \text{kJ mol}^{-1}$

(b) $\ln\dfrac{112.3}{100} = \dfrac{44800}{8.3145}\left(\dfrac{1}{T} - \dfrac{1}{373.15}\right)$

$$T = \frac{1}{\dfrac{8.3145}{44800}\ln\dfrac{112.3}{100} + \dfrac{1}{373.15}} = 370.18\ \text{K} = 97.03\ °\text{C}$$

$$\text{(c) } T = \frac{1}{\dfrac{8.3145}{44800}\ln\dfrac{112.3}{101.325} + \dfrac{1}{373.15}} = 370.51\ \text{K} = 97.36\ °\text{C}$$

6.9 For uranium hexafluoride the vapor pressures (in Pa) for the solid and liquid are given by

$$\ln P_s = 29.411 - 5893.5/T$$

$$\ln P_l = 22.254 - 3479.9/T$$

Calculate the temperature and pressure of the triple point.

SOLUTION

$$29.411 - \frac{5893.5}{T} = 22.254 - \frac{3479.9}{T}$$

$T = 337.2\ \text{K} = 64.0\ °\text{C}$

$P = e^{29.411\ -\ 5893.5/337.2} = 152.2\ \text{kPa}$

6.10 The heats of vaporization and of fusion of water are 2490 J g^{-1} and 33.5 J g^{-1} at 0 °C. The vapor pressure of water at 0 °C is 611 Pa. Calculate the sublimation pressure of ice at -15 °C, assuming that the enthalpy changes are independent of temperature.

SOLUTION

```
                 ΔfusH
liquid  <--------  solid
  |                  |
ΔvapH              ΔsubH
  ↓                  |
vapor  <-------------┘
```

$\Delta_{sub}H = \Delta_{fus}H + \Delta_{vap}H$

$= 33.5 + 2490$

$= 2824 \text{ J g}^{-1}$

$$\ln \frac{P_2}{P_1} = \frac{\Delta_{sub}H\,(T_2 - T_1)}{RT_1T_2}$$

$$P_2 = P_1 \exp\left[\frac{\Delta_{sub}H(T_2 - T_1)}{(RT_1T_2)}\right]$$

$$= (611 \text{ Pa}) \exp\left[\frac{(2824 \times 18 \text{ J mol}^{-1})(-15 \text{ K})}{(8.314 \text{ J K}^{-1} \text{ mol}^{-1})(273.15 \text{ K})(258.15 \text{ K})}\right]$$

$$= 166 \text{ Pa}$$

6.11 The sublimation pressures of solid Cl_2 are 352 Pa at -112 °C and 35 Pa at -126.5 °C. The vapor pressures of liquid Cl_2 are 1590 Pa at -100 °C and 7830 Pa at -80 °C. Calculate (a) $\Delta_{sub}H$, (b) $\Delta_{vap}H$, (c) $\Delta_{fus}H$, and (d) the triple point.

SOLUTION

(a) $$\Delta_{sub}H = \frac{RT_1T_2}{T_2 - T_1} \ln \frac{P_2}{P_1}$$

$$= \frac{(8.314 \text{ J K}^{-1} \text{ mol}^{-1})(161.15 \text{ K})(146.65 \text{ K})}{14.5 \text{ K}} \ln \frac{352}{35}$$

$$= 31.4 \text{ kJ mol}^{-1}$$

(b) $$\Delta_{vap}H = \frac{(8.314 \text{ J K}^{-1} \text{ mol}^{-1})(163.15 \text{ K})(193.15 \text{ K})}{30 \text{ K}} \ln \frac{7830}{1590}$$

$$= 13.9 \text{ kJ mol}^{-1}$$

(c) $\Delta_{fus}H = \Delta_{sub}H - \Delta_{vap}H = 3.14 - 13.9 = 17.5 \text{ kJ mol}^{-1}$

(d) For the solid

$$\ln P = \ln 352 + \frac{31400}{8.314} (\frac{1}{161.15} - \frac{1}{T}) = 29.300 - \frac{3777}{T}$$

For the liquid

$$\ln P = \ln 1590 + \frac{17500}{8.314} (\frac{1}{173.15} - \frac{1}{T}) = 20.273 - \frac{2105}{T}$$

At the triple point

$$29.300 - \frac{3777}{T} = 20,273 - \frac{2105}{T} \qquad T = \frac{1672}{9.027} = 185 \text{ K}$$

$$T = 185 \text{ K} - 273.15 \text{ K} = -80 \text{ °C}$$

6.12 The vapor pressure of solid benzene, C_6H_6, is 299 Pa at -30 °C and 3270 Pa at 0 °C, and the vapor pressure of liquid C_6H_6 is 6170 Pa at 10 °C and 15,800 Pa at 30 °C. From these data, calculate (a) the triple point of C_6H_6, and (b) the enthalpy fusion of C_6H_6.

SOLUTION

Calculate the enthalpy of sublimation of C_6H_6.

$$\Delta_{sub}H = \frac{8.314\,(253.15)(223.15)}{30} \ln \frac{3270}{299}$$

$$= 44{,}030 \text{ J mol}$$

Express sublimation pressures as a function of T.

$$\ln P_{sub} = -\frac{\Delta_{sub}H}{RT} + \frac{\Delta_{sub}S}{R}$$

$$\text{At 273.15 K, } \ln 3270 = -\frac{44030}{RT} + \frac{\Delta_{sub}S}{R}$$

$$\ln P_{sub} = -\frac{44030}{RT} + 27.481$$

Calculate the enthalpy of vaporization of C_6H_6

$$\Delta_{vap}H = \frac{(8.314)(283.15)(303.15)}{20} \ln \frac{15800}{6170}$$

$$= 33{,}550 \text{ J mol}^{-1}$$

Express vapor pressures as a function of T.

$$\ln P_{vap} = -\frac{\Delta_{vap}H}{RT} + \frac{\Delta_{vap}S}{R}$$

$$\text{At 303.15 K, } \ln 15{,}800 = -\frac{33550}{8.314(303.15)} + \frac{\Delta_{vap}S}{R}$$

$$\ln P_{vap} = -\frac{33550}{8.314\ T} + 22.979$$

(a) At the triple point, $\ln P_{sub} = \ln P_{vap}$

$$-\frac{44030}{RT} + 27.481 = -\frac{33550}{RT} + 22.979$$

T = 279.99 K = 6.85 °C

$$P = \exp\left(-\frac{44030}{8.314 \times 279.99} + 27.481\right) = 5249 \text{ Pa}$$

(b) $\Delta_{fus}H = \Delta_{sub}H - \Delta_{vap}H = 44.03 - 33.55 = 10.48 \text{ kJ mol}^{-1}$

6.13 The boiling point of n-hexane at 1 bar is 68.6 °C. Estimate (a) its molar heat of vaporization, and (b) its vapor pressure at 60 °C.

SOLUTION

(a) $\Delta_{vap}H = (88 \text{ J K}^{-1} \text{ mol}^{-1})(69 + 273)\text{K} = 30.1 \text{ kJ mol}^{-1}$

(b) $$\ln \frac{P_2}{P_1} = \frac{\Delta_{vap}H}{R}\left(\frac{1}{T_1} - \frac{1}{T_2}\right)$$

$$\ln \frac{1 \text{ bar}}{P} = \frac{30100 \text{ J mol}^{-1}}{8.314 \text{ J K}^{-1} \text{ mol}^{-1}}\left(\frac{1}{333 \text{ K}} - \frac{1}{341.8 \text{ K}}\right)$$

P = 0.755 bar

6.14 From tables giving $\Delta_f G^o$, $\Delta_f H^o$, and $\overline{C}_P^o$ for $H_2O(l)$ and $H_2O(g)$ at 298 K, calculate (a) the vapor pressure of $H_2O(l)$ at 25 °C, and (b) the boiling point at 1 atm.

SOLUTION

(a) $H_2O(1) = H_2O(g)$

$\Delta G^o = -228.572 - (-237.129) = 8.557 \text{ kJ mol}^{-1}$

$\Delta G^o = -RT \ln (\frac{P}{P^o})$

$\frac{P}{P^o} = \exp\frac{08557 \text{ J mol}^{-1}}{(8.314 \text{ J K}^{-1} \text{ mol}^{-1})(298.15 \text{ K})} = 3.168 \times 10^{-2}$

$P = (3.168 \times 10^{-2})(10^5 \text{ Pa}) = 3.17 \times 10^3 \text{ Pa}$

(b) ΔH^o (298.15 K) = -214.818 - (-285.830) = 44.012 kJ mol^{-1}

In the absence of data on the dependence of $\bar{C}_P$ on T we will calculate $\Delta H^o(T)$ from

$$\Delta H^o(T) = 44{,}012 + \int_{298.15}^{T} (33.577 - 75.291)\, dT$$

$$= 44{,}012 - 41.714\ (T - 298.15)$$

$$\frac{\partial(\frac{\Delta G^o}{T})}{\partial/T}) = -\frac{\Delta H^o(T)}{T^2}$$

$$\frac{\Delta G^o}{T} = -\int [\frac{-44012}{T^2} - \frac{41.714}{T} + \frac{(41.714)(298.15)}{T^2}]dT$$

$$= \frac{44.012}{T} + 41.714 \ln T + \frac{(41.714)(298.15)}{T} + I$$

where I is an integration constant which can be evaluated from $\Delta G^o(298.15 \text{ K})$.

$$\frac{8590}{298.15} = \frac{44.012}{298.15} + 41.714 \ln 298.15 + \frac{(41.714)(298.15)}{(298.15)} + I$$

$I = -398.191$

At the boiling point $\Delta G^o = 0$ and so we need to solve the following equation by successive approximations.

$$0 = \frac{44012}{T} + 41.714 \ln T + \frac{(41.714)(298.15)}{T} - 398.191$$

Trying T = 373 K RHS = 0.163

Trying T = 374 K RHS = -0.130

Therefore, the standard boiling point calculated in this way is close to 373.5 K.

6.15 What is the maximum number of phases that can be in equilibrium in one-, two-, and three-component systems?

SOLUTION

$F = c - p + 2$

$0 = 1 - p + 2, p = 3$

$0 = 2 - p + 2, p = 4$

$0 = 3 - p + 2, p = 5$

6.16 How many degrees of freedom are there for the following systems, and how might they be chosen?

(a) $CuSO_4{\cdot}5H_2O$(cr) in equilibrium with $CuSO_4$(cr) and H_2O(g).

(b) N_2O_4 in equilibrium with NO_2 in the gas phase.

(c) CO_2, CO, H_2O, and H_2 in equilibrium in the gas phase.

(d) The system described in (c) is made up of specified amount of CO and H_2.

SOLUTION

(a) $c = s - r = 3 - 1 = 2$

$F = c - p + 2 = 2 - 3 + 2 = 1$

Only the temperature or pressure may be fixed.

(b) $c = s - r = 2 - 1 = 1$

$F = c - p + 2 = 2 - 3 + 2 = 1$

Temperature and pressure may be fixed.

(c) $c = 4 - 1 = 3$

$F = c - p + 2 = 3 - 1 + 2 = 4$

Temperature, pressure, and two mole fractions may be fixed.

(d) $c = 2 - r - m = 4 - 1 - 2 = 1$

$F = c - p + 2 = 1 - 1 + 2 = 2$

Only temperature and pressure may be fixed if the elemental composition is fixed.

6.17 Graphite is in equilibrium with gaseous H_2O, CO, CO_2, H_2, and CH_4. How many degrees of freedom are there? What degrees of freedom might be chosen for an equilibrium calculation?

SOLUTION

$c = s - R = 5 - 2 = 3$

$F = c - p + 2 = 3 - 1 + 2 = 4$

The degrees of freedom chosen might be T, P, H/C, O/C.

6.18 A gaseous system contains CO, CO_2, H_2, H_2O, and C_6H_6 in chemical equilibrium. How many components are there? How many independent reactions? How many degrees of freedom are there?

SOLUTION

	CO	CO_2	H_2	H_2O	C_6H_6
C	1	1	0	0	6
O	1	2	0	1	0
H	0	0	2	2	6

Subtract the first row from the second and divide the third row by 2 to obtain

1	1	0	0	6
0	1	0	1	-6
0	0	1	1	3

Subtract the second row from the first

1	0	0	-1	12
0	1	0	1	-6
0	0	1	1	3

The rank of this matrix is 3, and so there are 3 independent components. The stoichiometric numbers of 2 independent reactions are given by the last 2 columns.

$$CO_2 + H_2 = H_2O + CO$$

$$12CO + 3H_2 = C_6H_6 + 6CO_2$$

If the species are arranged in a different order in the matrix, a different pair of independent equations will be obtained.

$$F = c - p + 2$$

If the numbers of moles of 3 components and the temperature are specified, the system is completely determined; that is, the numbers of moles of the other two species and the pressure can be calculated - given values of the equilibrium constant. Alternatively, the mole fractions of 2 species and the temperature and pressure may be specified. We then have two mathematical expressions between 5 mole fractions, 4 of which are independent since $\Sigma y_i = 1$. If 2 mole fractions are known, the other two can be calculated from the two simultaneous equations.

6.19 The vapor pressure of water at 25 °C is 23.756 mm Hg. What is the vapor pressure of water when it is in a container with an air pressure of 100 bar, assuming the dissolved gases do not affect the vapor pressure. The density of water is 0.99707 g/m^3.

<u>SOLUTION</u>

$RT \ln (P/P_o) = \nabla_L(P-P_o)$

$(8.3145 \text{ J K}^{-1} \text{ mol}^{-1})(298.15 \text{ K}) \ln (P/23.756)$

$= (18.015 \text{ g mol}^{-1}/0.99707 \text{ g m}^{-3})(10^{-3} \text{ m/cm})^3)(100 \times 10^5 \text{ Pa})$

$P/23.756 = \exp 0.07289$

$P = 25.552 \text{ mm Hg} = 3406 \text{ Pa}$

6.20 The freezing point is lowered to -5 °C under the skates, and so the answer is yes.

6.21 6.29×10^{-3} bar

6.22 2.66×10^{-6} bar

6.23 4.9 K

6.24 53.0 kJ mol^{-1}, 38.4 kJ mol^{-1}, 158.6 J K^{-1} mol^{-1}

6.25 383.38 K, -0.4 °C

6.26 96 °C

6.27 27.57 kJ mol^{-1}

6.28 0.0773 Pa

6.29 (a) 0.12314 bar, (b) 0.125 110 bar

6.31 (a) 38.1 kJ mol^{-1}, (b) 3.78 kPa

6.32 (a) 50.91 kJ mol^{-1}, (b) 50.14 Pa K^{-1}, 44.20 Pa K^{-1}, (c) P_{ice} = 260 Pa, P_{liq} = 290 Pa

6.33 (a) -0.46 °C, (b) -0.59 °C

6.34 -8.59 kJ mol^{-1}

6.35 4

6.36 (a) 1, (b) 2

6.37 (a) 4; T, P, y_{H_2}, y_{O_2}; (b) 3; T,P,H/O

6.38 166.5 Pa

CHAPTER 7 Phase Equilibrium: Two- and Three-Component Systems

7.1 What is the difference in chemical potential of benzene (component 1) in toluene (component 2) at $x_1 = 0.5$ and $x_1 = 0.1$ at 25 °C, assuming ideal solutions?

SOLUTION

$$\mu_1 = \mu_1^* + RT \ln 0.5$$

$$\mu_1 = \mu_1^* + RT \ln 0.1$$

$$\Delta\mu_1 = (8.314\ \mathrm{J\ K^{-1}\ mol^{-1}})(298\ \mathrm{K}) \ln \frac{0.5}{0.1}$$

$$= 3988\ \mathrm{J\ mol^{-1}}$$

7.2 A binary liquid mixture of A and B is in equilibrium with its vapor at constant temperature and pressure. Prove that $\mu_A(g) = \mu_A(\ell)$ and $\mu_B(g) = \mu_B(\ell)$ by starting with

$$G = G(g) + G(\ell)$$

and the fact that dG = 0 when infinitesimal amounts of A and B are simultaneously transferred from the liquid to the vapor.

SOLUTION

$$dG = dG(\ell) + dG(g) = 0 \qquad (1)$$

since the transfer is made at equilibrium

$$dG(\ell) = \mu_A(\ell\ dn_A(\ell) + \mu_B(\ell)\ dn_B(\ell) \qquad (2)$$

$$dG(g) = \mu_A(g)\ dn_A(g) + \mu_B(g)\ dn_B(g) \qquad (3)$$

since $dn_A(\ell) = -\ dn_A(g)$ and $dn_B(\ell) = -dn_B(g)$, substituting equation 2 and 3 into equation 1 yields

$dG = [\mu_A(g) - \mu_A(\ell)]\ dn_A(g) + [\mu_B(g) - \mu_A(\ell)]\ dn_B(g) = 0$

Since $dn_A(g)$ and $dn_B(g)$ are independently variable, both terms in equation 4 have to be zero for dG to be zero. This yields the desired expression for components A and B.

7.3 Ethanol and methanol form very nearly ideal solutions. At 20 °C, the vapor pressure of ethanol is 5.93 kPa, and that of methanol is 11.83 kPa. (a) Calculate the mole fractions of methanol and ethanol in a solution obtained by mixing 100 g of each. (b) Calculate the partial pressures and the toal vapor pressure of the solution. (c) Calculate the mole fraction of methanol in the vapor.

SOLUTION

(a) $x_{CH_3OH} = \dfrac{100/32}{100/46 + 100/32} = 0.590$

$x_{C_2H_5OH} = \dfrac{100/46}{100/46 + 100/32} = 0.410$

(b) $P_{CH_3OH} = x_{CH_3OH}\, P^*_{CH_3OH}$

$= (0.590)(22{,}830 \text{ Pa}) = 6980 \text{ Pa}$

$P_{C_2H_5OH} = x_{C_2H_5OH}\, P^*_{C_2H_5OH}$

$= (0.410)(5930 \text{ Pa}) = 2430 \text{ Pa}$

$P_{total} = 2430 \text{ Pa} + 6980 \text{ Pa} = 9410 \text{ Pa}$

(c) $y_{CH_3OH,vapor} = \dfrac{6980 \text{ Pa}}{9410 \text{ Pa}} = 0.742$

7.4 One mole of benzene (component 1) is mixed with two moles of toluene (component 2). At 60 °C the vapor pressures of benzene and toluene are 51.3 and 1815 kPa, respectively. (a) As the pressure is reduced, at what pressure will boiling begin? (b) What will be the composition of the first bubble of vapor?

SOLUTION

(a) $P = P_2^* + (P_1^* - P_2^*)x_1$

$= 18.5\text{ kPa} + (51.3\text{ kPa} - 18.5\text{ kPa})(0.333)$

$= 29.4\text{ kPa}$

(b) $y_1 = \dfrac{x_1 P_1^*}{P_2^* + (P_1 - P_2)x_1}$

$= \dfrac{(0.333)(51.3\text{ kPa})}{18.5\text{ kPa} + 932.8\text{ kPa}}(0.333))$

$= 0.581$

7.5 The vapor pressures of benzene and toluene have the following values in the temperature range between their boiling points at 1 bar:

t/°C	79.4	88	94	100	110.0
$P^*_{C_6H_6}$/bar	1.000	1.285	1.526	1.801	
$P^*_{C_7H_8}$/bar		0.508	0.616	0.742	1.000

(a) Calculate the compositions of the vapor and liquid phases at each temperature and plot the boiling point diagram. (b) If a solution contaning 0.5 mole fraction benzene and 0.5 mole fraction toluene is heated, at what temperature will the first bubble of vapor appear and what will be its composition?

<u>SOLUTION</u>

(a) At 88 °C $x_B(1.285) + (1 - x_B)\ 0.508 = 1$

$x_B = 0.633$

$y_B = (0.633)(1.285)/1 = 0.814$

At 94 °C $x_B(1.526) + (1 - x_B)\ 0.616 = 1$

$x_B = 0.422$

$y_B = (0.4222)(1.526)/1 = 0.644$

At 100 °C $x_B(1.801) + (1 - x_B)\ 0.742 = 1$

$x_B = 0.244$

$y_B = (0.244)(1.801)/1 = 0.439$

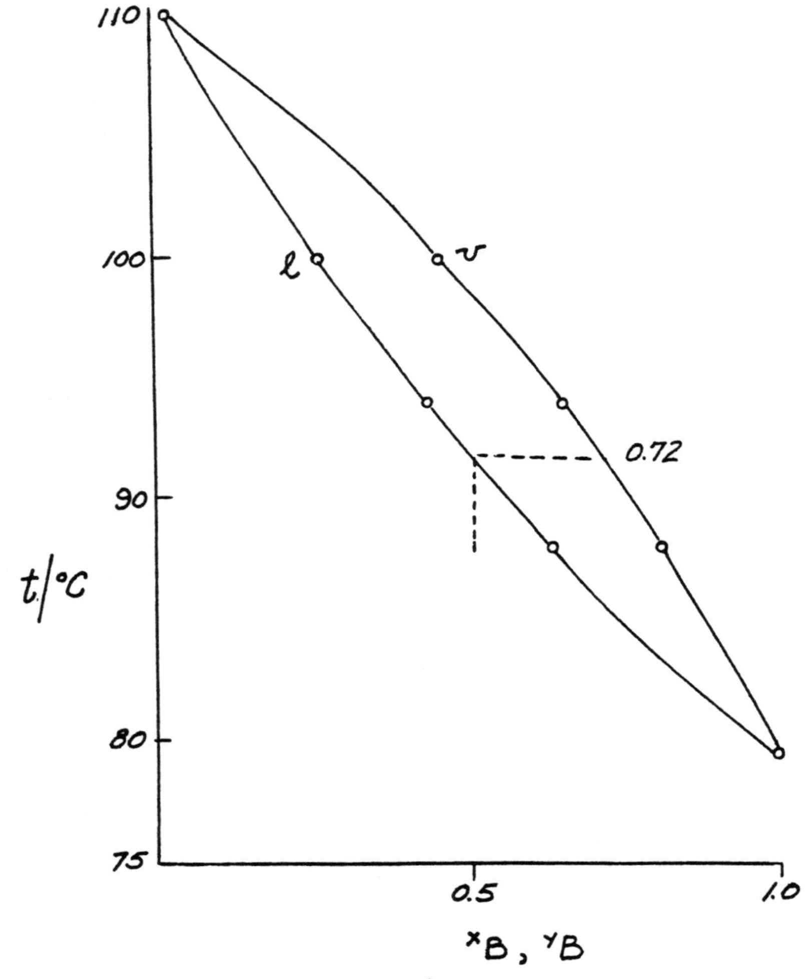

Bubble point = 92 °C y_B = 0.72

7.6 At 1.013 bar pressure propane boils at -42.1 °C and n-butane boils at -0.5 °C; the following vapor-pressure data are available.

t/°C	-31.2	-16.3
P/kPa(propane)	160.0	298.6
P/kPa(n-butane)	26.7	53.3

Assuming that these substances form ideal binary solutions with each other, (a) calculate the mole fractions of propane at which the solution will boil at 1.013 bar pressures at -31.2 and -16.3 °C. (b) Calculate the mole fractions of propane in the equilibrium vapor at these temperatures. (c) Plot the temperature-mole fraction diagram at 1.013 bar, using these data, and label the regions.

SOLUTION

(a) At -31.2 °C, $P_P = x_P 160.0$ kPa, $P_B = (1 - x_P)\ 26.7$ kPa

$$P = 101.3 \text{ kPa} = x_P\ 160.0 + (1 - x_P)\ 26.7$$

$$= 26.7 + 133.3\ x_P$$

$$x_P = 0.560$$

At -16.3 °C, $P_P = x_P\ 298.6$ kPa, $P_B = (1 - x_P)\ 53.3$ kPa

$$P = 101.3 \text{ kPa} = x_P\ 298.6 + (1 - x_P)\ 53.3$$

$$= 53.3 + 245.3\ x_P$$

$$x_P = 0.196$$

(b) At -31.2 °C,

$$y_P = \frac{(0.560)(160.0)}{(0.560)(160.0) + (0.440)(26.7)} = 0.884$$

At -16.3 °C

$$y_P = \frac{(0.196)(298.6)}{(0.196)(298.6) + (0.804)(53.3)} = 0.577$$

(c)

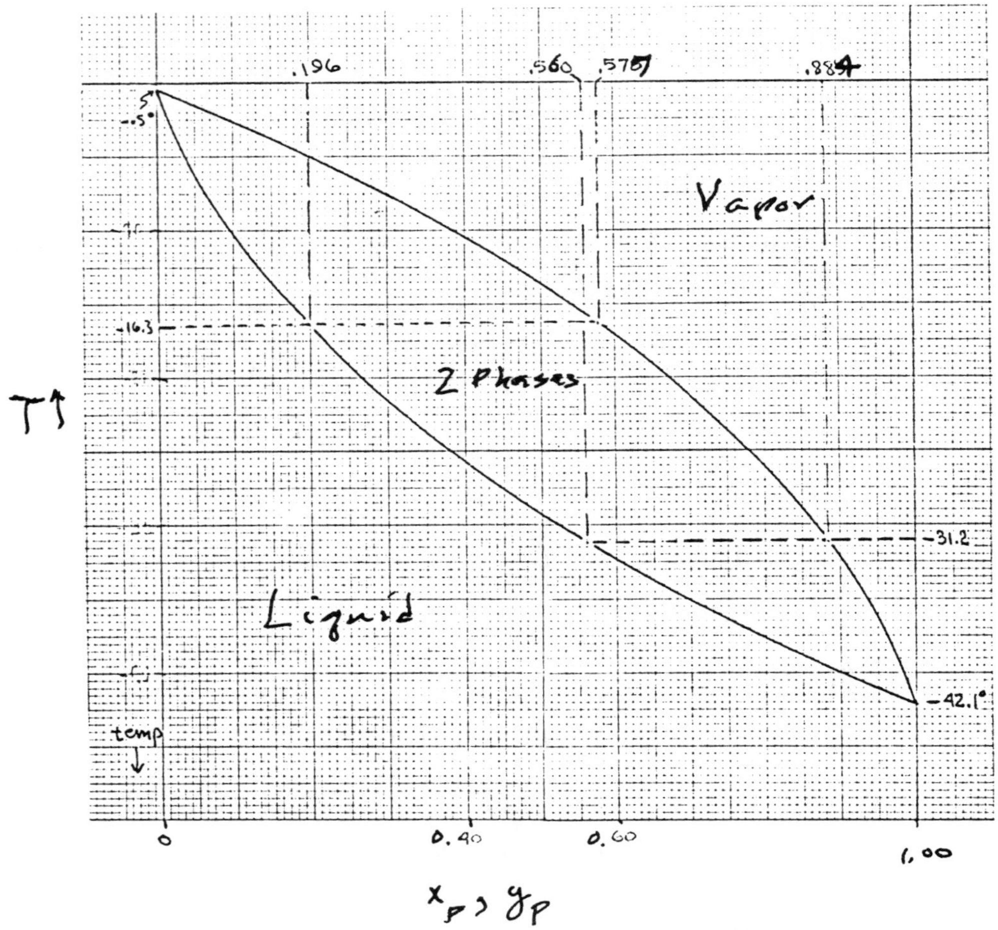

.196
.560
.5787
.8854
-.5°
-16.3
Vapor
2 Phases
T↑
-31.2
Liquid
-42.1°
temp ↓
0
0.40
0.60
1.00
x_p, y_p

7.7 The following table gives mole percent acetic acid in aqueous solutions and in the equilibrium vapor at the boiling point of the solution at 1.013 bar.

B.P., °C	118.1	113.8	107.5	104.4	102.1	100.0
Mole % of acetic acid						
In liquid	100	90.0	70.0	50.0	30.0	0
In vapor	100	83.3	57.5	37.4	18.5	0

Calculate the minimum number of theoretical plates for the column required to produce an initial distillate of 28 mole percent acetic acid from a solution of 80 mole percent acetic acid.

SOLUTION

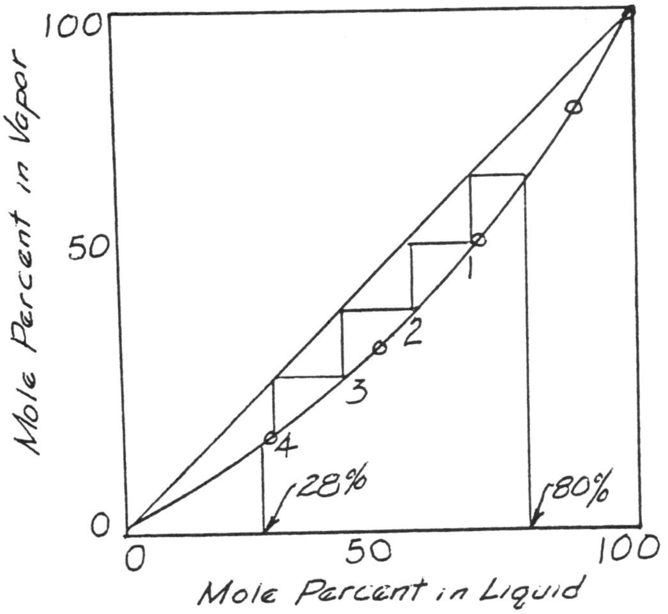

Since there are four steps, three theoretical plates are required in the column. The distilling pot counts as one plate.

7.8 If two liquids (1 and 2) are completely immiscible, the mixture will boil when the sum of the two partial pressures exceeds the applied pressure: $P = P_1^* + P_2^*$. In the vapor phase the ratio of the mole fractions of the two components is equal to the ratio of their vapor pressures.

$$\frac{P_1^*}{P_2^*} = \frac{x_1}{x_2} = \frac{m_1 M_2}{m_2 M_1}$$

where m_1 and m_2 are the masses of components 1 and 2 in the vapor phase, and M_1 and M_2 are their molar masses. The boiling point of the immiscible liquid system naphthalene-water is 98 °C under a pressure of 97.7 kPa. The vapor pressure of water at 98 °C is 94.3 kPa. Calculate the weight percent of naphthalene in the distillate.

SOLUTION

$$\frac{m_1}{m_2} = \frac{P_1^* M_1}{P_2^* M_2} = \frac{(97.7 - 94.3)(128)}{(94.3)(18)} = 0.261$$

$$\text{Weight percent naphthalene} = \frac{0.261}{1.261} = 0.207 \text{ or } 20.7\%$$

7.9 A regular binary solution is defined as one for which

$$\mu_1 = \mu_1^o + RT \ln x_1 + w x_2^2$$

$$\mu_2 = \mu_2^o + RT \ln x_2 + w x_1^2$$

Derive $\Delta_{mix}G$, $\Delta_{mix}S$, $\Delta_{mix}H$, and $\Delta_{mix}V$ for the mixing of x_1 moles of component 1 with x_2 moles of component 2. Assume that the coefficient w is independent of temperature.

SOLUTION

$$G = \sum_i n_i\mu_i = x_1\mu_1^o + x_1 RT \ln x_1 + wx_1x_2^2$$

$$+ x_2\mu_2^o + x_2 RT \ln x_2 + wx_2x_1^2$$

$$G^o = \sum_i n_i\mu^o = x_1\mu_1^o + x_2\mu_2^o$$

$$\Delta_{mix}G = RT(x_1\ln x_1 + x_2\ln x_2) + wx_1x_2$$

$$\Delta_{mix}S = -(\partial\Delta_{mix}G/\partial T)_P = -R(x_1\ln x_1 + x_2\ln x_2)$$

$$\Delta_{mix}H = \Delta_{mix}G + T\Delta_{mix}S = wx_1x_2$$

$$\Delta_{mix}V = (\partial\Delta_{mix}G/\partial P)_T = 0$$

7.10 From the data given in the following table construct a complete temperature-composition diagram for the system ethanol-ethyl acetate for 1.013 bar. A solution containing 0.8 mole fraction of ethanol, EtOH, is distilled completely at 1.013 bar. (a) What is the composition of the first vapor to come off? (b) That of the last drop of liquid to evaporate? (c) What would be the values of these quantities if the distillation were carried out in a cylinder provided with a piston so that none of the vapor escape?

x_{EtOH}	y_{EtOH}	B.P., °C	x_{EtOH}	y_{EtOH}	B.P., °C
0	0	77.15	0.563	0.507	72.0
0.025	0.070	76.7	0.710	0.600	72.8
0.100	0164	75.0	0.833	0.735	74.2
0.240	0.295	72.6	0.942	0.880	76.4
0.360	0.398	71.8	0.982	0.965	77.7
0.462	0.462	71.6	1.000	1.000	78.3

SOLUTION

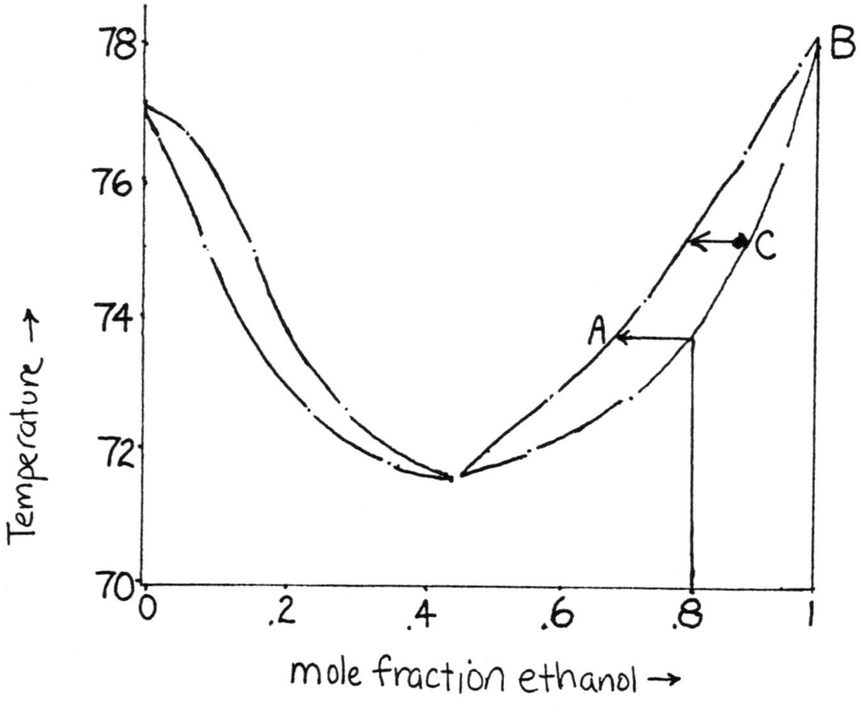

(a) The liquid will start to boil at 73.7 °C. The vapor in equilibrium with it is 0.69 mole fraction in ethanol (point A).

(b) As the boiling continues, more and more of the EtAc will be removed. As the mole fraction of EtOH increases, the boiling point will also increase. Finally, the very last vapors will be pure EtOH, as indicated by point B.

(c) The first vapor to come off will be the same as in part (a). However, when all of the mixture is in vapor state, the mole fraction of the vapor state will be 0.800, simply because that is the total mole fraction of EtOH in the entire mixture. Thus, to find the composition of the last drop to evaporate into the cylinder, simply note that liquid (y_{EtOH} = 0.9) is in equilibrium with vapor which has 0.800 EtOH in it. This is given by point C.

7.11 The Henry law constants for oxygen and nitrogen in water at 0 °C are 2.54 x 10^4 bar and 5.45 x 10^4 bar, respectively. Calculate the lowering of the freezing point of water by dissolved air with 80% N_2 and 20% O_2 by volume at 1 bar pressure.

<u>SOLUTION</u>

$$x_{N_2} = \frac{P_{N_2}}{K_{N_2}} = \frac{0.8 \text{ br}}{5.45 \times 10^4 \text{ bar}} = 1.47 \times 10^{-5}$$

$$x_{O_2} = \frac{P_{O_2}}{K_{O_2}} = \frac{0.2 \text{ bar}}{2.54 \times 10^4 \text{ bar}} = 7.87 \times 10^{-6}$$

The mole fraction dissolved air is $2.26 \times 10^{-5} = \dfrac{m}{m + \dfrac{1000}{18.02}}$

$m_{air} = 1.25 \times 10^{-3}$ mol/1000 g solvent

$\Delta T_{fp} = k_{fp}\, m = -(1.86)(1.25 \times 10^{-3}) = -0.00233$ K

7.12 Use the Gibbs-Duhem equation to show that if one component of a binary liquid solution follows Raoult's law, the other component will too.

SOLUTION

$x_1 d\mu_1 + x_2 d\mu_2 = 0$

If $\mu_1 = \mu_1^o + RT \ln x_1$

$d\mu_1 = \frac{RT}{x_1} dx_1$

Using the first equation

$d\mu_2 = -\frac{x_1}{x_2} d\mu_1 = -\frac{RT}{x_2} dx_1$

Since $x_1 + x_2 = 1$, $dx_2 = -dx_1$

$d\mu_2 = \frac{RT}{x_2} dx_2 = RT\, d \ln x_2$

$\mu_2 = \text{const} + RT \ln x_2$

If $x_2 = 1$, const $= \mu_2^o$

$\mu_2 = \mu_2^o + RT \ln x_2$

7.13 The following data on ethanol-chloroform solutions at 35 °C were obtained by G. Scatchard and C. L. Raymond [*J. Am. Chem. Soc.* **60**, 1278 (1938)]:

$x_{EtOH,liq}$	0	0.2	0.4	0.6	0.8	1.0
$y_{EtOH,vap}$	0.0000	0.1382	0.1864	0.2554	0.4246	1.0000

Total P, kPa 39.345 40.559 38.690 34.387 25.357 13.703

Calculate the activity coefficients of ethanol and chloroform based on the derivations from Raoult's law.

SOLUTION

$x_{EtOH} =$ 0.2 $\quad \gamma_{EtOH} = \dfrac{y_{EtOH}P}{x_{EtOH}P^*_{EtOH}}$

$$= \frac{(0.1382)(40.559)}{(0.2)(13.703)} = 2.045$$

$x_{EtOH} = 0.4 \quad \gamma_{EtOH} = \dfrac{(0.1864)(38.690)}{(0.4)(13.703)} = 1.316$

$x_{EtOH} = 0.6 \quad \gamma_{EtOH} = \dfrac{(0.2554)(34.287)}{(0.6)(13.703)} = 1.065$

$x_{EtOH} = 0.8 \quad \gamma_{EtOH} = \dfrac{(0.4246)(25.357)}{(0.8)(13.703)} = 0.982$

$x_{EtOH} = 1.0 \quad \gamma_{EtOH} = \dfrac{(1.000)(13.703)}{(1.000)(13.703)} = 1.000$

$$\gamma_{CHCl3} = \frac{y_{CHCl3}P}{x_{CHCl3}P^*_{CHCl_3}}$$

$x_{CHCl3} = 1 \quad \gamma_{CHCl3} = \dfrac{(1)(39.345)}{(1)(39.345)} = 1$

$\gamma_{CHCl3} = 0.8 \quad \gamma_{CHCl3} = \dfrac{(0.8618)(40.559)}{(0.8)(39.345)} = 1.111$

$\gamma_{CHCl3} = 0.6 \quad \gamma_{CHCl3} = \dfrac{(0.8136)(38.690)}{(0.6)(39.345)} = 1.333$

$\gamma_{CHCl_3} \quad = 0.4 \quad \gamma_{CHCl_3} \quad = \dfrac{(0.7446)(34.387)}{(0.4)(39.345)} \quad = 1.627$

$x_{CHCl_3} \quad = 0.2 \quad \gamma_{CHCl_3} \quad = \dfrac{(0.5754)(25.357)}{(0.2)(39.345)} \quad = 1.854$

7.14 Show that the equations for the bubble-point line and dew-point line for nonideal solutions are given by

$$x_1 = \frac{P - \gamma_2 P^*_2}{\gamma_1 P x^*_1 - \gamma_2 P^*_2}$$

$$y_1 = \frac{P\gamma_1 P^*_1 - \gamma_1 \gamma_2 P^*_1 P^*_2}{P\gamma_1 P^*_1 - P\gamma_2 P^*_2}$$

<u>SOLUTION</u>

$$P = P_1 + P_2 = x_1 \gamma_1 P^*_1 + (1 - x_1)\gamma_2 P^*_2$$

$$= \gamma_2 P^*_2 + x_1(\gamma_1 P^*_1 - \gamma_2 P^*_2)$$

$$x_1 = \frac{P - \gamma_2 P^*_2}{\gamma_1 P^*_1 - \gamma_2 P^*_2}$$

$$y_1 = \frac{P_1}{P_1 + P_2} = \frac{x_1 \gamma_1 P_1^*}{\gamma_2 P_2^* + x_1(\gamma_1 P_1^* - \gamma_2 P_2^*)}$$

Substituting the expression for x_1 yields the desired expression for y_1.

7.15 A regular binary solution is defined as one for which

$$\mu_1 = \mu_1^o + RT \ln x_1 + wx_2^2$$

$$\mu_2 = \mu_2^o + RT \ln x_2 + wx_1^2$$

Derive the expressions for the activity coefficients of γ_1 and $\gamma_2$in terms of w.

<u>SOLUTION</u>

$$\mu_1 = \mu_1^o + RT \ln \gamma_1 x_1$$

$$\mu_1 = \mu_1^o + RT \ln x_1 + wx_2^2$$

$$= \mu_1^o + RT \ln x_1 + RT \ln e^{wx_2^2/RT}$$

$$= \mu_1^o + RT \ln \left(e^{wx_2^2/RT} \right) x_1$$

Therefore

$$\gamma_1 = e^{wx_2^2/RT} \qquad \gamma_2 = e^{wx_1^2/RT}$$

7.16 The expressions for the activity coefficients of the components of a regular binary solution were derived in the preceding problem. Derive the expression for γ_1 in terms of the experimentally measured total pressure P, the vapor pressures of the two components, and the composition of the solution for the case that the deviations from ideality are small.

<u>SOLUTION</u>

$$\gamma_1 = \exp(wx_2^2/RT) \qquad \gamma_2 = \exp(wx_1^2/RT)$$

$$P = P_1 + P_2 = \gamma_1 x_1 P_1^* + \gamma_2 x_2 P_2^*$$

$$= x_1 P_1^* \exp\left(\frac{wx_2^2}{RT}\right) + x_2 P_2^* \exp\left(\frac{wx_1^2}{RT}\right)$$

If the exponentials are not far from unity, the exponentials may be expanded to obtain

$$P = x_1 P_1^* \left(1 + \frac{wx_2^2}{RT}\right) + x_2 P_2^* \left(1 + \frac{wx_1^2}{RT}\right)$$

$$\frac{w}{RT} = \frac{P - (x_1 P_1^* + x_2 P_2^*)}{x_1 x_2 \left[P_1^* + x_1 (P_2^* - P_1^*)\right]}$$

Thus from a measurement of the total pressure of a mixture of the two components, the activity coefficients of the components can be calculated over the entire concentration range using the first two equations given above. This is a particular example of the general situation that γ_1 and γ_2 can be calculated from the total pressure as a function of x_1.

7.17 Using the data in Problem 7.39, calculate the activity coefficients of water (1) and n-propanol (2) at 0.20, 0.40, 0.60, and 0.80 mole fraction of n-propanol, based on deviations from Henry's law and considering water to be the solvent.

SOLUTION

The data at the lowest mole fractions of n-propanol are used to calculate the Henry law constant for n-propanol in water.

x_2	$K_2 = P_2/x_2$
0.02	33.5
0.05	28.8
0.10	17.6

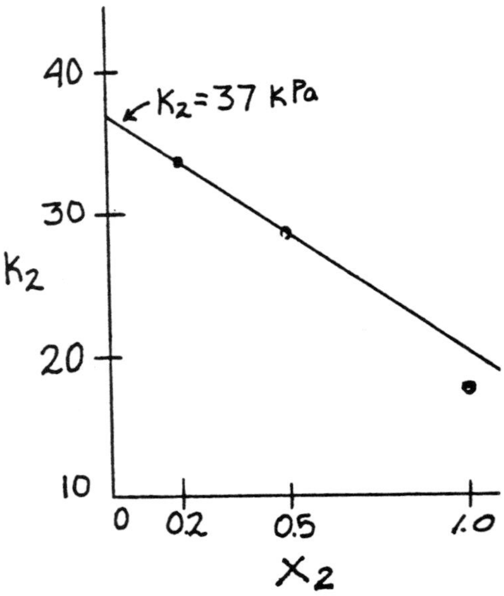

$$\gamma' = \frac{P_2}{x_2 K_2}$$

At

$x_2 = 0.2$ $\quad \gamma_2' = \frac{1.81}{0.2(37)} = 0.24$

$x_2 = 0.4$ $\quad \gamma_2' = \frac{1.89}{0.4(37)} = 0.13$

$x_2 = 0.6$ $\quad \gamma_2' = \frac{2.07}{0.6(37)} = 0.093$

$x_2 = 0.8$ $\quad \gamma_2' = \frac{2.37}{0.8(37)} = 0.080$

$$\gamma_1 = \frac{P_1}{x_1 P_1^*}$$

$x_2 = 0.2$ $\quad \gamma_1 = \frac{2.91}{(0.8)(3.17)} = 1.15$

$x_2 = 0.4$ $\quad \gamma_1 = \frac{2.89}{(0.6)(3.17)} = 1.52$

$x_2 = 0.6$ $\quad \gamma_1 = \frac{2.65}{(0.4)(3.17)} = 2.09$

$x_2 = 0.8$ $\quad \gamma_1 = \frac{1.79}{(0.2)(3.17)} = 2.82$

7.18 If 68.4 g of sucrose ($M = 342$ g mol^{-1}) is dissolved in 1000 g of water: (a) What is the vapor pressure at 20 °C? (b) What is the freezing point? The vapor pressure of water at 20 °C is 2.3149 kPa.

SOLUTION

(a) $$x_2 = \frac{\frac{68.4}{342}}{\frac{68.4}{342} + \frac{1000}{18}} = 3.59 \times 10^{-3}$$

$$\frac{P_1^0 - P_1}{P_1^0} = x_2 \qquad \frac{2.3149 - P_1}{2.3149} = 3.59 \times 10^{-3}$$

$$P_1 = 2.3149(1 - 3.59 \times 10^{-3}) = 2.3006 \text{ kPa}$$

(b) $$\Delta T_f = K_f m = 1.86(0.2) = -0.372$$

$$T_f = -0.372 \text{ °C}$$

7.19 The protein human plasma albumin has a molar mass of 69,000 g mol^{-1}. Calculate the osmotic pressure of a solution of this protein containing 2 g per 100 cm^3 at 25 °C in (a) pascals, and (b) millimeters of water. The experiment is carried out using a salt solution for solvent and a membrane permeable to salt as well as water.

SOLUTION

$$\pi = \frac{cRT}{M} = \frac{(20 \times 10^{-3} \text{ kg L}^{-1})(10^3 \text{ mL}^{-3})(8.314 \text{ J K}^{-1} \text{ mol}^{-1})(298.15 \text{ K})}{69 \text{ kg mol}^{-1}}$$

$$= 719 \text{ Pa}$$

$$h = \frac{\pi}{dg} = \frac{719 \text{ Pa}}{(1 \text{ g cm}^{-3})(10^{-3} \text{ kg g}^{-1})(10^{2} \text{ cm m}^{-1})(9.8 \text{ m s}^{-2})}$$

$= 7.34 \times 10^{-2}$ m = 73.4 mm of water

7.20 The following osmotic pressures were measured for solutions of a sample of polyisobutylene in benzene at 25 °C.

$c/\text{kg m}^{-3}$	5	10	15	20
Π/Pa	49.5	101	155	211

Calculate the number average molar mass from the value of π/c extrapolated to zero concentration of the polymer.

SOLUTION

$c/\text{kg m}^{-3}$	5	10	15	20
$(\pi/c)/(\text{Pa/kg m}^{-3})$	9.90	10.1	10.3	10.6

$\Delta = -0.2$ $\Delta = -0.2$

$$\frac{\pi}{c} = \frac{RT}{M} + Bc$$

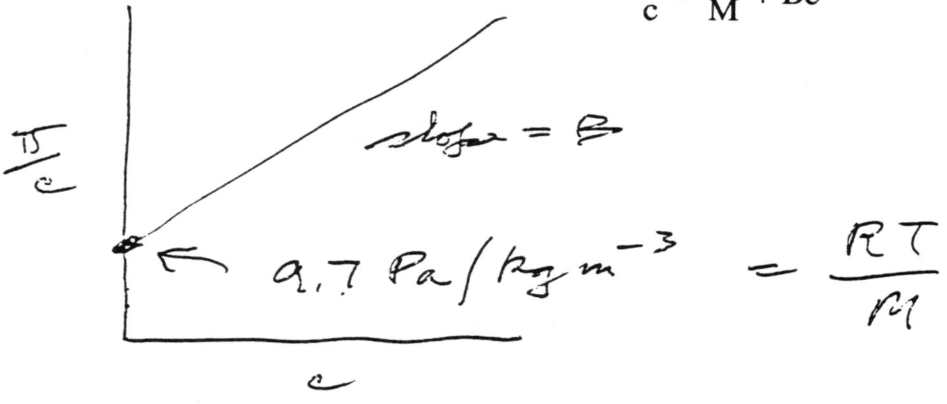

$$M = \frac{RT}{9.7 \text{ Pa/kg m}^{-3}} = \frac{(8.314 \text{ J K}^{-1} \text{ mol}^{-1})(298 \text{ K})}{9.7 \text{ Pa m}^{3} \text{ kg}^{-1}}$$

$= 255 \text{ kg mol}^{-1}$

7.21 A sample of polymer contains 0.50 mole fraction with molar mass 100,000 g mol^{-1} and 0.50 mole fraction with molar mass 200,000 g mol^{-1}. Calculate (a) M_n, and (b) M_m.

SOLUTION

(a) $M_n = \dfrac{\Sigma n_i M_i}{\Sigma n_i} = \Sigma x_i M_i$ where x_i is mole fraction of the ith component

$$= 0.5(10^5) + 0.5(2 \times 10^5) = 150{,}000 \text{ g mol}^{-1}$$

(b) $M_m = \dfrac{\Sigma n_i M_i^2}{\Sigma n_i M_i}$

Dividing numerator and denominator by Σn_i

$$M_m = \frac{\Sigma x_i M_i^2}{\Sigma x_i M_i} = \frac{0.5(10^{10}) + 0.5(4 \times 10^{10})}{0.5(10^5) + 0.5(2 \times 10^5)} = 167{,}000 \text{ g mol}^{-1}$$

7.22 Calculate the osmotic pressure of a 1 mol L^{-1} sucrose solution in water from the fact that at 30 °C the vapor pressure of the solution is 4.1606 kPa. The vapor pressure of water at 30 °C is 4.2429 kPa. The density of pure water at this temperature (0.99564 g cm^{-3}) may be used to estimate V_1 for a dilute solution. To do this problem, Raoult's law is introduced into equation 7.56.

SOLUTION

$\bar{V}_1 \pi = -RT \ln x_1$

Substituting Raoult's law $P_1 = x_1 P_1^{o}$

$$\pi = -\frac{RT}{\bar{V}_1} \ln \frac{P_1}{P_1^o}$$

$$\bar{V}_1 = \frac{18.02 \text{ g mol}^{-1}}{0.99564 \text{ g cm}^{-3}} = 1810 \text{ cm}^3 \text{ mol}^{-1} = 0.01810 \text{ L mol}^{-1}$$

$$\pi = \frac{(0.08314 \text{ L bar K}^{-1} \text{ mol}^{-1})(303.15 \text{ K})}{0.0180 \text{ L mol}^{-1}} \ln \frac{4.2429 \text{ kPa}}{4.1606 \text{ kPa}}$$

$$= 27.3 \text{ bar}$$

7.23 Calculate the solubility of p-dibromobenzene in benzene at 20 °C and 40 °C assuming ideal solutions are formed. The enthalpy of fusion of p-dibromobenzene is 13.22 kJ mol^{-1} at its melting point (86.9 °C).

<u>SOLUTION</u>

At 20 °C $\ln x_2 = \frac{\Delta_{fus}H_2(T - T_{2f})}{RTT_{2f}}$

$$= \frac{(13220 \text{ J mol}^{-1})(-66.9 \text{ K})}{(8.314 \text{ J K}^{-1} \text{ mol}^{-1})(360.1 \text{ K})(293.2 \text{ K})}$$

$$x_2 = 0.365$$

At 40 °C $\ln x_2$ $= \frac{(13220 \text{ J mol}^{-1})(-46.9 \text{ K})}{(8.314 \text{ J K}^{-1} \text{ mol}^{-1})(360.1 \text{ K})(313.2 \text{ K})}$

$$x_2 = 0.516$$

7.24 Calculate the solubility of naphthalene at 25 °C in any solvent in which it forms an ideal solution. The melting point of naphthalene is 80 °C, and the enthalpy of fusion is 19.19 kJ

mol^{-1}. The measured solubility of naphthalene in benzene is $x_1 = 0.296$.

SOLUTION

$$- \ln x_1 = \Delta_{fus}H_1(T_{0,1} - T)/RTT_{0,1}$$

$$\ln x_1 = \frac{-(19290 \text{ J mol}^{-1})(353 \text{ K} - 298 \text{ K})}{(8.314 \text{ J K}^{-1} \text{ mol}^{-1})(353 \text{ K})(298 \text{ K})}$$

$$x_1 = 0.297$$

7.25 *The NBS Tables of Chemical Thermodynamic Properties* list $\Delta_f G^o$ for I_2 in C_6H_6:x as 7.1 kJ mol. The x indicates that the standard state for I_2 in C_6H_6 is on the mole fraction scale. What is the solubility of I_2 in C_6H_6 at 298 K on the mole fraction scale? A chemical handbook lists the solubility as 16.46 g I_2 in 100 cm^3 of C_6H_6. Are these solubilities consistent?

SOLUTION

I_2 (cr) + $[C_6H_6]$ = I_2 in C_6H_6

$$\Delta G^o = -RT \ln x$$

$$7.1 = -(8.314 \times 10^{-3})(298) \ln x$$

$x = 0.0569$ where x is the equilibrium mole fraction of I_2

The mole fraction calculated from the chemical handbook is

$$\frac{\frac{16.46}{253.8}}{\frac{16.46}{253.8} + \frac{(100)(0.8787)}{78.12}} = 0.0545$$

so the values are consistent.

7.26 The following cooling curves have been found for the system antimony-cadmium.

Cd. Wt. %	0	20	37.5	47.5	50	58	70	93	100
First break in curve, °C	--	550	461	--	419	--	400	--	--
Continuing constant temperature, °C	630	410	410	410	410	439	295	295	321

Construct a phase diagram, assuming that no breaks other than these actually occur in any cooling curve. label the diagram completely and give the formula of any compound formed. How many degrees of freedom are there for each area and at each eutectic point?

SOLUTION

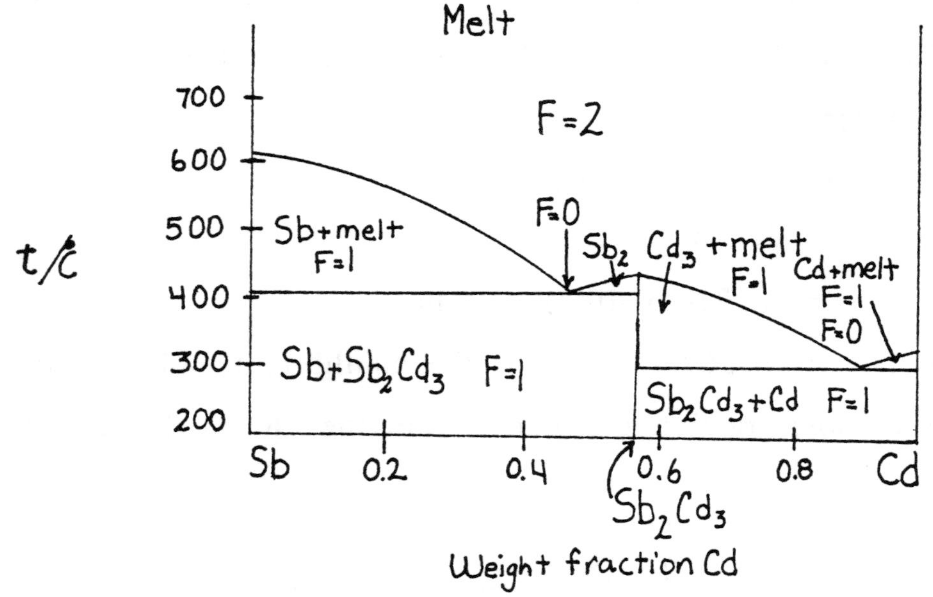

7.27 The phase diagram for magnesium-copper at constant pressure shows that two compounds are formed: $MgCu_2$ that melts t 800 °C, and Mg_2Cu that melts at 580 °C. Copper melts at 1085 °C, and Mg at 648 °C. The three eutectics are at 9.4% by weight Mg

(680 °C), 34% by weight Mg (560 °C), and 65% by weight Mg (380 °C). Construct the phase diagram. How many degrees of freedom are there for each area and at each eutectic point?

SOLUTION

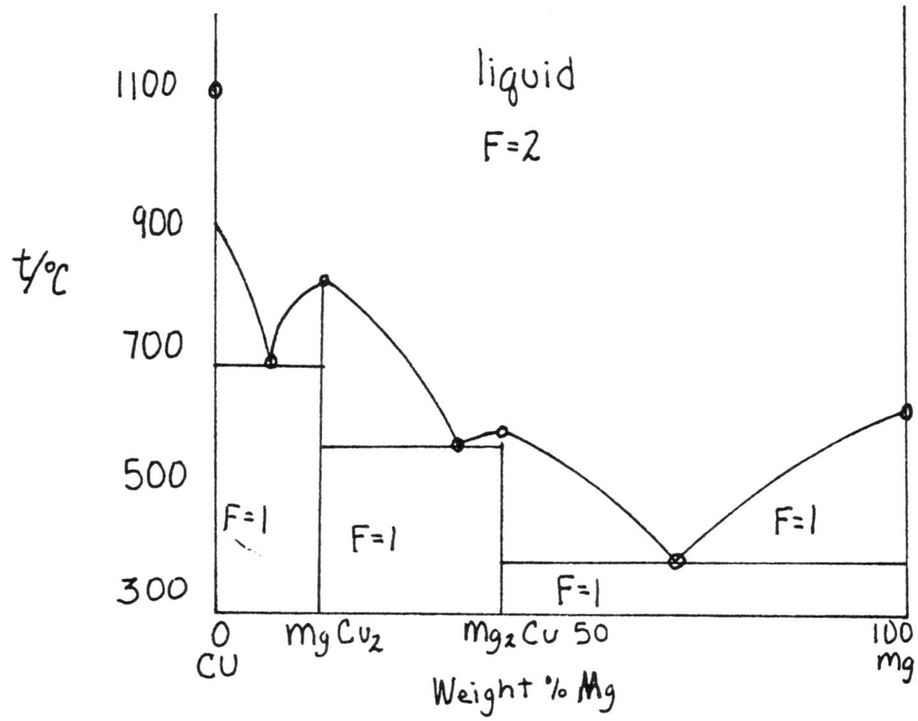

F = 2 - p + 1 In the liquid region F = 2; in the two-phase regions F = 1; and at the eutectic point F = 0.

7.28 For the ternary system benzene-isobutanol-water at 25 °C and 1 bar the following compositions have been obtained for the two phases in equilibrium.

Water-Rich Phase		Benzene-Rich Phase	
Isobutanol, wt.%	Water, wt.%	Isobutanol, wt. %	Benzene, wt.%
2.33	97.39	3.61	96.20
4.30	95.44	19.87	79.07
5.23	94.59	39.57	57.09
6.04	93.83	59.48	33.98
7.32	92.64	76.51	11.39

Plot these data on a triangular graph, indicating the tie lines. (a) Estimate the compositions of the phases that will be produced from a mixture of 20% isobutanol, 55% water, and 25% benzene. (b) What will be the composition of the principal phase when the first drop of the second phase separates when water is added to a solution of 80% isobutyl alcohol in benzene?

SOLUTION

(a) H_2O layer: 5.23% isobutanol and 94.5% H_2O
Benzene layer: 39.57% isobutanol and 57.09% benzene

(b) 10% H_2O
72% isobutanol
18% benzene

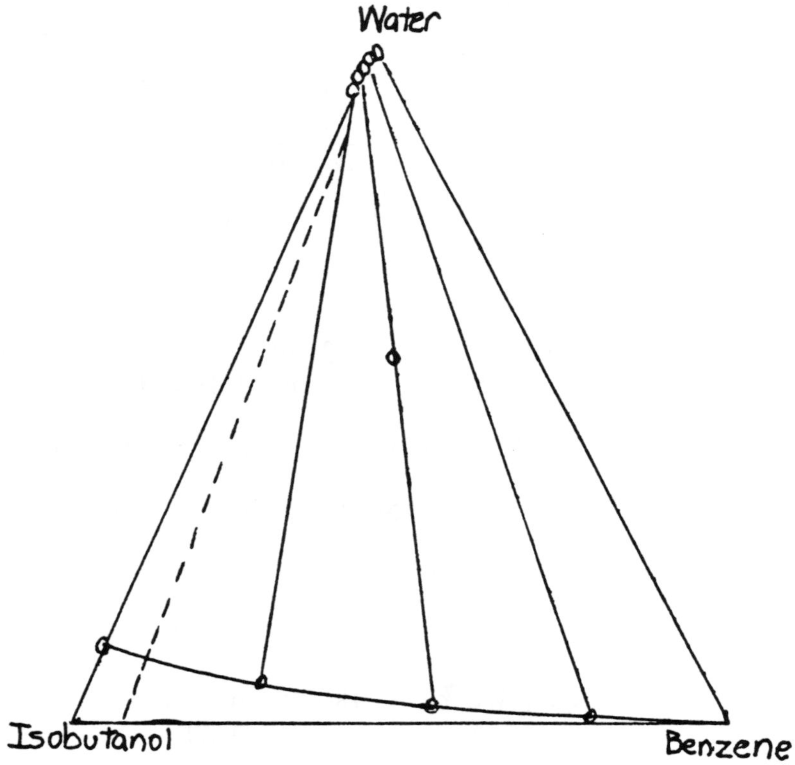

7.29 The following data are available from the system nickel sulfate-sulfuric acid-water at 25 °C. Sketch the phase diagram on triangular coordinate paper, and draw appropriate tie lines.

Liquid Phase		Solid Phase
$NiSO_4$, wt.%	H_2SO_4, wt.%	
28.13	0	$NiSO_4{\cdot}7H_2O$
26.34	1.79	$NiSO_4{\cdot}7H_2O$
27.16	3.86	$NiSO_4{\cdot}7H_2O$
26.15	4.92	$NiSO_4{\cdot}6H_2O$
15.64	19.34	$NiSO_4{\cdot}6H_2O$
10.56	44.68	$NiSO_4{\cdot}6H_2O$
9.65	48.46	$NiSO_4{\cdot}H_2O$
2.67	63.73	$NiSO_4{\cdot}H_2O$
0.12	91.38	$NiSO_4{\cdot}H_2O$
0.11	93.74	$NiSO_4$
0.08	96.80	$NiSO_4$

SOLUTION

	Wt.% $NiSO_4$
$NiSO_4{\cdot}H_2O$	89.6
$NiSO_4{\cdot}6H_2O$	58.8
$NiSO_4{\cdot}7H_2O$	54.9

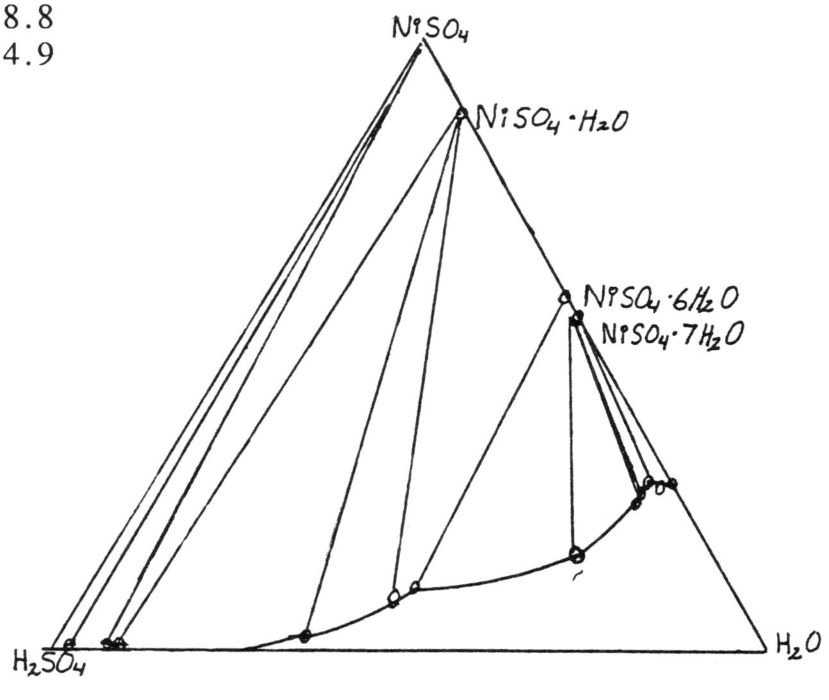

7.30 The Gibbs-Duhem equation in the form

$$\left(\frac{\partial M}{\partial T}\right)_{P,x} dT + \left(\frac{\partial M}{\partial P}\right)_{T,x} dP - \sum (x_i dM_i) = 0$$

applies to any molar thermodynamic property M in a homogeneous phase. If this is applied to G^E, it may be shown* that if the vapor is an ideal gas,

$$x_1 \frac{d \ln (y_1 P)}{dx_1} + x_2 \frac{d \ln (y_2 P)}{dx_1} = 0 \text{ (constant T)}$$

Show that this can be rearranged to the coexistence equation

$$\frac{dP}{dy_1} = \frac{P(y_1 - x_1)}{y_1(1 - y_1)}$$

Thus if P versus is measured, there is no need for measurements of x_1.

SOLUTION

$$x_1 d \ln y_1 = x_1 d \ln P = x_2 d \ln y_2 + x_2 d \ln P = 0$$

$$d \ln P = -\frac{x_1}{y_1} dy_1 = \frac{x_2}{y_2} dy_2 = \left(-\frac{x_1}{y_1} + \frac{x_2}{y_2}\right) dy_1$$

$$= \frac{x_2 y_1 - x_1 y_2}{y_1 y_2} dy_1 = \frac{y_1 - x_1}{y_1(1 - y_1)} dy_1$$

---

*J. M. Smith and H. C. Van Ness, *Introduction to Chemical Engineering Thermodynamics*, McGraw-Hill, New York, 1975, p. 346.

7.31 Use the activity coefficients for acetone-ether solutions at 30 °C from Table 7.3 to calculate G^E/RT at the various mole fractions of acetone.

<u>SOLUTION</u>

$$G^E/RT = x_1 \ln \gamma_1 + x_2 \ln \gamma_2$$

$$= 0.8 \ln 1.04 + 0.2 \ln 1.60 = 0.125$$

$$= 0.6 \ln 1.14 + 0.4 \ln 1.31 = 0.186$$

$$= 0.5 \ln 1.21 + 0.5 \ln 1.19 = 0.182$$

$$= 0.4 \ln 1.28 + 0.6 \ln 1.12 = 0.166$$

$$= 0.8 \ln 1.56 + 0.8 \ln 1.04 = 0.120$$

7.32 (a) $y_{EtBr_2} = 0.802$; (b) $x_{EtBr_2} = 0.425$

7.33 (a) $y_{CHCl_3} = 0.635$; (b) 20.91 kPa

7.34 (a) $y_B = 0.722$, $P = 69.55$ kPa; (b) 0.536

7.35 (a)$x_{ClB} = 0.591$; (b) $y_{ClB} = 0.731$; (c) 1.081 bar

7.36 $x_B = 0.240$, $y_B = 0.434$

7.37 571 g of $H_2O(g)$

7.38 5.293 J K^{-1} mol^{-1}, -1577 J mol^{-1}

7.39 $y_{n\text{-}Pr} = 0.406$

7.40 $x_{C_6H_6} = 0.55$; Pure C_6H_6 can be obtained by distillation provided $x_{C_6H_6} > 70.55$

7.41 (a) $y_{Pr} = 0.37$, (b) $y_{Pr} = 0.59$

7.42 33.0% O_2, 67.0% N_2

7.43 $\gamma_{acetone} = 1.67$, $\gamma_{cs_2} = 1.38$

7.44

x_1	0	0.2	0.4	0.6	0.8	1
γ_1	-	0.58	0.70	0.84	0.96	1.00
γ_2	1.00	0.98	0.89	0.74	0.61,	-

7.45 11.17 kJ mol^{-1}

7.46 (a) K = 19.9 kPa (b) $\gamma'_{CHCl_3} = 1.13, 1.37, 1.65, 1.88, 1.96$

7.47 $\gamma_1 = 3.12, 1.63, 1.19, 1.02$

$\gamma'_2 = 0.314, 0.417, 0.571, 0.772$

7.49 92.4 g mol^{-1}

7.51 122 kg mol^{-1}

7.52 0.0182 bar

7.53 (a) 36.8 kg mol^{-1} (b) 48.5 kg mol^{-1}

7.54 $x_A = 0.108$. The solution is not ideal.

7.55 $x_{Cd} = 0.842$

7.57 90.7% aniline, 5.0% n-heptane, 4.3% methylcyclohexane.

CHAPTER 8: Electrochemical Equilibrium

8.1 How much work is required to bring two protons from an infinite distance of separation to 0.1 nm? Calculate the answer in joules using the protonic charge 1.602 x 10^{-19} C. What is the work in kJ mol^{-1} for a mole of proton pairs?

SOLUTION

$$\text{Potential } \phi = \frac{Q_2}{4\pi\varepsilon_o r}$$

$$= \frac{(1.602 \times 10^{-19}\ C)(0.89875 \times 10^{10}\ N\ m^2\ C^{-2})}{10^{-10}\ m}$$

$$= 14.398\ J\ C^{-1}$$

$$Q_1\phi = (1.602 \times 10^{-19}\ C)(14.398\ J\ C^{-1})$$

$$= 2.307 \times 10^{-18}\ J$$

$$= (2.307 \times 10^{-18}\ J)(6.022 \times 10^{23}\ mol^{-1})$$

$$= 1389.3\ kJ\ mol^{-1}$$

8.2 How much work in kJ mol^{-1} can in principle be obtained when an electron is brought to 0.5 nm from a proton?

SOLUTION

$$w = \frac{Q_1 Q_2}{4\pi e_o r}$$

$$= \frac{(0.8988 \times 10^{10}\ N\ m^2\ C^{-2})(1.602 \times 10^{-19}\ C)^2(6.022 \times 10^{23}\ mol^{-1})}{(b5 \times 10^{-10}\ m)(10^3\ J\ kJ^{-1})}$$

$$= 277.8\ kJ\ mol^{-1}$$

8.3 A small dry battery of zinc and ammonium chloride weighing 85 g will operate continuously through a 4-Ω resistance for 450 min before its voltage falls below 0.75 V. The initial voltage is 1.60, and the effective voltage over the whole life of the battery is taken to be 1.00. Theoretically, how many kilometers above the earth could this battery be raised by the energy delivered under these conditions?

SOLUTION

$$I = \frac{E}{R} = \frac{1\ \text{V}}{4\ \Omega} = 0.25\ \text{A}$$

$$\text{Power} = I^2R = (0.25\ \text{A})^2(4\ \Omega) = 0.25\ \text{W}$$

$$\text{Work} = (0.25\ \text{W})(450 \times 60\ \text{s}) = 6.75 \times 10^3\ \text{J}$$

$$= (0.085\ \text{kg})(9.80\ \text{m s}^{-2})h$$

$$h = \frac{6.75 \times 10^3\ \text{J}}{(0.085\ \text{kg})(9.80\ \text{m s}^{-2})} = 8103\ \text{m}$$

$$= \frac{(8103 \times 10^5\ \text{cm})}{(2.54\ \text{cm in}^{-1})(12\ \text{in ft}^{-1})(5280\ \text{ft mile}^{-1})} = 5.04\ \text{miles}$$

8.4 (a) The mean ionic activity coefficient of 0.1 molar HCl(aq) at 25 °C is 0.796. What is the activity of HCl in this solution? (b) The mean ionic activity coefficient of 0.1 molar H_2SO_4 is 0.265. What is the activity of H_2SO_4 in this solution?

SOLUTION

$$a_{A_{\nu+}B_{\nu-}} = \gamma_{\pm}^{\ \nu_+\nu_-}\ m^{\nu_+\nu_-}\ (\nu_+^{\ \nu_+}\nu_-^{\ \nu_-})$$

$$\text{(a)}\ a_{HCl} = (0.796)^2(0.1)^2 = 0.00634$$

$$\text{(b)}\ a_{H_2SO_4} = (0.265)^3(0.1)^3\ 2^2 1^1 = 7.44 \times 10^{-5}$$

8.5 The solubility of Ag_2CrO_4 in water is 8.00 x 10^{-5} mol kg^{-1} at 25 °C, and its solubility in 0.04 mol kg^{-1} NaN_3 is 8.84 x 10^{-5} mol kg^{-1}. What is the mean ionic activity coefficient of Ag_2CrO_4 in 0.04 mol kg^{-1} $NaNO_3$?

SOLUTION

$$Ag_2CrO_4(s) = 2Ag^+ + CrO_4^{2-}$$

$$K_{sp} = a^2_{Ag^+} a_{CrO_4^{2-}} = 4\gamma^3_{\pm} m^3$$

When Ag_2CrO_4 is dissolved in H_2O, the ionic strength is essentially zero, and $\gamma_{\pm} \approx 1$.

$$K_{sp} = 4(8 \times 10^{-5})^3 = 2.048 \times 10^{-12}$$

In 0.04 mol L^{-1} $NaNO_3$,

$$K_{sp} = 2.048 \times 10^{-12} = 4(8.84 \times 10^{-5})^3 \gamma^3_{\pm}$$

$$\gamma_{\pm} = 0.905$$

8.6 A solution of NaCl has an ionic strength of 0.24 mol kg^{-1}. (a) What is its molality? (b) What molality of Na_2SO_4 would have the same ionic strength? (c) What molality of $MgSO_4$?

(a) $I = \frac{1}{2} (m_1 z_1^2 + m_2 z_2^2)$

$$0.24 = \frac{1}{2}(1^2 + 1^2)\ m \qquad m = 0.24\ \text{mol kg}^{-1}$$

(b) $$0.24 = \frac{1}{2}(2m \times 1^2 + m \times 2^2) \qquad m = 0.08\ \text{mol kg}^{-1}$$

(c) $$0.24 = \frac{1}{2}(2^2 + 2^2)\ m \qquad m = 0.06\ \text{mol kg}^{-1}$$

8.7 Using the limiting law calculate the mean ionic activity coefficients at 25 °C in water of the following electrolytes at 10^{-3} m: (a) NaCl, (b) $CaCl_2$, (c) $LaCl_3$.

SOLUTION

(a) $$I = 10^{-3}$$

$$\gamma_\pm = 10^{Az_+z_-I^{1/2}}$$

$$= 10^{-0.509(10^{-3})^{1/2}} = 0.964$$

(b) $$I = \frac{1}{2}[0.001 \times 2^2 + 0.002] = 3 \times 10^{-3}$$

$$\gamma_\pm = 10^{-2(0.909)(3 \times 10^{-3})^{1/2}} = 0.880$$

(c) $$I = \frac{1}{2}[0.001 \times 3^2 + 0.003] = 6 \times 10^{-3}$$

$$\gamma_\pm = 10^{-3(0.509)(6 \times 10^{-3})^{1/2}} = 0.762$$

8.8 The cell Pt|H_2(1 bar)|HBr(m)|AgBr|Ag has been studied by H. S. Harned, A. S. Keston, and J. G. Donelson [*J. Cm. Chem. Soc.* **58,** 989 (1936)]. The following table gives the electromotive forces obtained at 24 °C.

m/mol kg^{-1}	0.01	0.02	0.05	0.10
E/V	0.3127	0.2786	0.2340	0.2005

Calculate (a) E^o and (b) the activity coefficient for a 0.10 mol kg^{-1} solution of hydrogen bromide.

SOLUTION

(a) Plot $E + 0.1183 \log m - 0.0602\sqrt{m}$ versus m.

Intercept at m = 0 is $E^o = 0.0710$ V.

(b) $E = 0.0710 \text{ V} - (0.05915 \text{ V}) \log \gamma_{\pm}^{2} m^2$

$$0.2005 \text{ V} = 0.0710 \text{ V} - (0.1183 \text{ V}) \log \gamma_{\pm} - (0.1183 \text{ V}) \log 0.1$$

$$\log \gamma_{\pm} = -\frac{0.2005 - 0.1183 - 0.0710}{0.1183}$$

$$\gamma_{\pm} = 0.804$$

8.9 Design cells without liquid junction that could be used to determine the activity coefficients of aqueous solutions of (a)

NaOH and (b) H_2SO_4. Give the equations relating electromotive force to the mean ionic activity coefficient at 25 °C.

SOLUTION

(a) $Na|NaOH(m)|H_2|Pt$

$$H_2O + e^- = \frac{1}{2} H_2 + OH^- \quad E_R^o$$

$$Na^+ + e^- = Na \quad E_L^o$$

$$H_2O + Na = NaOH + \frac{1}{2} H_2 \quad E^o = E_R^o - E_L^o$$

At 25 °C, $E = E^o - 0.0591 \log (m^2\gamma_\pm^2 P_{H_2}^{1/2})$

(b) $Pt|H_2|H_2SO_4(m)|Ag_2SO_4|Ag$

$$Ag_2SO_4 + 2e^- = SO_4^{2-} + 2Ag \quad E_R^o$$

$$2H^+ + 2e^- = H_2 \quad E_L^o$$

$$H_2 + Ag_2SO_4 = H_2SO_4 + 2\ Ag \qquad E^o = E_R^o - E_L^o$$

At 25°C

$$E = E^o - \frac{0.0591}{2} \log \left(\frac{4m^3\gamma_\pm^3}{P_{H_2}}\right)$$

8.10 The electromotive force of the cell

$Pb(s)|PbSO_4(s)|Na_2SO_4 \cdot 10H_2O(sat)|Hg_2SO_4(s)|Hg(1)$

is 0.9647 V at 25 °C. The temperature coefficient is 1.74×10^{-4} $V\ K^{-1}$. (a) What is the cell reaction? (b) What are the values of $\Delta_r G$, $\Delta_r S$, and $\Delta_r H$?

SOLUTION

(a) $Pb(s) + Hg_2SO_4(s) = PbSO_4(s) + 2Hg(1)$

(b) $\Delta_r G = -nFE = -(2)(96{,}485\ C\ mol^{-1})(0.9647\ V) = -186.16\ kJ\ mol^{-1}$

$$\Delta_r S = nF\left(\frac{\partial E}{\partial T}\right)_P = (2)(96.485\ C\ mol^{-1})(1.74 \times 10^{-4}\ V\ K^{-1})$$

$$= 33.58\ J\ K^{-1}\ mol^{-1}$$

$$\Delta_r H = -nFE + nFT\left(\frac{\partial E}{\partial T}\right)_P$$

$$= -2(96{,}485\ C\ mol^{-1})(0.9647\ V) + 2(96{,}485\ C\ mol^{-1})$$

$$\times (298.15\ K)(1.74 \times 10^{-4}\ V\ K^{-1})$$

$$= -176.15\ kJ\ mol^{-1}$$

8.11 For the galvanic cell

$H_2(1\ bar)|HCl(ai)|Cl_2(1\ bar)$

the standard electromotive force at 298.15 K is 1.3604 V and $(\partial E^o/\Delta\partial T)_P = -1.247 \times 10^{-3}\ V\ K^{-1}$. (a) For the cell reaction, what are the values of $\Delta_r G^o$, $\Delta_r H^o$, $\Delta_r S^o$? (b) For $Cl^-(ao)$ what are the values of and $\Delta_f G^o$, $\Delta_f H^o$, and $\bar{S}^o$?

SOLUTION

(a) $\frac{1}{2} Cl_2(g) + e^- = Cl^-(aq)$

$H^+ (aq) + e^- = \frac{1}{2} H_2(g)$

$\frac{1}{2} Cl_2 + \frac{1}{2} H_2 = H^+(aq) + Cl^-(aq)$

$\Delta_r G^o = -nFE^o = -(96{,}485 \text{ C mol}^{-1})(1.3604 \text{ V})$

$= -131.260 \text{ kJ mol}^{-1}$

$\Delta_r H^o = \Delta_r G^o + T\Delta_r S^o$

$= -131.2604 + (298.15)(-120.3 \times 10^{-3} \text{ kJ K}^{-1} \text{ mol}^{-1})$

$= -167.127 \text{ kJ mol}^{-1}$

$\Delta_r S^o = nF(\frac{\partial E^o}{\partial T})_P = (96{,}485 \text{ C mol}^{-1})(-1.237 \times 10^{-3} \text{ V K}^{-1})$

$= -120.3 \text{ J K}^{-1} \text{ mol}^{-1}$

(b) Since $\Delta_f G^o[H^+(ao)] = 0$,

$\Delta_f G^o[Cl^-(ao)] = -131.260 \text{ kJ mol}^{-1}$

Since $\Delta_f H^o[H^+(ao)] = 0$,

$\Delta_f H^o[Cl^-(ao)] = -167.127 \text{ kJ mol}^{-1}$

Since $\bar{S}^o[H^+(ao)] = 0$

$$\Delta_r S^o = \bar{S}^o[Cl^-(ao)] - \frac{1}{2}\bar{S}^o[H_2(g)] - \frac{1}{2}\bar{S}^o[Cl_2(g)]$$

$$-120.3 = \bar{S}^o[Cl^-(ao)] - \frac{1}{2}(130.684) - \frac{1}{2}(223.066)$$

$$\bar{S}^o[Cl^-(ao)] = 56.6 \text{ J K}^{-1} \text{ mol}^{-1}$$

8.12 In problem 4.8 two equations were derived for calculating ΔG at another temperature if it is known at one. Compare the values of $\Delta_r G^o$ (323 K) and K_w calculated with these equations for

$$H_2O(1) = H^+(ao) + OH^-(ao)$$

<u>SOLUTION</u>

	$\Delta_f G^o$(298 K)	$\Delta_f H^o$ (298 K)	C_P^o(298 K)
$H_2O(1)$	-237.129	-285.830	75.291
$H^+(ao)$	0	0	0
OH^-	-157.144	-229.994	-148.5

$\Delta_r G^o = -157.244 + 237.129 = 79.885$ kJ mol^{-1}

$\Delta_r H^o = -229.994 + 285.830 = 55.836$ kJ mol^{-1}

$\Delta_r C_P = -148.5 - 75.291 = -223.8 \text{ J K}^{-1} \text{ mol}^{-1}$

(a) $\Delta_r G_2^o = \Delta_r G_1^o \, T_2/T_1 + \Delta_r H(1 - T_2/T_1)$

$$\Delta_r G^o(323) = \frac{(79.885)(323.15)}{298.15} + 55.836\left(1 - \frac{323.15}{298.15}\right)$$

$$= 81.902 \text{ kJ mol}^{-1}$$

$$K_w = 5.69 \times 10^{-14}$$

(b) $\Delta_r G_2 = \Delta_r G_1 \, T_2/T_1 + (\Delta_r H_1 - T_1 \Delta_r C_P)(1 - T_2/T_1)$

$$+ T_2 \Delta_r C_P \ln (T_2/T_1)$$

$$= \frac{(79.885)(323.15)}{298.15} + [55.836 + (298.15)(-0.2238)]$$

$$\times \left(1 - \frac{323.15}{298.15}\right) + (323.15)(-0.2238) \ln \left(\frac{323.15}{298.15}\right)$$

$$= 81.673 \text{ kJ mol}^{-1}$$

$$K_w = 6.28 \times 10^{-14}$$

8.13 Calculate E^o for the half cell $OH^- | H_2 | Pt$ at 25 °C using the value of the ion product for water, which is 1.006×10^{-14} (Section 9.1).

SOLUTION

For the cell $Pt|H_2|OH^-|H^+|H_2|Pt$

$$H^+ + e^- = \frac{1}{2} H_2 \qquad E_R^o = 0$$

$$H_2O + e^- = OH^- + \frac{1}{2} H_2 \qquad E_L^o$$

$$H^+ + OH^- = H_2O \qquad E^o = E_R^o - E_L^o = -E_L^o$$

$$E = -E_L^o = \frac{RT}{F} \ln \frac{1}{a_{H^+}a_{OH^-}}$$

$$E_L^o = \frac{RT}{F} \ln K_w$$

$$= \frac{(8.3145 \text{ K}^{-1} \text{ mol}^{-1})(298.15 \text{ K})}{96485 \text{ C mol}^{-1}} \ln (1.006 \times 10^{-14})$$

$$= 0.828 \text{ V}$$

8.14 What are the values of $\Delta_r G^o$ and K for the following reactions at 298 K from Appendix C.1?

(a) $Cu(s) + Zn^{2+}(ao) = Cu^{2+}(ao) + Zn(s)$

(b) $H_2(g) + Cl_2(g) = 2HCl(ai)$

(c) $Ca^{2+}(ao) + CO_3^{2-}(ao) = CaCO_3(s, \text{calcite})$

(d) $\frac{1}{2}Cl_2(g) + Br^-(ao) = \frac{1}{2}Br_2(g) + Cl^-(ao)$

(e) $Ag^+(ao) + Fe^{2+}(ao) = Fe^{3+}(ao) + Ag(s)$

<u>SOLUTION</u>

(a) $\Delta_r G^o = 65.49 + 147.06 = 212.55$ kJ mol^{-1}

$K = 5.79 \times 10^{-38}$

(b) $\Delta_r G^o = 2(-131.228) = -262.46$

$K = 9.56 \times 1045$

(c) $\Delta_r G^o = -1128.79 + 553.58 + 527.81 = -47.40$

$K = 2.01 \times 10^8$

(d) $\Delta_r G^o = \frac{1}{2}(3.110) - 131.228 + 103.96 = -25.71$

$K = 3.20 \times 10^4$

(e) $\Delta_r G^o = -4.7 - 77.107 + 78.90 = -2.9$

$K = 3.23$

8.15 For the electrochemical cell

$$Zn(s)|Zn^{2+}(aq)||Cu^{2+}(aq)|Cu(s)$$

what are (a) the half cell reactions, (b) cell reaction, (c) the standard electromotive force, (d) the equilibrium constant for the cell reaction, and (e) equilibrium constant expression?

SOLUTION

(a) R $Cu^{2+}(aq) + 2e^- = Cu(s)$ $E^o = 0.337$ V

L $Zn^{2+}(aq) + 2e^- = Zn(s)$ $E^o = -0.763$ V

(b) $Zn(s) + Cu^{2+}(aq) = Zn^{2+}(aq) + Cu(s)$

(c) $E^o = 0.337 - (-0.763) = +1.100$ V

(d) $K = e^{2(96,485)(1.1)/(8.314)(298)} = 1.62 \times 10^{37}$

(e) $$K = \frac{a_{ZnCl_2}}{a_{CuCl_2}} = \frac{4\, m^3_{ZnCl_2}\, \gamma^3_{\pm(ZnCl_2)}}{4\, m^3_{CuCl_2}\, \gamma^3_{\pm(CuCl_2)}}$$

If the $CuCl_2$ solution and $ZnCl_2$ solution have the same concentration, the activity coefficients will be approximately the same so that

$$\frac{m_{ZnCl_2}}{m_{CuCl_2}} \approx (1.62 \times 10^{37})^{1/3}$$

8.16 From standard electrode potentials in Table 8.2 what are the standard Gibbs energies of formation at 25 °C for $Cl^-(ao)$, $OH^-(ao)$, and $Na^+(ao)$?

SOLUTION

For $H_2(g)|HCl(ai)|Cl_2(g)$

R $\frac{1}{2} Cl_2 + e^- = Cl^-$ $E^o = 1.3604$ V

L $H^+ + e^- = \frac{1}{2} H_2$ $E^o = 0$

$\frac{1}{2} Cl_2 + \frac{1}{2} H_2 = H^+(ao) + Cl^-(ao)$ $E^o = 1.3604$ V

$\Delta_r G^o = -nFE^o = -(96{,}485 \text{ C mol}^{-1})(1.3604 \text{ V}) = -131.258 \text{ kJ mol}^{-1}$

This is $\Delta_f G^o[Cl^-(ao)]$ since $\Delta_f G^o [H^+(ao)]$ is zero by convention. Appendix C.1 gives $-131.228 \text{ kJ mol}^{-1}$.

For $H_2(g)|H^+(ao)||OH^-(ao)|O_2(g)$

R $\frac{1}{4} O_2 + \frac{1}{2} H_2O(1) + e^- = OH^-(ao)$ $E^o = 0.401$ V

L $H^+(ao) + e^- = \frac{1}{2} H_2$ $E^o = 0$

$\frac{1}{4} O_2 + \frac{1}{2} H_2 + \frac{1}{2} H_2O = H^+(ao) + OH^-(ao)$ $E^o = 0.401$

$\Delta_r G^o = -(96{,}485 \text{ C mol}^{-1})(0.401 \text{ V}) = -38.69 \text{ kJ mol}^{-1}$

$= \Delta_f G^o[OH^-(ao)] - \frac{1}{2}(237.129) = -157.26 \text{ kJ mol}^{-1}$

Appendix C.1 gives -157.244 kJ mol^{-1}.

For $Na(cr)|Na^+(ao)||H^+(ao)|H_2(g)$

R $H^+ + e^- = \frac{1}{2} H_2$ $E^o = 0$

L $Na^+ + e^- = Na$ $E^o = -2.714$ V

$Na(cr) + H^+(ao) = Na^+(ao) + \frac{1}{2} H_2(g)$ $E^o = 2.714$ V

$\Delta_r G^o = -(96{,}485 \text{ C mol}^{-1})(2.174 \text{ V}) = -261.86 \text{ kJ mol}^{-1} = \Delta_f G^o[Na^+(ao)]$

Appendix C.1 gives -261.905 kJ mol^{-1}.

8.17 According to Table 8.2 what are the equilibrium constants for the following reactions at 25 °C?

(a) $H^+(ao) + Li(s) = Li^+(ao) + \frac{1}{2} H_2(g)$

(b) $2H^+(ao) + Pb(s) = Pb^{2+}(ao) + H_2(g)$

(c) $3H^+(ao) + Au(s) = Au^{3+}(ao) + \frac{3}{2} H_2(g)$

SOLUTION

(a) R $H^+ + e^- = \frac{1}{2} H_2$ $E^o = 0$

L $Li^+ + e^- = Li$ $E^o = -3.045$ V

$H^+ + Li = Li^+ + \frac{1}{2} H_2$ $E^o = 3.045$ V

$$K = e^{nFE^{o}/RT}$$

$$= e^{(96{,}485\ \text{C mol}^{-1})(3.045\ \text{V})/(8.314\ \text{J K}^{-1}\ \text{mol}^{-1})(298\ \text{K})}$$

$$= 3.1 \times 10^{51}$$

(b) R $2H^{+} + 2e^{-} = H_2$ $\quad E^{o} = 0$

L $Pb^{2+} + 2e^{-} = Pb$ $\quad E^{o} = -0.126$ V

$2H^{+} + Pb = Pb^{2+} + H_2$ $\quad E^{o} = 0.126$

$$K = e^{2(96{,}485)(0.126)/(8.314)(298)} = 1.8 \times 10^{4}$$

(c) R $3H^{+} + 3e^{-} = \frac{3}{2} H_2$ $\quad E^{o} = 0$

L $Au^{3+} + 3e^{-} = Au$ $\quad E^{o} = 1.50$ V

$3H^{+} + Au = Au^{3+} + \frac{3}{2} H_2$ $\quad E^{o} = -1.50$ V

$$K = e^{3(96{,}485)(-1.50)/(8.314)(298)} = 7.8 \times 10^{-77}$$

The striking resistance of gold to corrosion by acid is evident.

8.18 At 25 °C the standard electrode potential for the $Ag^{+}|Ag$ electrode is 0.7991 V, and the solubility product for AgI is 8.2 x 10^{-17}. What is the standard electrode potential for $I^{-}|AgI|Ag$?

SOLUTION

$Ag|Ag^{+}||I^{-}|AgI|Ag$

$$AgI + e^- = Ag + I^- \qquad E^o_R$$

$$Ag^+ + e^- = Ag \qquad E^o_L = 0.7991\ V$$

$$AgI = Ag^+ + I^- \ K_{sp} \quad E^o = E^o_R - 0.7991\ V$$

$$E^o_R - 0.7991 = \frac{RT}{F} \ln K_{sp}$$

$$= \frac{(8.314)(298)}{96485} \ln 8.2 \times 10^{-17}$$

$E^o_R = -0.152$ V, the E^o for $I^-|AgI|Ag$

8.19 Using data from Appendix C.1 calculate the solubility of AgCl(cr) in water at 298.15 K. The salt is completely dissociated in the aqueous phase.

SOLUTION

$$AgCl(s) = Ag^+(ao) + Cl^-(ao)$$

$$\Delta_r G^o = 77.107 - 131.228 + 109.789 = 55.668 \text{ kJ mol}^{-1}$$

$$K = \exp(-55{,}668/8.3145 \times 298.15) = 1.767 \times 10^{-10}$$

$$= (a_{Ag^+})(a_{Cl^-}) = m^2 \gamma_\pm^2 \approx m^2$$

$$m = 1.33 \times 10^{-5} \text{ mol kg}^{-1}$$

8.20 Calculate standard electrode potentials at 25 °C for the following electrodes using Appendix C.1.

(a) Li^{+} (ao)|Li(s)

(b) F^{-}(ao)|F2(g)

(c) Pb^{2+}(ao)|PbO_2(s)|Pb

<u>SOLUTION</u>

$Li^{+} + e^{-} = Li$

$\Delta_r G^o$ = -(293.31) = 293.31 kJ mol^{-1}

$E^o = -\Delta_r G^o/nF$ = -(293,310 J mol^{-1})/(96,485 C mol^{-1}) = -3.040 V

$\frac{1}{2} F_2(g) + e^{-} = F^{-}$

$\Delta_r G^o$ = -278.79 kJ mol^{-1}

E^o = (278,790 J mol^{-1})/(96.485 C mol–^{1}) = 2.889 V

$\frac{1}{2} PbO_2 + 2H^{+} + e^{-} = \frac{1}{2} Pb^{2+} + H_2O$

$\Delta_r G^o = -237.129 + \frac{1}{2}(-24.43) - \frac{1}{2}(-217.33) = -140.68$ kJ mol^{-1}

E^o = (140,680 J mol^{-1})/(96,485 C mol^{-1}) = 1.458 V

8.21 Using Appendix C.1 calculate the values of $\Delta_r G^o$, $\Delta_r H^o$, $\Delta_r S^o$, and $\Delta_r C_P^o$ at 25 °C for the electrode reaction for the $Na^+ | Na$ electrode.

<u>SOLUTION</u>

The shorthand notation for the electrode reaction is $Na^+ + e^- = Na$, but we must remember that we are really talking about the cell for which the cell reaction is

$$Na^+ + \frac{1}{2} H_2(g) = Na(s) + H^+$$

$\Delta_r G^o$ = -(-261.905) = 261.905 kJ mol^{-1}

$\Delta_r H^o$ = -(-240.12) = 240.12 k J mol^{-1}

$\Delta_r S^o = 51.21 - 59.0 - \frac{1}{2}(130.684) = -73.132$ J K^{-1} mol^{-1}

$\Delta_r C_P^o = 28.24 - 46.4 - \frac{1}{2}(28.824) = -32.572$ J K^{-1} mol^{-1}

According to Table 8.2

$\Delta_r G^o = -nFE^o$ = -(96,485 C mol^{-1})(-2.714 V) = 261.860 kJ mol^{-1}

8.22 The standard electrode potentials E^o in the earlier literature are based on a standard state pressure of 1 atm. Show that when the bar is used as the standard state pressure, standard electrode potentials E^o(atm) need to be corrected to E^o(bar) using

$$E^o(\text{bar}) = E^o(\text{atm}) + (0.000169\ \text{V})\Delta\nu$$

where $\Delta\nu$ is the increase in the number of gaseous molecules as the cell reaction (including hydrogen) proceeds as written.

<u>SOLUTION</u>

$$H_2(g) + Ox = 2H^+ + Red$$

We have already seen (problem 5.53)

$$\Delta G^o(\text{bar}) = \Delta G^*(\text{atm}) - [RT \ln (P^*/P^o)]\ \Delta\nu$$

$$E^o(\text{bar}) = -\Delta G^o(\text{bar})\ nF = E^*(\text{atm}) + [RTuF\ln(P^*/P^o)]\Delta\nu$$

8.23 Calculate the standard Gibbs energy of formation of $Na^+(ao)$ from its $\Delta_f H^o$ and $\bar{S}^o$ using data from Appendix C.1 and equation 8.78.

<u>SOLUTION</u>

The formation reaction is

$$Na + H^+ = Na^+ + \frac{1}{2} H_2$$

Therefore, the standard Gibbs energy of formation is given by

$$\Delta_f G^o = \Delta_f H^o - T(\bar{S}^o_{Na^+} + \frac{1}{2} \bar{S}^o_{H_2} - \bar{S}^o_{Na})$$

$$= -240.12 - (298.15)(10^{-3})[59.0 - 51.21 + \frac{1}{2} (130.684)]$$

$$= -261.92 \text{ k J mol}^{-1}$$

8.24 When a hydrogen electrode and a normal calomel electrode are immersed in a solution at 25 °C a potential of 0.664 V is obtained. Calculate (a) the pH, and (b) the hydrogen-ion activity.

SOLUTION

(a) $$E = E^o - \frac{RT}{nF} \ln a_{H^+} = E^o + 0.0591 \text{ pH}$$

$$\text{pH} = \frac{E - E^o}{0.0591} = \frac{0.664 - 0.2802}{0.0591} = 6.49$$

(b) $$a_{H^+} = 10^{-6.49} = 3.24 \times 10^{-7}$$

8.25 Calculate the equilibrium constant at 25 °C for the reaction

$$2H^+ + D_2(g) = H_2(g) + 2D^+$$

from the electrode potential for $D^+|D_2|Pt$, which is -3.4 mV at 25 °C.

SOLUTION

$$Pt|D_2, D^+||H^+, H_2|Pt$$

$E \quad = 0 - (-0.0034) = 0.0034 \text{ V}$

$K \quad = e^{nFE^o/RT}$

$$= \exp\left[\frac{2(96500 \text{ C mol}^{-1})(0.0034 \text{ V})}{(8.314 \text{ J K}^{-1} \text{ mol}^{-1})(298 \text{ K})}\right] = 1.30$$

8.26 A water electrolysis cell operated at 25 °C consumes 25 kWh/lb of hydrogen produced. Calculate the cell efficiency using Δ_rG^o for the decomposition of water.

SOLUTION

$$H_2O(l) \quad = H_2(g) + \frac{1}{2} O_2(g)$$

$\Delta_rG^o = -237.129 \text{ kJ mol}^{-1}$

The electrical energy used per mole of H_2 produced is

$(25\ \text{kwh/lb})(10^3\ \text{w kw}^{-1})(3600\ \text{s hr}^{-1})(454\ \text{g lb})^{-1}(2\text{g mol}^{-1})$

$= 396{,}475\ \text{J mol}^{-1}$

$$\text{Efficiency} = \frac{237.129}{396.475} = 0.60$$

8.27 (a) When methane is oxidized completely to $CO_2(g)$ and $H_2O(1)$ at 25 °C, how much electrical energy can be produced using a fuel cell, assuming no electrical losses? What is the electromotive force of the fuel cell? (b) When one mole of methane is oxidized completely in a Carnot engine that operates between 500 K and 300 K, how much electrical energy can be produced, assuming that the mechanical energy can be converted completely into electrical energy?

SOLUTION

(a) R $2\ O_2 + 8\ e^- + 8\ H^+ = 4\ H_2O$

L $CO_2 + 8\ H^+ + 8\ e^- = CH_4 + 2\ H_2O$

$2\ O_2 + CH_4 = CO_2 + 2\ H_2O$

$\Delta_rG^o = -394.359 + 2(-237.129) - (50.72) = -817.90\ \text{kJ mol}^{-1}$

$= -nFE^o$

$$E^o = \frac{817.90 \times 10^3 \text{ J mol}^{-1}}{8(96485 \text{ C mol}^{-1})} = 1.0596 \text{ V}$$

(b) $\Delta_r H^o = -393.509 + 2(-285.830) - (-74.81)$

$= -890.4 \text{ kJ mol}^{-1}$

$$|w| = |q|\frac{(T_1 - T_2)}{T_1} = 890.4\,\frac{200}{500} = 356 \text{ kJ mol}^{-1}$$

Thus the fuel cell would produce over twice as much electrical energy.

8.28 Calculate the electromotive force of Li(l)|LiCl(l)|Cl_2(g) at 900 K for $P_{Cl_2} = 1$ bar. This high-temperature battery is attractive because of its high electromotive force and low atomic masses. Lithium chloride melts at 883 K and lithium at 453.69 K. [The $\Delta_f G^o$ for LiCl(l) at 900 K in JANAF Thermochemical Tables is -335.140 kJ mol^{-1}.]

SOLUTION

$$\text{Li}(l) + \frac{1}{2}\text{Cl}_2(\text{g}) = \text{LiCl}(l)$$

$\Delta_r G^o = -335{,}140 \text{ J mol}^{-1} = -FE^o$

$$E^o = \frac{-335140 \text{ J mol}^{-1}}{-96485 \text{ C mol}^{-1}} = 3.474 \text{ V}$$

8.29 A membrane permeable only by Na^+ is used to separate the following two solutions:

α	0.10 mol kg^{-1} NaCl	0.05 mol kg^{-1} KCl
β	0.05 mol kg^{-1} NaCl	0.10 mol kg^{-1} KCl

What is the membrane potential at 25 °C and which solution has the highest positive potential?

SOLUTION

$$\phi^\beta - \phi^\alpha = \frac{RT}{z_iF} \ln \frac{a_i^\beta}{a_i^\alpha}$$

$$= -\frac{(8.314 \text{ J K}^{-1} \text{ mol}^{-1})(298 \text{ K})}{96485 \text{ C mol}^{-1}} \ln \frac{0.05}{0.10} = 0.018 \text{ V}$$

The β phase is more positive because of the diffusion of Na^+ from α to β. Since the ionic strengths of the two solutions are the same, the activity coefficients of Na^+ in α and β are very nearly the same.

8.30 5.759×10^9 V m^{-1}

8.31 (a) 463 kJ mol^{-1} (b) 46.3 kJ mol^{-1} (c) 5.78 kJ mol^{-1}

8.32 $a = 4\ m^3\gamma_{\pm}^3$

8.33 (a) $\frac{a_{LICl}^{1/2}}{m}$ (b) $\frac{a_{AlCl_3}^{1/4}}{27^{1/4}\ m}$ (c) $\frac{a_{MgSO_4}^{1/2}}{m}$

8.34 (a) 0.1 (b) 0.3 (c) 0.4 (d) 0.4

8.35 0.834

8.36 $K = a(HCl)/(P_{H_2}/P^o)^{1/2} = 5.701 \times 10^3$

8.37 $pK = 1.018\ \sqrt{I} + \log \frac{m_1}{m_2} + \frac{(E - E^o)F}{2.303\ RT} + \log m_3$

8.38 (a) $2AgCl + Zn = 2Ag + ZnCl_2(0.555\ mol\ kg^{-1})$

(b) -195.9 kJ mol^{-1} (c) -77.6 J K^{-1} mol^{-1}

(d) -219.0 kJ mol^{-1}

8.39 -130.318 kJ mol^{-1} -125 J K^{-1} mol^{-1} -167.580 kJ mol^{-1}

8.40 (a) The 10.02% electrode is negative.

(b) $\Delta H = -2913$ J mol^{-1}

(c) 0.030462 V

8.41 791.885 kJ mol^{-1}, 55.835 kJ mol^{-1}, -80.668 J K^{-1} mol^{-1}, 1.008×10^{-14}

8.42 -109.805 kJ mol^{-1}

8.44 (a) 4.405, (b) 4.400 V

8.45 $H_2O(l) = H^+(ao) + OH^-(ao)$

$K_w = 1.003 \times 10^{-14}$

8.46 Ag|Ag+||Br|AgBr|Ag, $K = 10^{-11.85}$

8.47 1.34×10^{-5} mol L^{-1}

8.48 -0.17 mV, -0.34 mV

8.49 -0.4009 V

8.50 -108.86 kJ mol^{-1}

8.51 -744.49 kJ mol^{-1}

8.52 1.239 V

8.53 (a) 2.62 (b) 2.399×10^{-3}

8.54 (a) 237.129 kJ mol^{-1} (b) 285.83 kJ mol^{-1} (c) 50.8 K

8.55 1.172 V

8.56 15.3

CHAPTER 9: Ionic Equilibria and Biochemical Reactions

9.1 Show that the slope of the titration curve of a monobasic weak acid is given by

$$\frac{d\alpha}{dpH} = \frac{2.303\ K[H^+]}{(K + [H^+])^2}$$

where α is the degree of neutralization.

SOLUTION

	HA	$= H^+ + A^-$	
Initial	1		
eq	$1-\alpha$	$[H^+]$	α

$$K = \frac{[H^+]\alpha}{1 - \alpha}$$

$$\alpha = \frac{K}{(K + [H^+])}$$

$$\frac{d\alpha}{d[H^+]} = -\frac{K}{(K + [H^+])^2}$$

$$\frac{dpH}{d[H^+]} = -\frac{1}{2.303}\frac{d\ln[H^+]}{d[H^+]} = \frac{-1}{2.303[H^+]}$$

$$\frac{d\alpha}{dpH} = \frac{d\alpha}{d[H^+]}\frac{d[H^+]}{dpH} = +\frac{2.303\ K[H^+]}{(K + [H^+])^2}$$

9.2 Using the Debye-Hückel theory, estimate the apparent pK for acetic acid at 0.01 mol L^{-1} ionic strength. At 25 °C the thermodynamic pK value is 4.756. It is assumed that the activity coefficient for undissociated acetic acid is unity at this value of the ionic strength.

SOLUTION

Using equation 9.14

$$K_{app} = \frac{K}{\gamma_-}$$

From equation 8.48

$$\log \gamma_i = -0.509\sqrt{0.01}$$

$$\gamma_i = 0.889$$

$$K_{app} = \frac{10^{-4.756}}{0.889} = 1.973 \times 10^{-5}$$

$$pK_{app} = 4.70$$

9.3 According to Appendix C.1 what are the values of $\Delta_r G^o$, $\Delta_r H^o$, and $\Delta_r S^o$ at 298 K for

$$H_2O(1) \quad = H^+(ao) \quad + OH^-(ao)$$

Show that the same value of $\Delta_r S^o$ is obtained from $\Delta_r G^o$ and $\Delta_r H^o$ as by using $\Delta_r S^o = \Sigma \nu_i S_i^o$. Calculate K_w at 298 K.

SOLUTION

$\Delta_r G^o$ = -157.244 + 2237.129 = 79.885 kJ mol^{-1}

$\Delta_r H^o$ = -229.994 + 285.830 = 55.836 kJ mol^{-1}

$\Delta_r S^o$ = -10.75 - 69.91 = -80.66 J K^{-1} mol^{-1} or

$$\Delta_r S^o = \frac{\Delta H^o - \Delta G^o}{T} = \frac{(55836 - 79885)\ \text{J mol}^{-1}}{298.15\ \text{K}}$$

$$= -80.66\ \text{J K}^{-1}\ \text{mol}^{-1}$$

$K_w = \exp[-79.885/(8.3143 \times 10^{-3})(298.15)]$

$= 1.010 \times 10^{-14}$

Note the use of the value of R used in constructing Appendix C.1.

9.4 For the acid dissociation of acetic acid, $\Delta_r H^o$ is approximately zero at room temperature in H_2O. For the acidic form of aniline, which is approximately as strong an acid as acetic acid, $\Delta_r H^o$ is approximately 21 kJ mol^{-1}. Calculate $\Delta_r S^o$ for each of the following reactions.

$CH_3CO_2H = H^+ + CH_3CO_2^-$ pK = 4.75

$C_6H_5NH_3^+ = H^+ + C_6H_5NH_2$ pK = 4.63

How do you interpret these entropy changes? What compensates for the increase in entropy expected from the increase in number of molecules in the balanced chemical reaction?

<u>SOLUTION</u>

For acetic acid

$\Delta_r G^o = -RT \ln K$

$= RT\ 2.303\ pK$

$= (8.314\ J\ K^{-1}\ mol^{-1})(298\ K)(2.303)(4.75)$

$= 27.1\ kJ\ mol^{-1}$

$\Delta_r S^o = (\Delta_r H^o - \Delta_r G^o)/T = -(27.1 \times 10^3\ J\ mol^{-1})/(298\ K)$

$= -91\ J\ K^{-1}\ mol^{-1}$

This increase in order is due to the hydration of the ions that are formed.

For aniline

$$\Delta_r G^o = (8.314 \text{ J K}^{-1} \text{ mol}^{-1})(298 \text{ K})(2.303)(4.63)$$

$$= 26.4 \text{ kJ mol}^{-1}$$

$$\Delta_r S^o = [(21 - 26.4) \times 10^3 \text{ J mol}^{-1}]/(298 \text{ K})$$

$$= -18 \text{ J K}^{-1} \text{ mol}^{-1}$$

The entropy change is much smaller than for acetic acid because there is no change in the number of ions.

9.5 Estimate pK_3 and pK_2 for H_3PO_4 at 25 °C and 0.1 mol L^{-1} ionic strength. The values at zero ionic strength are $pK_3 = 2.148$ and $pK_2 = 7.198$.

<u>SOLUTION</u>

$$pK_I = pK_{I=0} - \frac{(2n+1)AI^{1/2}}{1 + I^{1/2}}$$

A = 0.509 at 25 °C

n is defined by $HA^{-n} = H^+ + A^{-(n+1)}$

For pK_3 of H_3PO_4, n = 0

$$pK_3 = 2.148 - \frac{(0.509)(0.1)^{1/2}}{1 + (0.1)^{1/2}} = 2.148 - 0.122 = 2.026$$

For pK_2 of H_3PO_4, n=1

$$pK_2 = 7.198 - \frac{(3)(0.509)(0.1)^{1/2}}{1 + (0.1)^{1/2}}$$

$$= 7.198 - 0.369 = 6.831$$

9.6 In a strong acid solution, the amino acid histidine binds three protons. The acid dissociation constants numbered from the weakest acid dissociation are 6.92×10^{-10}, 1.00×10^{-6}, and 1.51×10^{-2} at 25 °C. Calculate the concentrations of the four forms of histidine (His^-, HisH, $HisH_2^+$, and $HisH_3^{2+}$) assuming these constants apply at the ionic strength of the solution.

SOLUTION

The equations derived for H_3PO_4 can be used.

$K_1 = 6.92 \times 10^{-10}$

$K_1K_2 = 6.92 \times 10^{-16}$

$K_1K_2K_3 = 1.045 \times 10^{-17}$

$1 + [H^+]/K_1 + [H^+]^2/K_1K_2 + [H^+]^3/K_1K_2K_3$

$= 1 + 144.5 + 14.45 + 9.6 \times 10^{-5} = 159.95$

$[His^-] = 0.1(1)/159.95 = 0.00063$ M

$[HisH] = 0.1(144.5)/159.95 = 0.0903$ M

$[HisH_2^+] = 0.1(14.45)/159.95 = 0.00903$ M

$[HisH_3^{2+}] = 0.1(9.6 \times 10^{-5})/159.95 = 6 \times 10^{-8}$ M

9.7 The pK for the dissociation of $CaATP^{2-}$ at 25 °C in 0.2 mol L^{-1} $(n\text{-propyl})_4NCl$ is 3.60. The pK for

$$HATP^{3-} = H^+ + ATP^{4-}$$

is 6.95. Calculate the apparent pK of this ATP ionization when ATP is titrated in a solution 0.1 mol L^{-1} $CaCl_2$. Assume that the

Ca^{2+} = concentration is much larger than the total ATP concentrations.

SOLUTION

$K_{app} = K_{ATP}(1 + [Ca^{2+}]/K)$

$pK_{app} = pK_{ATP} - \log (1 + [Ca^{2+}]/K)$

$= 6.95 - \log (1 + 0.1/10^{-36})$

$= 4.35$

9.8 Will 0.01 mol L^{-1} creatine phosphate react with 0.01 mol L^{-1} adenosine diphosphate to produce 0.04 mol L^{-1} creatine and 0.02 mol L^{-1} adenosine triphosphate at 25 °C, pH 7, pMg 4? What concentration of ATP can be formed if the other reactants are maintained at the indicated concentration?

SOLUTION

Creatine P + H_2O = Creatane + P $\Delta_r G^{o\prime} = -43.5$ kJ mol^{-1}

ADP + P = ATP + H_2O $\Delta_r G^{o\prime} = 39.8$ kJ mol^{-1}

Creatine P + ADP = Creatine + ATP $\Delta_r G^{o\prime} = -3.7$ kJ mol^{-1}

$$\Delta_r G' = \Delta_r G^{O\prime} + RT \ln \frac{[\text{Creatine}][\text{ATP}]}{[\text{Creatine P}][\text{ADP}]}$$

$$= -3700 + (8.314)(298) \ln \frac{(0.04)(0.02)}{(0.01)(0.01)}$$

$$= 1340 \text{ J mol}^{-1}$$

Therefore the answer to the first question is no.

$$K = e^{-\Delta_r G^{O\prime}/RT} = e^{3700/(8.1314)(298)} = 4.5$$

$$= \frac{[\text{Creatine}][\text{ATP}]}{[\text{Creatine P}][\text{ADP}]}$$

If the reactants are maintained at the indicated concentrations,

$$[\text{ATP}] = \frac{4.5\ [\text{Creatine P}][\text{ATP}]}{[\text{Creatine}]} = \frac{4.5\ (0.01)(0.01)}{0.04}$$

$$= 1.1 \times 10^{-2}\ \text{mol L}^{-1}$$

9.9 The cleavage of fructose 1,6-diphosphate (FDP) to dihydroxyacetone phosphate (CHP) and glyceraldehyde 3-phosphate (GAP) is one of a series of reactions most organisms use to obtain energy. At 37 °C and pH 7, $\Delta G^{o'}$ for the reaction FDP = DHP + GAP is 23.97 kJ mol^{-1}. What is $\Delta_r G^{o'}$ in an erythrocyte in which [FDP] = 3×10^{-6} mol L^{-1}, [DHP] = 138×10^{-6} mol L^{-1}, and [GAP] = 18.5×10^{-6} mol L^{-1}?

SOLUTION

FDP + H_2O = DHP + GAP

$$\Delta G' = \Delta G^{o'} + RT \ln \frac{[\text{DHP}][\text{GAP}]}{[\text{FDP}]}$$

$$= 23{,}970 + (8.314)(310) \ln \frac{(138 \times 10^{-6})(18.5 \times 10^{-6})}{(3 \times 10^{-6})}$$

$$= 5770\ \text{J mol}^{-1}$$

9.10 How many grams of ATP have to be hydrolyzed to ADP to lift 100 lb 100 ft if the available Gibbs energy can be converted into mechanical work with 100% efficiency? It is assumed that [ATP] = [ADP] = [P] = 0.01 mol L^{-1} and that $\Delta G^{o'}$ is -39.8 kJ mol^{-1} at 25 °C.

SOLUTION

$$w = mgh$$

$$= \left(\frac{100\ lb}{2.2\ lb/kg}\right)(9.8\ ms^{-2})(100\ ft)(12\ in/ft)(2.54\ cm/in)(0.01\ m/cm)$$

$$= 1.358 \times 10^4\ J$$

$$\Delta_r G' = \Delta_r G^{o'} = RT \ln \frac{[ADP][P]}{[ATP]}$$

for $ATP + H_2O = ADP + P$

$$\Delta_r G' = -39{,}800\ K + (8.314)(298)\ln[(0.01)^2/0.01]$$

$$= -51{,}210\ J\ mol^{-1}$$

$$= -(51{,}210\ J\ mol^{-1})/(507.2\ g\ mol^{-1})$$

$$= -101\ J\ g^{-1}$$

$$\text{Mass of ATP required} = \frac{1.358 \times 10^4\ J}{101\ J\ g^{-1}} = 135\ g$$

9.11 Biochemistry textbooks give $\Delta_r G^{o'}$ = -20.1 kJ mol^{-1} for the hydrolysis of ethyl acetate at pH 7 and 25 °C. Experiments in acid solution show that

$$\frac{[CH_3CH_2OH][CH_3CO_2H]}{[CH_3CO_2CH_2CH_3]} = 14$$

where the equilibrium concentrations are in moles per liter. What is the value of $\Delta_r G^{o'}$ obtained from this equilibrium constant? The pK of acetic acid = 4.60 at 25 °C.

SOLUTION

$$K'' = \frac{[CH_3CH_2OH][([CH_3CO_2H] + [CH_3CO_2^-])}{[CH_3CO_2CH_2CH_3]}$$

$$= \frac{[CH_3CH_2OH][CH_3CO_2H]}{[CH_3CO_2CH_2CH_3]} (1 + \frac{[CH_3CO_2^-]}{[CH_3CO_2H]})$$

$$= 14 \text{ mol L}^{-1}(1 + \frac{K_{CH_3CO_2H}}{[H^+]})$$

At pH 7

$$K' = 14 \ (1 + \frac{10^{-4.6}}{10^{-7}}) = 3530$$

$$\Delta_r G^{o\prime} = -RT \ln K = -(8.314 \text{ J K}^{-1} \text{ mol}^{-1})(298.15 \text{ K}) \ln 3540$$

$$= -20.3 \text{ kJ mol}^{-1}$$

9.12 Fumarase catalyzes the reaction fumarate + H_2O = L-malate. At 25 °C and pH 7, K' = 4.4 = [L-malate]/[fumarate]. What is the value of K' at pH 4? Given:

For fumaric acid $K_1 = 10^{-4.18}$

For L-malic acid $K_1 = 10^{-4.73}$

<u>SOLUTION</u>

$$\frac{[\text{L-malate}]}{[\text{fumarate}]} = \frac{[M] + [HM]}{[F] + [HF]} = \frac{[M][1 + [H^+]/K_{1M}]}{[F][1 + [H^+]/K_{1F}]}$$

$$= 4.4 \frac{[1 + 10^{-4}/10^{-4.73}]}{[1 + 10^{-4}/10^{-4.18}]} = 4.4 \frac{6.37}{2.51} = 11.2$$

9.13 Given $\Delta_r G^o = 49.4$ kJ mol^{-1} for

$$ATP^{4-} + H_2O = AMP^{2-} + P_2O_4^{7} + 2H^+$$

calculate $\Delta G^{o\prime}$ at pH 7 and 25 °C and 0.2 mol L^{-1} ionic strength. See Table 9.3.

<u>SOLUTION</u>

ATP^{4-} +	H_2O	AMP^{2-}	+ $P_2O_7^{4-}$	+ $2H^+$
$10^{-6.95}$		$10^{-6.45}$	$10^{-8.95}$	
$HATP^{3-}$		$HAMP^{1-}$	$HP_2O_7^{3-}$	
			$10^{-6.12}$	
			$H_2P_2O_7^{2-}$	

$$K' = \frac{([AMP^{2-}] + [HAMP^-])([P_2O_7^{4-}] + [HP_2O_7^{3-}] + [H_2P_2O_7^{2-}])}{([ATP^{4-}] + [HATP^{3-}])}$$

$$= \frac{[AMP^{2-}][P_2O_7^{4-}](1 + \frac{[HAMP^-]}{[AMP^{2-}]})(1 + \frac{[HP_2O_7^{3-}]}{[P_2O_7^{4-}]} + \frac{H_2P_2O_7^{2-}]}{[P_2O_7^{4-}]})}{[ATP^{4-}](1 + \frac{[HATP^{3-}]}{[ATP^{4-}]})}$$

$$= \frac{[AMP^{2-}][P_2O_7^{4-}][H^+]^2}{[ATP^{4-}]} \cdot \frac{(1 + \frac{[H^+]}{K_{1AMP}})(1 + (\frac{[H^+]}{K_{1PP}}) + \frac{[H^+]^2}{K_{1PP}K_{2PP}})}{[H^+]^2 (1 + \frac{[H^+]}{K_{1ATP}})}$$

$$= e^{-49,400/(8.314)(298)} \left(\frac{1 + \frac{10^{-7}}{10^{-6.45}} (1 + \frac{10^{7}}{10^{-8.95}}) + (\frac{(10^{-7})^2}{10^{-8.95}\ 10^{-6.12}})}{(10^{7})^2 (1 + \frac{10^{-7}}{10^{-6.95}})} \right)$$

$= 1.53 \times 10^7$

$\Delta_r G^{o\prime} = -RT \ln K' = -(8.314 \text{ J K}^{-1} \text{ mol}^{-1})(298 \text{ K}) \ln (1.53 \times 10^7)$

$= -41.0 \text{ kJ mol}^{-1}$

9.14 Ethyl acetate is hydrlyzed in an aqueous buffer at pH 7 in a calorimeter. The enthalpy of hydrolysis as measured in the calorimeter does not correspond with what is calculated from the following standard enthalpies of formation from a chemical thermodynamic table.

	$\Delta_f H^o(298 \text{ K})/\text{kJ mol}^{-1}$
acetic acid (1)	-484.5
ethanol (1)	-277.0
ethyl acetate (1)	-479.0
H_20 (1)	-285.8

Please explain why. What additional information would you need to calculate the heat absorbed in this experiment?

SOLUTION

ethyl acetate (1) + H_2O(1) = ethanol (1) + acetatic acid (1)

$\Delta_r H^o$ = 3.3 kJ mol^{-1}

This heat absorption would be obtained in acidic solution (where the acetic acid is undissociated) if the reactants and products form ideal solutions. At pH 7 H^+ i produced anda reacts with the buffer. Therefore, it is necessary to know the heat of acid dissociation of the buffer. To calculate an accurate value for the heat absorbed, $\Delta_f H^o$ is needed for

acetic acid	dissolved in water
ethanol	dissolved in water
ethyl acetate	dissolved in water

9.15 If n molecules of a ligand A combine with a molecule of protein to form PA_n without intermediate steps, derive the relation between the fractional saturation Y and the concentration of A.

<u>SOLUTION</u>

$$P + nA = PA_n$$

$$K = \frac{[P][A]^n}{[PA_n]} \qquad [P]_o = [P] + [PA_n]$$

$$\frac{([P]_o - [PA_n])[A]^n}{[PA_n]} = K \qquad [P]_o [A]^n = [PA_n] (K = [A]^n)$$

$$Y = \frac{[PA_n]}{[P]_o} = \frac{1}{1 + K/[A]^n} = \frac{[A]^n}{1 + [A]^n/K}$$

This equlibrium represents a cooperative effect in that as soon as one ligand molecule is bound, the other (n - 1) ligand molecules are also bound.

9.16 Since it is difficult to determine the values of the four dissociation constants in equation 9.52, the empirical Hill equation

$$Y = \frac{1}{1 + \frac{K_h}{P_{O_2}^h}}$$

is frequently used to characterize binding. Show that the Hill coefficient h may be obtained by plotting log [Y/(1 - Y)] versus P_{O_2}.

SOLUTION

$$\frac{Y}{1 - Y} = \frac{P_{O_2}^h}{K_h} \qquad \log \frac{Y}{1-Y} = -\log K_h + h \log P_{O_2}.$$

For a variety of binding systems relatively linear Hill plots are obtained for values of Y in the range 0.1 to 0.9, but deviations usually occur at the extremes unless h = 1. At the extremes the plot usually approaches a slope of unity.

9.17 The percent saturation of a sample of human hemoglobin was measured at a series of oxygen partial pressures at 20 °C, pH 7.1, 0.3 mol L^{-1} phosphate buffer, and 3 x 10^{-4} mol L^{-1} heme.

P_{O_2}/P	Percent Saturation
393	4.8
787	20
1183	45
2510	78
2990	90

Calculate the values of h and K_h in the Hill equation. (See problem 9.16.)

SOLUTION

$\log P$	$\log\left(\frac{Y}{1 - Y}\right)$
2.594	-1.31
2.896	-0.602
3.074	-0.087
3.400	+0.550
3.476	+0.954

$$\log\left(\frac{Y}{1 - Y}\right) = -\log K_h + n \log P$$

n = slope = 2.4 $\qquad\qquad$ $K_h = 4 \times 10^7$

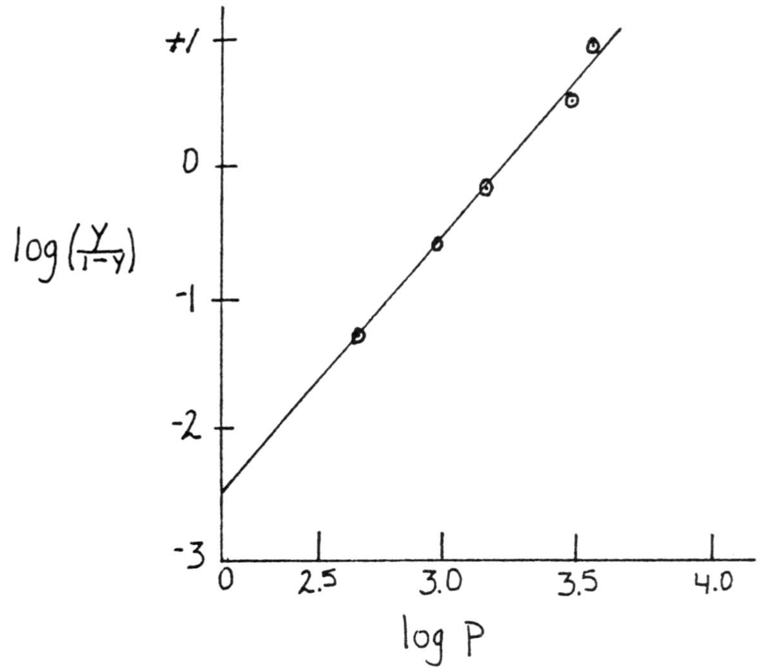

9.18	5.301×10^{-14}, 6.638
9.19	-10.75 J K^{-1} mol^{-1}
9.20	-0.25, 27.15 kJ mol^{-1}, -92.1 J K^{-1} mol^{-1}, 1.752×10^{-5}
9.22	$P_o = 1/D$, $P_A = K_A[AD/D$, $P_B = K_B[B]/D$, and $P_{AB} = K_AK_B[A][B]/D$.
9.23	3.14
9.24	(a) $K_1 = K/3$, $K_2 = K$, $K_3 = 3K$ (b) $F_{-3} = 1/8$, $F_{-2} = 3/8$, $F_{-1} = 3/8$, $F_o = 1/8$
9.25	1.3
9.26	0.53
9.27	-42.3 kJ mol^{-1}
9.28	0.038 mol L^{-1}
9.29	-29 kJ mol^{-1}
9.30	-29.8 kJ mol^{-1}
9.31	(a) $K' = K_1(1 + K_{CH}/[H^+])/(1 + K_{AH}/[H^+])$ (b) $K_2K_{CH}/K_1K_{AH} = 1$
9.32	0.158 mol L^{-1}
9.34	64 ppm

CHAPTER 10: Quantum Mechanics

10.1 If the exitance, M_λ [equation 10.5] is integrated over all wavelengths, the total radiant energy emitted per second per square meter, I, is equal to σT^4. The constant σ is called the Stefan-Boltzmann constant and is equal to 5.67×10^{-8} J s^{-1} $m^{-2} K^{-4}$.

a) What is the total radiant energy emitted per second per m^2 at 2000 K?

b) If all the radiant energy coming through a 1 cm^2 pinhole from a cavity at 2000 K were to be absorbed by 1 g of water at 373.15 K, how long will it take to vaporize the water?

SOLUTION:

(a) $I = 5.67 \times 10^{-8} \times (2000)^4$ J s^{-1} m^{-2}

$= 9.07 \times 10^{-5}$ J s^{-1} m^{-2}

(b) Energy/sec $= 9.07 \times 10^{5}$ J s^{-1} $m^{-2} \times 10^{-4}$ $m^2 = 9.07$ J s^{-1} for 1 g H_2O

$\Delta H_{vap} = 2259$ J $g^{-1} \times 1$ g $= 2259$ J

$$t = \frac{2259}{90.7} \text{ s} = 24.9 \text{ s}$$

10.2 (a) The distribution of wavelengths from a certain star peaks in the visible at $\lambda = 600$ nm. Assuming the distribution obeys the Planck distribution law, use Wien's displacement law to estimate the temperature of the star.

(b) A metal bar is heated to "red-heat" so that its radiation peaks at λ = 800 nm. Estimate the temperature of the bar.

SOLUTION:

(a) $$T = \frac{2.898 \times 10^{-3}\ \text{K m}}{600 \times 10^{-9}\ \text{m}} = 4830\ \text{K}$$

(b) $$T = \frac{2.898 \times 10^{-3}\ \text{K m}}{800 \times 10^{-9}\ \text{m}} = 3623\ \text{K}$$

10.3 Calculate the energy per photon and the number of photons emitted per second from

(a) a 100-W yellow light bulb (λ = 550 nm)

(b) a 1-kW microwave source (λ = 1 cm)

SOLUTION:

(a) $$\varepsilon_{photon} = \frac{hc}{\lambda} = \frac{(6.626 \times 10^{-34}\ \text{J s})(2.998 \times 10^{8}\ \text{m s}^{-1})}{550 \times 10^{-9}\ \text{m}}$$

$$= 3.61 \times 10^{-19}\ \text{J}$$

$$\text{\#photons/s} = \frac{100\text{W}}{3.61 \times 10^{-19}\ \text{J}} = 2.77 \times 10^{20}\ \text{photons/s}$$

(b) $$\varepsilon_{photon} = \frac{hc}{\lambda} = \frac{(6.626 \times 10^{-34}\ \text{J s})(2.998 \times 10^{8}\ \text{m s}^{-1})}{10^{-2}\ \text{m}}$$

$$= 1.99 \times 10^{-23}\ \text{J}$$

$$\text{\#photons/s} = \frac{1000\ \text{W}}{1.99 \times 10^{-23}\ \text{J}} = 5.03 \times 10^{25}\ \text{photons/s}$$

10.4 (a) Derive the value of the constant in the Wien displacement law (equation 10.3) in terms of h, c, and k.

(b) If, from experiment, the values of h and c were measured to be 6.6×10^{-34} J s^{-1} and 3.0×10^{8} m s^{-1}, and the value of the constant in the Wien displacement law were measured to be 2.9×10^{-3} K/m, find the value of k from (a). Since R is measured to be 8.3 J K^{-1} mol^{-1}, you can also calculate the Avogadro's constant.

<u>SOLUTION</u>:

(a) $$\rho(\lambda,t) = \frac{8\pi hc}{\lambda^5} \; \frac{1}{e^{hc/\lambda kT} - 1}$$

differentiate w.r.t. λ , and set = 0:

($x = hc/\lambda kT$)

$$5 = \frac{x_{max} e^{x_{max}}}{e^{x_{max}} - 1} \quad \frac{x_{max}}{1 - e^{-x_{max}}}$$

Solve by successive approximation: $x_{max} = 4.965$

Thus $$\lambda_{max} T = \frac{hc}{(4.965)k}$$

(b) $$k = \frac{hc}{(4.965)(2.9 \times 10^{-3} \text{ K/m})} = \frac{(6.6 \times 10^{-34})(3 \times 10^{8})}{(4.965)(2.9 \times 10^{-3})}$$

$k = 1.4 \times 10^{-23}$ J K^{-1}

$$N_A = R/k = \frac{8.3 \text{ J K}^{-1} \text{ mol}^{-1}}{1.4 \times 10^{-23} \text{ J K}^{-1}}$$

$= 5.9 \times 10^{23}$ mol^{-1}

10.5 Calculate the wavelengths (in micrometers) of the first three lines of the Paschen series for atomic hydrogen.

SOLUTION:

$$\lambda = \frac{1}{R}\frac{n_1^2 n_2^2}{(n_2^2 - n_1^2)} = \frac{1}{R}\frac{9\ n_2^2}{(n_2^2 - 9)}$$

For $n_2 = 4$ $\quad \lambda = \dfrac{(9)(16)}{(109677.58\ \mathrm{cm^{-1}})(7)} = 1.8756\ \mu\mathrm{m}$

For $n_2 = 5$ $\quad \lambda = \dfrac{(9)(25)}{(109677.58\ \mathrm{cm^{-1}})(16)} = 1.2822\ \mu\mathrm{m}$

For $n_2 = 6$ $\quad \lambda = \dfrac{(9)(36)}{(109677.58\ \mathrm{cm^{-1}})(27)} = 1.0941\ \mu\mathrm{m}$

10.6 Calculate the wavelength of light emitted when an electron falls from the n = 100 orbit to the n = 99 orbit of the hydrogen atom. Such species are known as high Rydberg atoms. They are detected in astronomy and are more and more studied in the laboratory.

SOLUTION:

$$E_{100} - E_{99} = -(1.0967 \times 10^7\ \mathrm{m^{-1}})(10^{-2}\ \mathrm{m\ cm^{-1}})\left(\frac{1}{100^2} - \frac{1}{99^2}\right)$$

$$= 0.2227\ \mathrm{cm^{-1}}$$

$$\lambda = 4.49\ \mathrm{cm}$$

10.7 In the photoelectric effect an electron is emitted from a metal as the result of absorption of a photon of light. Part of the energy of the photon is required to release the electron from the metal; this energy ϕ is called the work function or binding energy. The kinetic energy of the ejected electron is given by

$$\frac{1}{2}mv^2 = h\nu - \phi$$

where m and v are the mass and velocity of the electron. For the 100 face of silver metal (see Chapter 23) the velocity of electrons emitted using 200 nm photons is 7.42×10^5 m s^{-1}. Calculate the work function of this face in eV.

SOLUTION:

$$\frac{1}{2}mv^2 = \frac{1}{2}(9.10 \times 10^{-31} \text{ kg})(7.42 \times 10^5 \text{ m s}^{-1})^2$$

$$= 2.50 \times 10^{-19} \text{ J}$$

$$\phi = \frac{(6.626 \times 10^{-34} \text{ J s})(2.998 \times 10^8 \text{ m s}^{-1})}{200 \times 10^{-9}\text{m}} - 2.50 \times 10^{-19} \text{ J}$$

$$\phi = 7.432 \times 10^{-19} \text{ J}$$

$$= \frac{7.432 \times 10^{-19} \text{ J}}{1.602 \times 10^{-19} \text{ J/eV}} = 4.64 \text{ eV}$$

10.8 Photoelectron spectroscopy utilizes the photoelectron effect to measure the binding energy of electrons in molecules and solids, by measuring the kinetic energy of the emitted electrons and using the relation in problem 10.7 between kinetic energy, wavelength and binding energy. One variant of this photoelectron spectroscopy is X-ray Photoelectron Spectroscopy (XPS). If the X-ray wavelength is 0.2 nm, calculate the velocity of electrons emitted from molecules in which the binding energies are 10 eV, 100 eV and 500 eV.

SOLUTION:

$$\frac{1}{2}\,mv^2 = \frac{hc}{\lambda} - \phi$$

$$v = \{\frac{2}{m}(\frac{hc}{x} - \phi)\}^{1/2}$$

$$v_{10eV} = 4.668 \times 10^7 \text{ m s}^{-1}$$

$$v_{100eV} = 4.634 \times 10^7 \text{ m s}^{-1}$$

$$v_{500eV} = 4.473 \times 10^7 \text{ m s}^{-1}$$

10.9 Electrons are accelerated by a 1000-V potential drop. (a) Calculate the de Broglie wavelength. (b) Calculate the wavelength of the X-rays that could be produced when these electrons strike a solid.

SOLUTION:

(a) $$Ee = \frac{1}{2}mv^2$$

$$v = \left(\frac{2Ee}{m}\right)^{1/2} = \left[\frac{2(1000\ V)(1.602 \times 10^{-19}\ C)}{9.110 \times 10^{-31}\ kg}\right]^{1/2}$$

$$= 1.875 \times 10^7\ m\ s^{-1}$$

$$\lambda = \frac{h}{mv} = \frac{6.626 \times 10^{-34}\ J\ s}{(9.110 \times 10^{-31}\ kg)(1.875 \times 10^7\ m\ s^{-1})} = 0.0387\ nm$$

(b) $$Ee = \frac{hc}{\lambda} \qquad \lambda = \frac{hc}{Ee}$$

$$\lambda = \frac{(6.626 \times 10^{-34}\ J\ s)(2.998 \times 10^8\ m\ s^{-1})}{(1000\ V)(1.602 \times 10^{-19}\ C)} = 1.24\ nm$$

10.10 An ultraviolet photon (λ = 58.4 nm) from a helium gas discharge tube is absorbed by a hydrogen molecule whish is at rest. Since momentum is conserved, what is the velocity of the hydrogen molecule after absorbing the photon? What is the translational energy of the hydrogen molecule in J mol^{-1}?

SOLUTION:

$$p = \frac{h}{\lambda} = \frac{6.626 \times 10^{-34}\ J\ s}{58.4 \times 10^{-9}\ m} = 1.135 \times 10^{-26}\ kg\ m\ s^{-1} = mv$$

$$= \frac{2(1.0079 \times 10^{-3} \text{ kg mol}^{-1})}{6.022 \times 10^{23} \text{ mol}^{-1}} v$$

$$v = \frac{(1.135 \times 10^{-26} \text{ kg m s}^{-1})(6.022 \times 10^{23} \text{ mol}^{-1})}{2(1.0079 \times 10^{-3} \text{ kg mol}^{-1})} = 3.39 \text{ m s}^{-1}$$

$$E = \frac{1}{2} mv^2 N_A = 0.012 \text{ J mol}^{-1}$$

10.11 What is the de Broglie wavelength of an oxygen molecule at room temperature? Compare this to the average distance between oxygen molecules in a gas at 1 bar at room temperature.

<u>SOLUTION</u>:

$$\frac{1}{2} mv^2 = \frac{3}{2} kT \qquad v = \left(\frac{3kT}{m}\right)^{1/2} = \frac{h}{mv}$$

$$\lambda = \frac{h}{(3mkT)^{1/2}}$$

$$= \frac{6.63 \times 10^{-34} \text{ J s}}{[(3)(5.31 \times 10^{-26} \text{ kg})(1.38 \times 10^{-23} \text{ J K}^{-1})(298 \text{ K})]^{1/2}}$$

$$= 0.0259 \text{ nm}$$

Since $m = \frac{(32 \times 10^{-3} \text{ kg mol}^{-1})}{(6.022 \times 10^{23} \text{ mol}^{-1})}$

$$= 5.31 \times 10^{-26} \text{ kg}$$

To calculate the average distance l between molecules we calculate the length of the side of the cube containing one molecule.

$$\bar{V} = \frac{RT}{P} = \frac{(8.314\ J\ K^{-1}\ mol^{-1})(298\ K)}{(10^5\ Pa)}$$

$$= 0.0247\ m^3\ mol^{-1}$$

$$l^3 = \frac{(0.0247\ m^3\ mol^{-1})}{(6.022 \times 10^{23}\ mol^{-1})}$$

$$l = 3.46\ nm$$

Since the de Broglie wavelength is much shorter than the average distance between molecules, translational motion can be treated classically.

10.12 What is the deBroglie wavelength of a thermal neutron at 300 K?

SOLUTION:

$$E = \frac{3}{2} kT = \frac{1}{2} m_n v^2 = \frac{p^2}{2m_n}$$

$$p = \sqrt{3kTm_n}$$

$$= \sqrt{3(1.381 \times 10^{-23}\ J\ K^{-1})(300\ K)(1.675 \times 10^{-27}\ kg)}$$

$$= 4.563 \times 10^{-24}\ kg\ m\ s^{-1}$$

$$\lambda = \frac{h}{p} = \frac{6.626 \times 10^{-34} \text{ J s}}{4.563 \times 10^{-24} \text{ kg m s}^{-1}} = 1.452 \times 10^{-10} \text{ m}$$

$$= 0.1452 \text{ nm}$$

10.13 Calculate the deBroglie wavelengths of

(a) a 1-g bullet with velocity 300 m s^{-1}

(b) a 10^{-6}-g particle with velocity 10^{-6} m s^{-1}

(c) a 10^{-10}-g particle with velocity 10^{-10} m s^{-1}

(d) a H_2 molecule with energy of (3/2) kT at T = 20 K.

SOLUTION:

$$\text{(a) } \lambda = \frac{h}{p} = \frac{6.626 \times 10^{-34} \text{ J s}}{10^{-3} \text{ kg} \times 300 \text{ m s}^{-1}} = 2.21 \times 10^{-33} \text{ m}$$

$$\text{(b) } \lambda = \frac{6.626 \times 10^{-34} \text{ J s}}{10^{-9} \text{ kg} \times 10^{-6} \text{ m/s}} = 6.626 \times 10^{-19} \text{ m}$$

$$\text{(c) } \lambda = \frac{6.626 \times 10^{-34} \text{ J s}}{10^{-13} \text{ kg} \times 10^{-10} \text{ m s}^{-1}} = 6.626 \times 10^{-10} \text{ m}$$

$$\text{(d) } m_{H_2} = \frac{2 \times 1.008 \times 10^{-3} \text{ kg mol}^{-1}}{N_A} = 3.348 \times 10^{-27} \text{kg}$$

$$\frac{1}{2} m_{H_2} v^2 = \frac{3}{2} kT$$

$$v = \left(\frac{3kT}{m_{H_2}}\right)^{1/2} = \left[\left(\frac{3(1.38 \times 10^{-23})(20)}{3.348 \times 10^{-27}}\right)\right]^{1/2} = 497.3 \text{ m s}^{-1}$$

$$\lambda = \frac{6.626 \times 10^{-34} \text{ J S}}{(3.348 \times 10^{-27} \text{ kg})(497.3 \text{ m s}^{-1})} = 3.980 \times 10^{-10} \text{ m}$$

10.14 The lifetime of a molecule in a certain electronic state is 10^{-10} s. What is the uncertainty in energy of this state? Give the answer in J and in J mol^{-1}.

SOLUTION:

$$\Delta E \quad \geq \frac{\hbar}{2\Delta t} \qquad = \frac{6.626 \times 10^{-34}\ \text{J s}}{4\pi(10^{-10}\ \text{s})} = 5.27 \times 10^{-25}\ \text{J}$$

$$= (5.27 \times 20^{-25}\ \text{J})(6.022 \times 10^{23}\ \text{mol}^{-1}) = 0.318\ \text{J mol}^{-1}$$

10.15 For a particle in a one-dimensional box, the ground state wave function is

$$\phi = (2/a)^{1/2} \sin(\pi x/a).$$

(a) What is the probability that the particle is in the right half of the box? (b) What is the probability that the particle is in the middle third of the box?

SOLUTION:

(a) $\frac{1}{2}$, by symmetry

(b) probability $= \int_{a/3}^{2a/3} \psi^2 dx = \frac{2}{a}\int_{a/3}^{2a/3} \sin^2(\frac{\pi x}{a})dx$

$$= \frac{2}{a}\left[\frac{1}{2}x - \frac{a}{4\pi}\sin(\frac{2\pi x}{a})\right]_{a/3}^{2a/3}$$

$$= \frac{2}{a}\left[\frac{a}{3} - \frac{a}{6} - \frac{a}{4\pi}\sin\frac{2\pi 2a}{3a} + \frac{a}{4\pi}\sin\frac{2\pi a}{3a}\right]$$

$$= \frac{2}{a}\left[\frac{a}{6} - \frac{a}{4\pi}\sin\frac{4\pi}{3} + \frac{a}{4\pi}\sin\frac{2\pi}{3}\right]$$

$$= \frac{1}{3} - \frac{1}{2\pi}\sin\frac{4\pi}{3} + \frac{1}{2\pi}\sin\frac{2\pi}{3}$$

$$= \frac{1}{3} - \frac{1}{2\pi}\left(-\frac{\sqrt{3}}{2}\right) + \frac{1}{2\pi}\left(\frac{\sqrt{3}}{2}\right) = \frac{1}{3} + \frac{\sqrt{3}}{2\pi} = 0.609$$

10.16 Show that the function $f = 8e^{5x}$ is an eigenfunction of the operator d/dx. What is the eigenvalue?

SOLUTION:

$\frac{d\psi}{dx} = \frac{d}{dx} 8e^{5x} = 40\ e^{5x} = 5\psi$. Thus the eigenvalue is 5.

10.17 What are the results of operating on the following functions with the operator d/dx and d^2/dx^2?

(a) e^{-ax^2} (b) cos bx (c) e^{ikx}

Which functions are eigenfunctions of these operators? What are the corresponding eigenvalues?

SOLUTION:

(a) $\frac{de^{-ax^2}}{dx} = -2axe^{-ax^2}$

$$\frac{d^2}{dx^2} e^{-ax^2} = 4ax^2e^{-ax^2} - 2ae^{-ax^2}$$

(b) $\frac{d}{dx} \cos bx = - b\sin bx$

$$\frac{d^2}{dx^2} \cos bx = - b^2\cos bx$$

(Thus cosbx is an eigenfunction of $\frac{d^2}{dx^2}$ with eigenvalue $-b^2$)

(c) $\frac{d}{dx} e^{ikx} = ik^2e^{ikx}$

$$\frac{d^2}{dx^2} e^{ikx} = - k^2e^{ikx}$$

(Thus e^{ikx} is an eigenfunction of both $\frac{d}{dx}$ and $\frac{d^2}{dx^2}$ with eigenvalues ik and $(ik)^2 = -k^2$, respectively)

10.18 Show that the operators for the x coordinate and for the momentum in the x direction p_x do not commute. Calculate the operator representing the commutator of x and p_x.

SOLUTION:

$\hat{x} = x$ and $\hat{p} = \frac{\hbar}{i}\frac{\partial}{\partial x}$

$$\hat{x}\hat{p}_x f(x) = x \cdot \frac{\hbar}{i} f'(x) \quad \text{where } f'(x) = \frac{\partial f(x)}{\partial x}.$$

$$\hat{p}_x\hat{x} f(x) = \frac{\hbar}{i}\frac{\partial}{\partial x}(xf(x)) = \frac{\hbar}{i}[f(x) + xf'(x)]$$

Since these operators do not commute, we cannot specify precise values of both x and p_x. The relationship between the uncertainties in these two variables is given by the Heisenberg uncertainty principle.

$$(\hat{x}\hat{p}_x - \hat{p}_x\hat{x})f(x) = x\frac{\hbar}{i} f'(x) - \frac{\hbar}{i}[f(x) + xf'(x)]$$
$$= -\frac{\hbar}{i} f(x)$$

$$(\hat{x}\hat{p}_x - \hat{p}_x\hat{x})f(x) = -\frac{\hbar}{i}$$

10.19 (a) Calculate the energy levels for n=1, 2, and 3 for an electron in a potential well of width 0.25 nm with infinite barriers on either side. The energies should be expressed in kJ mol^{-1}. (b) If an electron makes a transition from n = 2 to n = 1 what will be the wavelength of the emitted radiation ?

SOLUTION:

(a) $n = 1, E = \frac{n^2h^2}{8ma^2} = \frac{(6.626 \times 10^{-34})^2}{8(9.109 \times 10^{-31} kg)(2.5 \times 10^{-10}\ m)^2}$

$= 9.640 \times 10^{-19}$ J

$= 580.5$ kJ/mol

$n = 2, E = 2322$ kJ/mol

$n = 3, E = 5225$ kJ/mol

(b) $\Delta E = (4-1)(9.640 \times 10^{-19})J = 2.892 \times 10^{-18}$ J

$$\Delta E = \frac{hc}{\lambda}$$

$$\text{Thus } \lambda = \frac{(6.626 \times 10^{-34})(2.998 \times 10^{8})}{2.892 \times 10^{-18}}$$

$$m = 6.87 \times 10^{-8} \text{ m} = 68.7 \text{ m}$$

10.20 For a helium atom in a one-dimensional box calculate the value of the quantum number of the energy level for which the energy is equal to (3/2)kT at 25°C (a) for a box 1 nm long, (b) for a box 10^{-6} m long and (c) for a box 10^{-2} m long.

<u>SOLUTION</u>:

(a) $E = \frac{3}{2} kT = \frac{3}{2} (1.38 \times 10^{-23} \text{ J K}^{-1})(298\text{K}) = 6.17 \times 10^{-21} \text{ J}$ ·

$$n = \frac{a}{h} (8mE)^{1/2}$$

$$= \frac{10^{-9}\text{m}}{6.63 \times 10^{-34} \text{ J s}} \left[\frac{8(4.003 \times 10^{-3} \text{ kg/mol})(6.17 \times 10^{-21} \text{ J})}{6.02 \times 10^{23} \text{ mol}^{-1}}\right]^{1/2}$$

= 27.3 or approximately 27.

(b) n $= 2.7 \times 10^{4}$

(c) n $= 2.7 \times 10^{8}$

10.21 Show that the wave functions for a particle in a one-dimensional box are orthogonal using

$$\sin\alpha \sin\beta = \frac{1}{2}\cos(\alpha - \beta) - \frac{1}{2}\cos(\alpha + \beta)$$

<u>SOLUTION</u>:

$$\psi_n(x) = \left(\frac{2}{a}\right)^{1/2} \sin\frac{n\pi x}{a} \qquad n = 1,2,\ldots$$

$$\frac{2}{a}\int_0^a \sin\frac{n\pi x}{a}\sin\frac{m\pi x}{a}\,dx = \frac{1}{a}\int_0^a \cos\frac{(n-m)\pi x}{a}\,dx - \frac{1}{a}\int_0^a \cos\frac{(n+m)\pi x}{a}\,dx$$

$$= 0 \text{ if } m \neq n$$

because n - m and n + m are both integers. The integrals are over complete cycles of $\cos\frac{(\text{integer})\pi x}{a}$ which are equal to zero.

10.22 Calculate the degeneracies of the first three levels for a particle in a cubical box.

SOLUTION

$E = \frac{h^2}{8ma^2}(n_x^2 + n_y^2 + n_z^2)$ where n_x, n_y, n_z are 1, 2, 3 . . .

n_x	n_y	n_x	$E/\frac{h^2}{8ma^2}$	
1	1	1	3	Degeneracy = 1
2	1	1		
1	2	1	6	Degeneracy = 3
1	1	2		
2	2	1		
2	1	2	9	Degeneracy = 3
1	2	2		

10.23 (a) Later in Table 14.4, we will find that the following molecules have the indicated vibrational frequencies

$^{35}Cl_2(560\ cm^{-1})$, $^{39}K^{35}Cl(281\ cm^{-1})$, $^{1}H_2(4401\ cm^{-1})$.

What are the force constants for these molecules if we treat them as harmonic oscillators?

(b) Assuming the force constant for $^{37}Cl_2$ is the same as for $^{35}Cl_2$, predict the fundamental vibrational frequency of $^{37}Cl_2$.

<u>SOLUTION</u>:

(a) $k = \mu(2\pi\nu)^2 = \mu(2\pi c\tilde{\nu})^2$

$$^{35}Cl_2:\ k = \frac{17.50 \times 10^{-3}\ kg\ mol^{-1}}{6.022 \times 10^{23}\ mol^{-1}}(2\pi \times 2.998 \times 10^8\ m\ s^{-1} \times 5.60 \times 10^{+4}\ m^{-1})^2$$

$$= 323.3\ N\ m^{-1}$$

$$^{39}K\,^{35}Cl:\ k = \frac{26.33 \times 10^{-3}\ kg\ mol^{-1}}{6.022 \times 10^{23}\ mol^{-1}}(2\pi \times 2.998 \times 10^8\ m\ s^{-1} \times 2.81 \times 10^4\ m^{-1})^2$$

$$= 122.5\ N\ m^{-1}$$

$$^{1}H_2:\ k = \frac{0.5039 \times 10^{-3}}{6.022 \times 10^{23}}(2\pi \times 2.998 \times 10^8 \times 4.40 \times 10^5)^2$$

$$= 575.1\ N\ m^{-1}$$

(b) $$\frac{\tilde{\nu}_{37}}{\tilde{\nu}_{35}} = \sqrt{\left(\frac{\mu_{35}}{\mu_{37}}\right)} = \left(\frac{17.5}{18.5}\right)^{1/2} = 0.9726 \Rightarrow \tilde{\nu}_{37} = 545\ cm^{-1}$$

10.24 Check the normalizations of ψ_0 and ψ_1 for the harmonic oscillator and show they are orthogonal.

SOLUTION:

$$\int_{-\infty}^{\infty}) \psi_0^2 \, dx = (\frac{\alpha}{\pi})^{1/2} \int_{-\infty}^{\infty} e^{-\alpha x^2} \, dx = (\frac{\alpha}{\pi})^{1/2} \frac{\pi^{1/2}}{\alpha^{1/2}} = 1$$

$$\text{since} \int_{\infty}^{\infty} e^{-\alpha^2 x^2} \, dx = \frac{1}{2\alpha} \pi^{1/2}$$

$$\int_{-\infty}^{\infty} \psi_1^2 \, dx = 2(\frac{\alpha^3}{\pi})^{1/2} \int_{-\infty}^{\infty}) x^2 e^{-\alpha x^2} \, dx = 4(\frac{\alpha^3}{\pi})^{1/2} \int_{\infty}^{\infty} x^2 e^{-\alpha x^2} \, dx$$

$$= 4(\frac{\alpha^3}{\pi})^{1/2} \frac{\pi^{1/2}}{2^2 \alpha^{3/2}} = 1 \text{ since} \int_{\infty}^{\infty} x^2 e^{-\alpha \xi^2} \, dx = \frac{\pi^{1/2}}{4\alpha^{3/2}}$$

To check orthogonality

$$\int_{-\infty}^{\infty} \psi_0 \psi_1 dx = 2^{1/2} (\frac{\alpha}{\pi})^{1/2} \int_{-\infty}^{\infty} x \, e^{-\alpha x^2} \, dx = 0$$

since x is an odd function and $e^{-\alpha x^2}$ is an even function.

10.25 Substitute the v=1 eigenfunction for the harmonic oscillator into the Schrödinger equation for the harmonic oscillator, and obtain the expression for the eigenvalue (energy).

SOLUTION:

$$\psi_1 = c\, x\, e^{-ax^2} \qquad a = \frac{\pi}{h}\,(km)^{1/2}$$

$$\frac{d\psi_1}{dx} = c\{e^{-ax^2} - 2ax^2e^{-ax^2}\}$$

$$\frac{d^2\psi_1}{dx^2} = c\{-6axe^{-ax^2} + 4a^2x^3e^{-ax^2}\} = \{-6a + 4a^2x^2\}\,\psi_1$$

Substituting into the Schrödinger. equation

$$(-\frac{\hbar}{2m}\frac{d^2}{dx^2} + \frac{1}{2}k\,x^2)\psi_1 = E\psi_1$$

the terms proportional to x^2 cancel, so

$$E = \frac{\hbar^2}{2m}\cdot 6a = \frac{h^2}{8\pi^2 m}\cdot\frac{6\pi}{m}\,(km)^{1/2} = \frac{3}{4}\frac{h}{\pi}\,(\frac{k}{m})^{1/2}$$

$$= \frac{3}{2}\,h\Big[\frac{1}{2\pi}\,(\frac{k}{m})^{1/2}\Big] = \frac{3}{2}\,h\nu_0$$

10.26 In the vibrational motion of HI, the iodine atom essentially remains stationary because of its large mass. Assuming that the hydrogen atom undergoes harmonic motion and that the force constant k is 317 N m^{-1}, what is the fundamental vibrational frequency n_0? What is ν_0 if H is replaced by D?

SOLUTION:

$$m = \frac{1.0078 \times 10^{-3}\ \text{kg mol}^{-1}}{6.022 \times 10^{23}\ \text{mol}^{-1}} = 1.674 \times 10^{-27}\ \text{kg}$$

$$\nu_0 = \frac{1}{2\pi}\sqrt{\frac{k}{m}} = \frac{1}{2\pi}\sqrt{\frac{317\ \text{kg s}^{-2}}{1.674 \times 10^{-27}\text{kg}}} = 6.93 \times 10^{13}\ \text{s}^{-1}$$

If H is replaced by D:

$$\nu_0 = (\frac{1}{\sqrt{2}})\ (6.93 \times 10^{13}\ \text{s}^{-1}) = 4.52 \times 10^{13}\ \text{s}^{-1}$$

10.27 What are the expectation values for <x> and $<x^2>$ for a quantum mechanical harmonic oscillator in the v = 1 state? What is the standard deviation Δx?

<u>SOLUTION</u>:

$$\langle x \rangle = \int \psi_1^*(x)\, x\, \psi_1(x)\, dx$$

$$= (\frac{4\alpha^3}{\pi})^{1/2} \int_{-\infty}^{+\infty} x^3\, e^{-\alpha x^2}\, dx = 0 \quad \text{since the integrand is odd}$$

$$\langle x^2 \rangle = (\frac{4\alpha^3}{\pi})^{1/2} \int_{-\infty}^{+\infty} x^4\, e^{-\alpha x^2}\, dx = \frac{1}{\alpha}(\frac{4}{\pi})^{1/2} \int_{-\infty}^{+\infty} z^4 e^{-z^2}\, dz$$

$$= \frac{1}{\alpha}(\frac{4}{\pi})^{1/2}\frac{3\pi^{1/2}}{4} = \frac{3}{2\alpha} = \frac{3}{2}(\frac{\hbar^2}{k\mu})^{1/2}$$

$$\Delta x = [\langle x^2 \rangle - \langle x \rangle^2]^{1/2} = \left[\frac{3}{2}(\frac{\hbar^2}{k\mu})^{1/2}\right]^{1/2}$$

10.28 What are the reduced mass and moment of inertia of $^{23}Na^{35}Cl$? The equilibrium internuclear distance R_e is 236 pm. What are the values of E for the states with J = 1 and J = 2?

SOLUTION:

$$\mu = \frac{1}{\frac{1}{m_1} + \frac{1}{m_2}} = \frac{m_1 m_2}{m_1 + m_2}$$

$$= \frac{(22.98977 \times 10^{-3} \text{ kg mol}^{-1})(34.96885 \times 10^{-3} \text{ kg mol}^{-1})}{[(22.98977 + 34.96885) \times 10^{-3} \text{ kg mol}^{-1}](6.022137 \times 10^{23} \text{ mol}^{-1})}$$

$$= 2.30328 \times 10^{-26} \text{ kg}$$

$$I = \mu R_e^2 = (2.303 \times 10^{-26} \text{ kg})(236 \times 10^{-12} \text{ m})^2$$

$$= 1.283 \times 10^{-45} \text{ kg m}^2$$

$$E = \frac{\hbar^2}{2I} J(J+1) = \frac{(6.626 \times 10^{-34} \text{ J s})^2 (2)}{8\pi^2 (1.283 \times 10^{-45} \text{ kg m}^2)}$$

$$= 8.668 \times 10^{-24} \text{ J} \quad \text{for } J = 1$$

$$= 2.600 \times 10^{-23} \text{ J for } J = 2$$

10.29 The commutator of two operators, $\hat{A}$ and $\hat{B}$, is defined as the operator $\hat{A}\hat{B}-\hat{B}\hat{A}$. Using the definitions of $\hat{L}_x$, $\hat{L}_y$, and $\hat{L}_z$ given in equation 10.127, find the commutator of $\hat{L}_x$ with $\hat{L}_y$ and $\hat{L}_x$ with $\hat{L}_z$.

SOLUTION:

$$\hat{L}_x = -i\hbar \left(y\frac{\partial}{\partial z} - z\frac{\partial}{\partial y}\right)$$

$$\hat{L}_y = -i\hbar \left(z\frac{\partial}{\partial x} - x\frac{\partial}{\partial z}\right)$$

$$\hat{L}_z = -i\hbar \left(x\frac{\partial}{\partial y} - y\frac{\partial}{\partial x}\right)$$

$$\therefore \hat{L}_x \hat{L}_y = -\hbar^2\left(y\frac{\partial}{\partial z} - z\frac{\partial}{\partial y}\right)\left(z\frac{\partial}{\partial x} - x\frac{\partial}{\partial z}\right)$$

$$\hat{L}_y\hat{L}_x = -\hbar^2\left(z\frac{\partial}{\partial x} - x\frac{\partial}{\partial z}\right)\left(y\frac{\partial}{\partial z} - z\frac{\partial}{\partial y}\right)$$

so,

$$\hat{L}_x \hat{L}_y - \hat{L}_y\hat{L}_x = -\hbar^2\left\{y\frac{\partial}{\partial z}\left(z\frac{\partial}{\partial x}\right) + zx\frac{\partial}{\partial y}\frac{\partial}{\partial z} - zy\frac{\partial}{\partial x}\frac{\partial}{\partial z} x \frac{\partial}{\partial z}\left(z\frac{\partial}{\partial y}\right)\right\}$$

$$= -\hbar^2\left(y\frac{\partial}{\partial x} - x\frac{\partial}{\partial y}\right) = i\hbar\hat{L}_z$$

Similarly

$$\hat{L}_x \hat{L}_z - \hat{L}_z\hat{L}_x = -i\hbar\hat{L}_y$$

10.30 The $^{12}C^{16}O$ molecule has an equilibrium bond distance of 112.8 pm. Calculate (a) the reduced mass and (b) the moment of inertia. Calculate (c) the wavelength of the photon emitted when the molecule makes the transition from $\ell=1$ to $\ell = 0$ using equation 10.120 for the energy levels.

<u>SOLUTION</u>:

(a) reduced mass

$$\mu = \frac{m_1 m_2}{m_1+m_2} = \frac{(12)(15.995)}{(12 + 15.995)} \frac{10^{-3}}{6.022 \times 10^{23}}$$

$$\mu = 1.139 \times 10^{-26} \text{ kg}$$

(b) moment of inertia

$$I = (1.139 \times 10^{-26})(112.8 \times 10^{-12})^2 \text{ kg m}^2$$

$$= 1.449 \times 10^{-46}\ \text{kg m}^2$$

$$\text{(c) } \Delta E = \frac{\hbar^2}{2I}\ [1(1+1) - 0(0+1)] = \frac{\hbar^2}{I} = \frac{h^2}{4\pi^2 I}$$

$$= 7.675 \times 10^{-23}\ \text{J}$$

$$\lambda = \frac{hc}{\Delta E} = 2.58 \times 10^{-3}\ \text{m}$$

10.31 $\nu_{max} = (5.87 \times 10^{10}\ s^{-1}\ K^{-1})T$

10.32 (a) $1.76 \times 10^{14}\ s^{-1}$ (1700 nm), $5.88 \times 10^{14}\ s^{-1}$ (510 nm)

(b) 9.66, 290 nm

10.33 $2.740\ 03 \times 10^{14}\ s^{-1}$, $1.094\ 11 \times 10^{-4}$ cm

10.34 $4.052\ 27 \times 10^{-4}$, $2.625\ 87 \times 10^{-4}$ nm

10.35 (a) 0.0388 nm, (b) 9.052×10^{-4} m

10.36 0.1303 nm

10.37 $1.325 \times 10^{-27}\ kg\ m\ s^{-1}$, $0.0113\ m\ s^{-1}$

10.38 5.38×10^{-26} J, 1.6×10^{-5} %

10.40 (a) -6a, (b) $1/k_x$

10.41 $E = n^2h^2/8\ ml^2$

10.42 145.1, 580.5, 1306.1 kJ mol^{-1}

10.43 1,3,3,3

10.44 $(km)^{1/2}\ \pi/k$

10.45 (a) $P = 2 \int_{x_{class}}^{\infty} (\alpha/\pi)^{1/2}\ e^{-\alpha x^2}\ dx$

where $x_{class} = \hbar^{1/2}/(\mu k)^{1/4}$

$\alpha = (k\mu/\hbar^2)^{1/2}$

(b) 16% outside the classically allowed region

10.46 $E = (n_x + 1/2)\ h\nu_x + (h_y + 1/2)h\nu_y + (n_z + 1/2)H\nu_z$

$E_o = h\nu_x/2 + h\nu_y/2 + h\nu_z/2$

10.47 $E = (n_x + n_y + n_z + 3/2)\ h\nu$

The degeneracies are 1, 3, 6.

10.51 1.240×10^{-27} kg, 2.827×10^{-46} kg m^2.

$E_o = 0$, $E_1 = 3.92 \times 10^{-23}$ J, $E_2 = 11.76 \times 10^{-23}$ J

CHAPTER 11: ATOMIC STRUCTURE

11.1 Using data from Table C.2 at 0 K, what is the ionization energy of $H(g)$?

$$H(g) = H^+(g) + e^-$$

SOLUTION:

For $H(g)$ at 0 K $\Delta_f H^o = 216.037$ kJ mol^{-1}

$H^+(g)$ a 0 K $\Delta_f H^o = 1528.085$ kJ mol^{-1}

$e^-(g)$ at 0 K $\Delta_f H^o = 0$

$\Delta H^o = 1312.048$ J mol^{-1}

$$= \frac{1312.048 \text{ J mol}^{-1}}{(1.602\ 177.33 \times 10^{-19}\text{ C})(6.022\ 136\ 7 \times 10^{23}\text{ mol}^{-1})}$$

$= 13.698\ 42$ eV

11.2 How much energy in eV and kJ mol^{-1} is required to remove electrons from the following orbitals in an H atom (a) 3d; (b) 4f; (c) 4p; (d) 6s?

SOLUTION:

$$\Delta E = \frac{13.6058 \text{ eV}}{n^2} = \frac{(13.6058)(96.485)}{n^2} \frac{\text{kJ}}{\text{mol}}$$

(a) n=3 ΔE = 1.512 eV = 145.9 kJ/mol

(b) n=4 ΔE = 0.8504 eV = 82.05 kJ/mol

(c) n=4 same as (b)

(d) n=6 ΔE = 0.3779 eV = 36.47 kJ/mol

11.3 Calculate the ionization potentials for He^+, Li^{2+}, Be^{3+}, B^{4+}, and C^{5+}.

SOLUTION:

$E = (13.61\ V)Z^2$

	Z	E/V		Z	E/V
He^+	2	54.44	B^{4+}	5	340.25
Li^{2+}	3	122.49	C^{5+}	6	489.96
Be^{3+}	4	217.76			

11.4 Since H and D have different reduced masses, they also have slightly different energy levels. Calculate the (a) ionization potentials and (b) the wavelengths of the first line in the Balmer series for these atoms.

SOLUTION:

$$\mu_H = \frac{m_e m_p}{m_e + m_p} = 9.104431 \times 10^{-31}\ kg$$

$$\mu_D = \frac{m_e m_D}{m_e + M_D} = 9.106908 \times 10^{-31}\ kg$$

$$\text{(a) } IP_H = hcR_H = hcR_\infty \frac{\mu_H}{m_e} = 2.178718 \times 10^{-18}\ J$$

$$= 13.59848\ eV$$

$$IP_D = hcR_\infty \frac{\mu_D}{m_e} = 2.179311 \times 10^{-18} \text{ J}$$

$$= 13.60218 \text{ eV}$$

$$\% \text{ diff} = 2.7 \times 10^{-2} \ \%$$

(b) $$\frac{1}{\lambda_H} = R_H \left(\frac{1}{2^2} - \frac{1}{3^2}\right) = \frac{R_H}{7.2} \ ; \ \frac{1}{\lambda_D} = \frac{R_D}{7.2}$$

$$\lambda_H = 656.470 \text{ nm}$$

$$\lambda_D = 656.291 \text{ nm}$$

11.5 The muon is an elementary particle with a negative charge equal to the charge of the electron and a mass approximately 200 times the electron mass. The muonium atom is formed from a proton and a muon. Calculate the reduced mass, the Rydberg constant, and the formula for the energy levels for this atom. What is the most probable radius of the 1s orbital for this atom?

SOLUTION:

$$\mu_M = \mu_{muonium} = \frac{m_{muon} m_{prot}}{m_{muon} + m_{prot}} = \frac{200 \ m_e m_{prot}}{200 \ m_e + m_{prot}} = 1.646397 \times 10^{-28} \text{ kg}$$

$$R_M = R_\infty \frac{\mu_M}{m_e} = 1.98335 \times 10^{-9} \text{ m}^{-1}$$

$$E = -\frac{hcR_M}{n^2} = \frac{-3.9425 \times 10^{-16} \text{ J}}{n^2} = \frac{-2461}{n^2} \text{ eV}$$

$$a_M = a_0 \; m_e/\mu_M = \frac{(52.918 \text{ pm})(9.1094 \times 10^{-31} \text{ kg})}{1.6463 \times 10^{-28} \text{ kg}}$$

$$= 0.2928 \text{ pm}$$

11.6 A hydrogen-like atom has a series of spectral lines at λ = 26.2445 nm, λ = 19.4404 nm, λ = 17.3578 nm, and λ = 16.4028 nm. What is the atomic number Z of this atom? What is the formula for this spectral series (i.e. n_1 and n_2 in eq. 11.9)?

<u>SOLUTION</u>:

$$\lambda^{-1} = R_\infty Z^2 \left(\frac{1}{n_2^1} - \frac{1}{n_2^2}\right) \quad ; \; n_2 = n_1 + m$$

$$= 1.0973732 \times 10^7 \; Z^2 \left(\frac{1}{n_1^2} - \frac{1}{n_2^2}\right) \text{ m}^{-1} \left(10^{-9} \frac{\text{nm}^{-1}}{\text{m}^{-1}}\right)$$

$$= 0.010973732 \; Z^2 \left(\frac{1}{n_1^2} - \frac{1}{n_2^2}\right) \text{ nm}^{-1}$$

Expt: λ^{-1} = 0.0381032 nm^{-1}

= 0.0514393 nm^{-1}

= 0.0576110 nm^{-1}

= 0.06096752 nm^{-1}

By trial and error substitution, or by graphing these numbers vs $\frac{1}{(n_1+m)^2}$ for m = 1,2,3,··· and looking for the intercept for $n_1+m = \infty$ gives $\frac{Z^2}{n_1^2}$; the slope gives Z^2. This leads to Z=5, n_1 = 2 and

$$\lambda^{-1} = 25(0.010973732)\left(\frac{1}{4} - \frac{1}{n_2^2}\right)$$

$n_2 = 2 + m$, m = 1,2,···

11.7 What are the degeneracies of the following orbitals for hydrogen-like atoms? (a) n = 1, (b) n = 2, and (c) n = 3.

SOLUTION:

(a) n = 1 $l = 0$ m = 0 $m_s = \pm\frac{1}{2}$

degeneracy = 2

(b) n = 2 $l = 0$ m = 0 $m_s = \pm\frac{1}{2}$

$l = 1$ m = -1,0,+1 $m_s = \pm\frac{1}{2}$

degeneracy = 8

(c) n = 3 $l = 0$ m = 0 $m_s = \pm\frac{1}{2}$

$l = 1$ m = -1,0,+1 $m_s = \pm\frac{1}{2}$

$l = 2$ m = -2,-1,0,+1,+2 $m_s = \pm\frac{1}{2}$

degeneracy = 18

[Note that the degeneracy is $2n^2$.]

11.8 Show that for a 1s orbital of a hydrogen-like atom the most probable distance from proton to electron is $\frac{a_o}{Z}$. Find the numerical values for C^{5+} and B^{3+}.

SOLUTION:

$$\psi_{1s} \; \alpha \; e^{-Zr/a_o}$$

Probability density of $p(r) \; \alpha^2 r^2 \; \psi_{1s}^2 = r^2 \times e^{-2Zr/a_o}$.

Setting the derivative of the probability density equal to zero,

$$r^2 \, e^{-2Zr/a_o} \left(\frac{-2Z}{a_o}\right) + 2re^{-Zr/a_o} = 0 \quad , \text{or} \qquad r = \frac{a_o}{Z} \; .$$

$$r(C^{5+}) = \frac{a_o}{5} = 105.8 \text{ pm}$$

$$r(B^{3+}) = \frac{a_o}{3} = 176.4 \text{ pm}$$

11.9 Find the values of r for which the hydrogen-like atom 2s and 3s wave functions are equal to zero (these are the radial nodes). Compare to Fig. 11.1.

SOLUTION:

$$2s \quad \psi_{2s} = \frac{1}{4\sqrt{2\pi}}\left(\frac{Z}{a_o}\right)^{1/2} \left(2 - \frac{Zr}{a_o}\right) e^{-Zr/a_o}$$

$$r_{NODE} = \frac{2a_o}{Z}$$

$$3s \quad \psi_{3s} = \frac{1}{81\sqrt{3\pi}}\left(\frac{Z}{a_o}\right)^{1/2} \left(27 - 18\frac{Z}{a_o} + 2\left(\frac{Zr}{a_o}\right)^2\right)$$

$$\left(\frac{Zr}{a_o}\right)_{NODE} = \frac{1}{4}\{18 \pm \sqrt{(18)^2 - 8(27)}\ \}$$

$$r_{NODE} = 7.098\ \frac{a_o}{Z}\ ,\ 1.902\ \frac{a_o}{Z}$$

11.10 What is the average distance form an orbital electron to the nucleus for a 2s and 2p electron in (a) H and (b) Li^{2+}?

<u>SOLUTION</u>:

(a) $$\langle r\rangle_{n,l} = \frac{n^2 a_o}{Z}\left\{1 + \frac{1}{2}\left[1 - \frac{l(l+1)}{n^2}\right]\right\}$$

For H $$\langle r\rangle_{2,0} = \frac{2^2(52.92\ \text{pm})}{1} \cdot F(3,2) = 317.5\ \text{pm}$$

$$\langle r\rangle_{2,1} = \frac{2^2(52.92\ \text{pm})}{1}\left\{1 + \frac{1}{2}\left[1 - \frac{2}{4}\right]\right\} = 264.6\ \text{pm}$$

(b) For Li^{2+}

$$\langle r\rangle_{2,0} = \frac{2^2(52.92\ \text{pm})}{3} \bullet \frac{3}{2} = 105.8\ \text{pm}$$

$$\langle r\rangle_{2,1} = \frac{2^2(52.92\ \text{pm})}{3}\left\{1 + \frac{1}{2}\left[1 - \frac{2}{4}\right]\right\} = 88.2\ \text{pm}$$

11.11 Use equation 11.18 to derive the expression for the average distance between the electron and proton in a hydrogen-like atom in the 1s level.

SOLUTION:

$$\langle r \rangle = \int_0^\infty r\rho(r)dr = \int_0^\infty r4\pi r^2 \frac{Z^3}{\pi a_o^3} e^{-2Zr/a_o} dr = \frac{3}{2}\frac{a_o}{Z}$$

11.12 Calculate the expectation value of the distance between the nucleus and the electron of a hydrogen-like atom in the $2p_z$ state using equation 10.59. Show that the same result is obtained using equation 11.19.

SOLUTION:

$$\langle r \rangle = \int \psi_{21}^* \, r \, \psi_{21} d\tau = \int \psi_{21}^2 \, r d\tau$$

$$\psi_{21} = \frac{1}{\sqrt{4}\ \ 2\pi} \left(\frac{Z}{a_o}\right)^{3/2} \left(\frac{Zr}{a_o}\right) \cos\theta \ \exp(-Zr/2a_o)$$

$$\langle r \rangle = \frac{1}{16(2\pi)\left(\frac{Z}{a_o}\right)^5} \int_0^{2\pi}\int_0^{\pi}\int_0^{\infty} r^3\exp(-Zr/a_o)\cos^2\theta \ r^2\sin\theta \ d\theta d\phi dr$$

The element of volume in spherical polar coordinates is

$$d\tau = r^2\sin\theta \ d\theta d\phi dr$$

and the ranges of the coordinates are

$$0 < \theta < \pi \qquad 0 < \phi < 2\pi \qquad 0 < r < \infty$$

as may be seen in Figure 10.11.

When the integration is carried out using

$$\int_0^\infty x^n e^{-bx} dx = n!/b^{n+1}$$ we obtain $<r> = 5a_o/Z$

Equation 12.19 yields

$$<r> = <r> = \frac{4a_o}{Z}\left[1 + \frac{1}{2}\left(1 - \frac{1}{2}\right)\right] = \frac{5a_o}{Z}$$

11.13 It can be shown that a linear combination of two eigenfunctions belonging to the same degenerate level is also an eigenfunction of the hamiltonian with the same energy. In terms of mathematical formulas, if $\hat{H}\psi_1 = E_1\psi_1$ and $\hat{H}\psi_2 = E_1\psi_2$, then $\hat{H}(c_1\psi_1 + c_2\psi_2) = E_1(c_1\psi_1 + c_2\psi_2)$. The wave function $c_1\psi_1 + c_2\psi_2$ still needs to be normalized. Equation 11.5 yields the following expressions for the 2p eigenfunctions

$$\psi_{2p_{+1}} = be^{-Zr/2a}\, r \sin\theta\, e^{i\phi}$$
$$\psi_{2p_{-1}} = be^{-Zr/2a}\, r \sin\theta\, e^{-i\phi}$$
$$\psi_{2p_0} = ce^{-Zr/2a}\, r \cos\theta$$

Use this information to find the real functions for the 2p orbitals that are given in Table 11.1. Given:

$$e^{i\phi} = \cos\phi + i \sin\phi$$
$$e^{-i\phi} = \cos\phi - i \sin\phi$$

SOLUTION:
As indicated in the footnote in Section 11.2, we define ψ_{2p_x} and ψ_{2p_y} as the following linear combinations

$$\psi_{2p_x} = \frac{(\psi_{2p_{+1}} + \psi_{2p_{-1}})}{\sqrt{2}}$$

$$\psi_{2p_x} = \frac{(\psi_{2p_{+1}} + \psi_{2p_{-1}})}{i\sqrt{2}}$$

Substituting in these expressions we obtain

$$\psi_{2p_x} = \frac{2be^{-Zr/2a}\, r \sin\theta\, (e^{i\phi} + e^{-i\phi})}{\sqrt{2}}$$

$$= \frac{2be^{-Zr/2a}\, r \sin\theta \cos\theta}{\sqrt{2}}$$

$$\psi_{2p_y} = \frac{2be^{-Zr/2a}\, r \sin\theta\, (e^{i\phi} - e^{-i\phi})}{\sqrt{2}}$$

$\lvert L \rvert = \sqrt{l\,(l+1)}\,\hbar$	0	$\sqrt{2}\hbar$	$\sqrt{6}\hbar$
Radial modes	2	1	0
Angular modes	0	1	2

11.16 Using equation 11.40, calculate the difference in energy between the two spin angular momentum states of a hydrogen atom in the 1s orbital in a magnetic field of 1 Tesla. What is the wavelength of radiation emitted when the electron spin "flips"? In what region of the electromagnetic spectrum is this?

SOLUTION:

$$\Delta E = g_e \mu_B B = (2.0023)\left(\frac{e\hbar}{2m_e}\right)(1\ \text{T})$$

$= (2.0023)(9.274 \times 10^{-24}\ J\ T^{-1})(1\ T)$

$= 1.8569 \times 10^{-23}\ J$

$$\lambda = \frac{hc}{\Delta E} = \frac{6.626 \times 10^{-34} \times 2.998 \times 10^{8}}{1.8569 \times 10^{-23}}\ m = 1.0698 \times 10^{-2}\ m$$

$= 1.07$ cm, microwave

11.17 For the wave function

$$\psi = \begin{vmatrix} \psi_A(1) & \psi_A(2) \\ \psi_B(1) & \psi_B(2) \end{vmatrix}$$

show that (a) the interchange of two columns changes the sign of the wave function, (b) the interchange of two rows changes the sign of the wave function, and (c) the two electrons cannot have the same spin orbital.

SOLUTION:

(a) The interchange of two columns yields

$$\begin{vmatrix} \psi_A(2) & \psi_A(1) \\ \psi_B(2) & \psi_B(1) \end{vmatrix} = \psi_A(2)\ \psi_B(1) - \psi_A(1)\,\psi_B(2)$$

This is the wave function for the atom with the two electrons interchanged, and it is the negative of the function given in the problem, as required by the anti-symmetrization principle.

(b) The interchange of two rows yields

$$\psi = \begin{vmatrix} \psi_B(1) & \psi_B(2) \\ \psi_A(1) & \psi_A(2) \end{vmatrix} = \psi_A(2)\ \psi_B(1) - \psi_A(1)\,\psi_B(2)$$

which is the negative of the wave function given in the problem. Thus the interchange of two columns or two rows is equivalent to the interchange of two electrons between orbitals.

(c) If two electrons are presented by the same spin orbitals, the determinant

$$\begin{vmatrix} \psi_A(1) & \psi_A(2) \\ \psi_A(1) & \psi_A(2) \end{vmatrix} = 0$$

is in accord with the principle that two electrons in an atom cannot have four identical quantum numbers.

11.18 What are the electron configurations for H^-, Li^+, O^{2-}, F^-, Na^+, and Mg^{2+}?

SOLUTION:

$1s^2$, $1s^2$, $1s^22s^22p^6$, $1s^22s^22p^6$, $1s^22s^22p^6$, $1s^22s^22p^6$.

11.19 How many electrons can enter the following sets of atomic orbitals: 1s, 2s, 2p, 3s, 3p, 3d?

SOLUTION:

There are $2l + 1$ values of m for n given l and each of these orbitals may be occupied by 2 electrons so $2(2l+1)$ electrons can enter a set of orbitals.

1s: $l=0$, $2(2l+1) = 2$

2s: $l=0$, $2(2l+1) = 2$

2p: $l=1$, $2(2l+1) = 6$

3s: $l=0$, $2(2l+1) = 2$

3p: $l=1$, $2(2l+1) = 6$

3d: $l=2$, $2(2l+1) = 10$

11.20 3.40 V, 1.259

11.21 1.84, 2.26, 2.777

11.22 -2.903 E_h, -7.478 E_h, -14.668 E_h

11.24 0.7542 eV

11.25 ^3P_o

11.26 (b) and (d)

11.28 -2.626 kJ mol^{-1}. This is half the classical value.

11.29 $n_1 \gtrsim 10$

11.30 656.172 12, 656.163 58 nm

11.31 1.98 x 10^{-7} m

11.32 6 a_o

11.33 (a) 243 nm, (b) 6.8 eV

11.34 $\theta = 0, \pi$, $r = 2a_o$ are the most probable positions

11.35 $n \gtrsim 8$

11.38 17.639 pm

11.39 0.7144, 0.6615, 0.5557 nm

11.40 $\Delta E = 5.788 \times 10^{-4}$ eV

$kT = 0.0259$ eV

11.41

	4s	4p	4d	4f
$\lvert L \rvert/\hbar$	0	$\sqrt{2}$	$\sqrt{6}$	$\sqrt{12}$
radial nodes	3	2	1	0
angular nodes	0	1	2	3

11.45 Sc [Ar]$3d4s^2$

Sc^+ [Ar]$3d4s$

Sc^{2+} [Ar]$3d$

Sc^{3+} [Ar] $1s^2 2s^2 2p^6 3s^2 3p^6$

12.1 Given that the equilibrium distance in H_2^+ is 106 pm and that of H_2 is 74.1 pm, calculate the internuclear repulsion energy in both cases at R_e. Using $D_e(H_2^+) = 2.79$ eV and $D_e(H_2) = 4.78$ eV, calculate $E_{e\ell}$ at R_e for both.

SOLUTION:

$$H_2^+: \frac{e^2}{4\pi\varepsilon_o R_e} = \frac{(1.602 \times 10^{-19})^2}{4\pi(8.854 \times 10^{-12})(1.06 \times 10^{-10})} = 2.176 \times 10^{-18}\ \text{J}$$

$$= 13.6\ \text{eV}$$

$$H_2: \frac{e^2}{4\pi\varepsilon_o R_e} = 19.4\ \text{eV}$$

$$E_{H_2^+}(R_e) - E_{H_2^+}(\infty) = -\ 2.79\ \text{eV}$$

$$\therefore E_{H_2^+}(R_e) = E_{H_2^+}(\infty)\quad -2.79\ \text{eV}$$

$$E_{e\ell,H_2^+}(R_e) = -13.6\ \text{eV}$$

but,

$$E_{e\ell,H_2^+}(\infty) = -13.6\ \text{eV} + E_{e\ell,H_2^+}(\infty) - 2.79\ \text{eV} = -30.0\ \text{eV}$$

Similarly,

$$E_{e\ell,H_2}(R_e) = E_{H_2,}{}^{2}{}_{)}(\infty) - 4.78 - 19.4\ \text{eV} = -51.4\ \text{eV}$$

12.2 Given the equilibrium dissociation energy D_e for N_2 in Table 12.2 and the fundamental vibration frequency 2331 cm^{-1}, calculate the spectroscopic dissociation energy D_o in kJ mol^{-1}.

SOLUTION:

$$\frac{1}{2}h\nu_o = \frac{(6.626 \times 10^{-34}\text{ J s})(2.998 \times 10^{8}\text{ m s}^{-1})(2.331 \times 10^{5}\text{ m}^{-1})}{2(1000\text{ J kJ}^{-1})}$$

$$\times \frac{(6.022 \times 10^{23}\text{ mol}^{-1})}{1}$$

$$= 13.93\text{ kJ mol}^{-1}$$

$$D_o = D_2 - \frac{1}{2}h\nu_o = 955.42 - 13.93$$

$$= 941.49\text{ kJ mol}^{-1}$$

This is the thermodynamic dissociation energy at absolute zero. The standard change in enthalpy for dissociation at 0 K calculated from Table C.2 is 941.57 kJ mol^{-1}.

12.3 Derive the values of the normalization constants given in equations 12.23 and 12.24.

SOLUTION:

$$\psi_g = C(1s_A + 1s_B)$$

$$\int \psi_g^2\, d\tau = C^2 \int [1s_A^2 + 21(s_A 1s_B) + 1s_B^2\, d\tau]$$

$$1 = C^2\left[1 + 2\int 1s_A 1s_B d\tau + 1\right]$$

$$C = \frac{1}{(2 + 2S)^{1/2}} \text{ where } S = \int 1s_A 1s_B \, d\tau$$

$$\psi_u = C'(1s_A - 1s_B)$$

$$\int \psi_u d\tau = (C')^2 \int [1s_A^2 - 2(1s_A 1s_B) + 1s_B^2] d\tau$$

$$= (C')^2 [2 - 2S]$$

$$C' = \frac{1}{(2 - 2S)^{1/2}}$$

12.4 Plot ψ_g and ψ_u versus distance along the internuclear axis for H_2^+ in the ground state without worrying about normalization (R_e = 106 pm).

SOLUTION:

The values of these wave functions along the line through the two nuclei are given by

$$\psi_g = C\left(e^{-|z|a_o} + e^{-(z - R_e)/a_o}\right)$$
$$\psi_u = C'\left(e^{-|z|a_o} - e^{-(z - R_e)/a_o}\right)$$

where z is the distance measured from nucleus A toward nucleus B, and C and C′ are the normalization constants that we are going to ignore. These plots are given in Fig. 12.5.

12.5 Plot ψ_g^2 and ψ_7^2 versus distance along the internuclear axis for H_2^+ in the ground state without worrying about normalization (R_e = 106 nm).

SOLUTION

See Fig. 12.5

12.6 The overlap integral S for the H_2^+ molecule can be evaluated as

$$S = [1 + R/a_o + R^2/3a_o^2]\, e^{-R/a_o}$$

Plot this as a function of R/a_o. At what value of R is it a maximum? At what value of R is it a minimum? What is the value of S at the equilibrium separation, R = 106 pm?

SOLUTION:

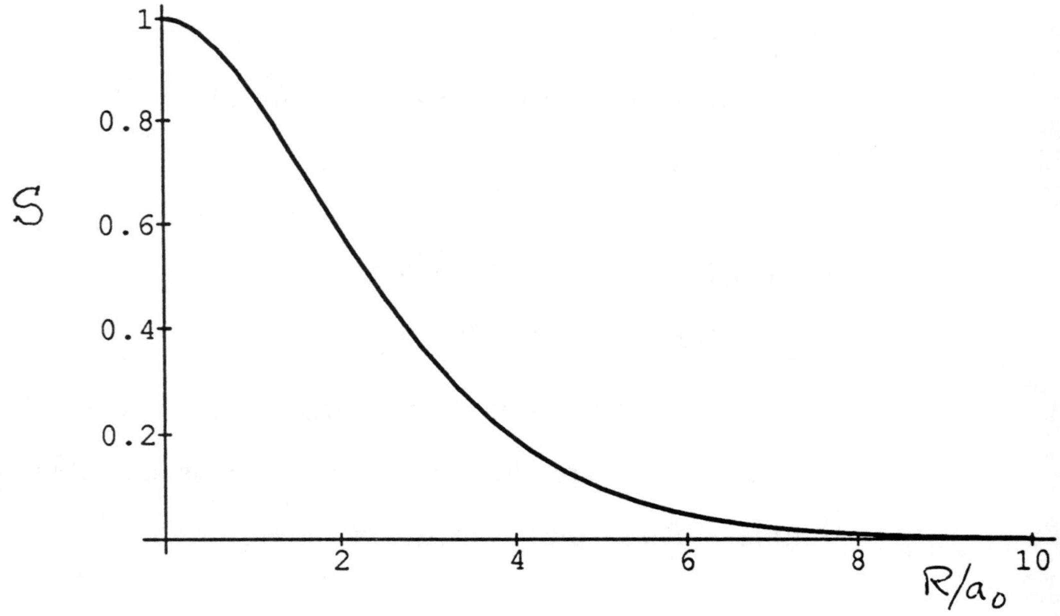

Maximum at R = 0
Minimum at R = ∞
S(106 pm) = 0.766

12.7 Express the four VB wave functions for H_2 as Slater determinants.

SOLUTION:

$$\psi_1 = \frac{1}{2^{1/2}}\begin{vmatrix} 1s_A(1)\alpha(1) & 1s_A(2)\alpha(2) \\ 1s_B(1)\beta(1) & 1s_B(2)\beta(2) \end{vmatrix}$$

$$- \frac{1}{2^{1/2}}\begin{vmatrix} 1s_A(1)\beta(1) & 1s_A(2)\beta(2) \\ 1s_B(1)\alpha(1) & 1s_B(2)\alpha(2) \end{vmatrix}$$

$$\psi_2 = \frac{1}{2^{1/2}}\begin{vmatrix} 1s_A(1)\alpha(1) & 1s_A(2)\alpha(2) \\ 1s_B(1)\alpha(1) & 1s_B(2)\alpha(2) \end{vmatrix}$$

ψ_3 is the same as ψ_2 with β replacing α. ψ_4 is obtained from ψ_1 by replacing the minus sign with a plus sign.

12.8 Show that the hybrid orbitals for tetravalent caarbon given by equations 12.49 to 12.52 are orthogonal.

SOLUTION:

$$\int \psi_{sp^3(i)}\, \psi_{sp^3(ii)}\, d\tau$$

$$= \frac{1}{4}\int (\psi_s + \psi_{px} + \psi_{py} + \psi_{pz})(\psi_s - \psi_{px} - \psi_{py} - \psi_{pz})\, d\tau$$

cross terms do not contribute since atomic orbitals are orthogonal

$$= \frac{1}{4} \int (\psi_s^2 + \psi_{px}^2 - \psi_{py}^2 - \psi_{pz}^2)\, d\tau$$

$$= \frac{1}{4} (1 + 1 - 1 - 1) = 0$$

since orbitals are normalized.

12.9 Using the hydrogen-like orbitals of Table 11.1, write out the sp^2 orbitals of equations 12.46-12.48. For $\psi_{sp2(i)}$, write the wave function as a function of r for $\theta = 0$ (along the positive z axis), $\theta = 90^o$ (in the xy plane), and $\theta = 180^0$ (along the negative z axis).

SOLUTION:

e.g $\psi_{sp^2(i)} = \dfrac{1}{4\sqrt{2\pi}} [(2-\sigma) + \sqrt{2}\, \sigma\cos\theta] \dfrac{e^{-\sigma}}{\sqrt{3}}$

$\theta = 0°$: $\quad \psi = \dfrac{1}{4\sqrt{6\pi}} [2 + 0.414\, \sigma]e^{-\sigma}$

$\theta = 90°$ $\quad \psi = \dfrac{1}{4\sqrt{6\pi}} [2 - \sigma]e^{-\sigma}$

$\theta = 180°$ $\quad \psi = \dfrac{1}{4\sqrt{6\pi}} [2 - 2.414\, \sigma]e^{-\sigma}$

12.10 Discuss the electronic structure of the methyl radical and the location of the unpaired electron.

SOLUTION:

The CH_3 radical is planar with the 3 two-electron bonds due to the overlapping of the sp^2 orbitals of C with the s orbitals of H. The odd electron is in the remaining unhybridized p orbital which is perpendicular to the plane.

12.11 How many electrons are involved in sigma and pi bonding orbitals in the following molecules: (a) ethylene, (b) ethane, (c) butadiene, (d) benzene ?

SOLUTION:

(a) 10σ, 2π (b) 14σ, 0π (c) 18σ, 4π (d) 24σ, 6π

[σ electron pairs not labeled]

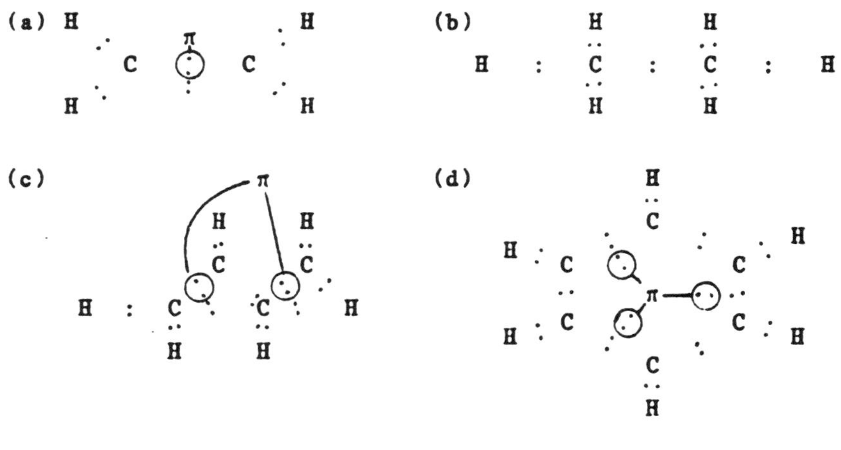

12.12 The heat of hydrogenation of cyclohexane is -121 kJ mol^{-1} and the heat of hydrogenation of benzene is -209 kJ mol^{-1}. What is the reduction in the energy due to the formation of the π bond system in benzene?

SOLUTION:

If the molecule hexatriene existed, its heat of hydrogenation would be expected to be 3(-121 kJ mol^{-1}) = -362 kJ mol^{-1}.

hexatriene   benzene

The resonance stabilization is equal to the difference in heats of hydrogenation.

$$-362 \text{ kJ mol}^{-1} - (-209 \text{ kJ mol}^{-1}) = -153 \text{ kJ mol}^{-1}$$

12.13 Consider the Hückel molecular orbitals for butadiene given in equation (12.64). Each ϕ_i is an atomic p_z orbital on carbon atom i, so there is a nodal plane in the xy plane for each molecular orbital. There are other nodes in these orbitals as we move from atom 1 to atom 4 since the orbitals change sign. How many nodes are there for $\psi 1$, $\psi 2$, $\psi 3$, and $\psi 4$? Where are they? Compare to Fig 12.22.

SOLUTION:

ψ_1	no nodes	
ψ_2	1 node:	between atoms 2 and 3
ψ_3	2 notes:	between 1 and 2 and between 3 and 4
ψ_4	3 nodes:	between 1 and 3, 2 and 3, and 3 and 4

12.14 In Hückel theory, the contribution to the electronic energy of a single π-bond (see the discussion regarding ethylene in Section 12.7) is 2β. For butadiene, the contribution for two conjugated bonds is 4.472β, while for benzene (three conjugated bonds) it is 8β. What is the extra stabilization in butadiene and benzene due to the conjugation (in terms of β)? Using the data of problem 12.12, calculate the value of β for benzene and predict the value of the extra stabilization for butadiene.

SOLUTION:

butadiene: $\Delta = 4.472\beta - 4\beta = 0.472\beta$

benzene: $\Delta = 8\beta - 6\beta = 2\beta$

$2\beta = -153$ kJ/mol from prob. 12.12

$\beta = -76.5$ kJ/mol

$\therefore \Delta_{butadiene} = 0.472(-76.5)$ kJ/mol $= -36.1$ kJ/mol

12.15 The Lennard-Jones parameters for nitrogen are $\varepsilon/k = 95.1$ K and $\sigma = 0.37$ pm. Plot the potential energy (expressed as V/k in K) for the interaction of two molecules of nitrogen.

SOLUTION:

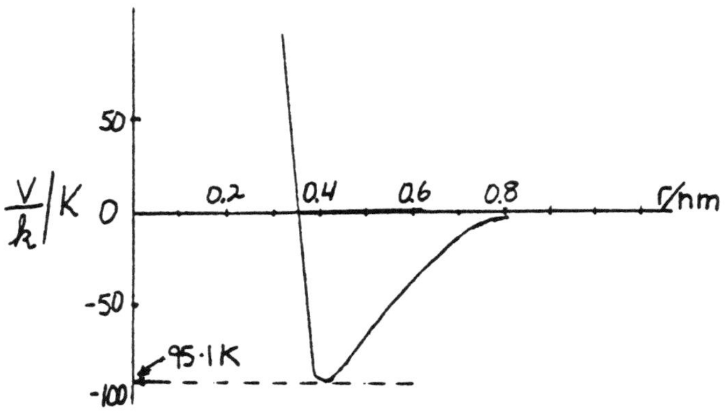

12.16 For Ne the parameters of the Lennard-Jones 6-12 potential are $\varepsilon/k = 35.6$ K and $\sigma = 275$ pm. Plot V in J mol^{-1} versus r and calculate the distance r_m where $dV/dr = 0$.

SOLUTION:

$r_m = 2^{1/6}\,\sigma = 2^{1/6}\,(0.275\text{ nm}) = 0.309\text{ nm}$

$\varepsilon = (8.314\text{ J K}^{-1}\text{ mol}^{-1})(35.6\text{ K}) = 296.0\text{ J mol}^{-1}$

$$V = 4\varepsilon\left[\left(\frac{\sigma}{r}\right)^{12} - \left(\frac{\sigma}{r}\right)^{6}\right]$$

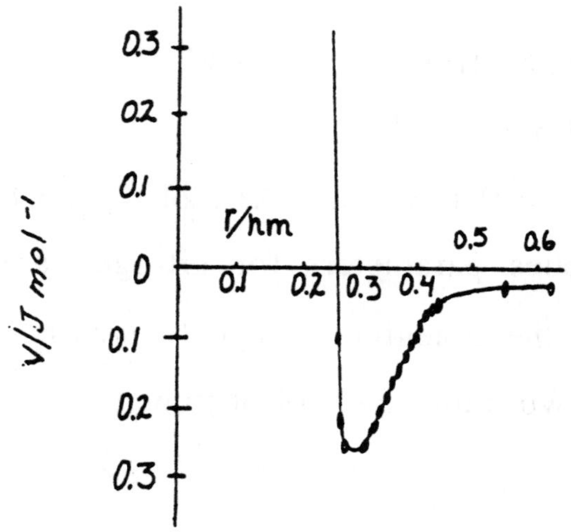

$$V = 4(296.0\text{ J mol}^{-1})\left[\left(\frac{2.75}{r}\right)^{12} - \left(\frac{2.75}{r}\right)^{6}\right]$$

12.17 For KF(g) the dissociation constant D_e is 5.18 eV and the dipole moment is 28.7 x 10^{-30} C m. Estimate these values assuming that the bonding is entirely ionic. The ionization potential K(g) is 4.34 eV, and the electron affinity of F(g) is 3.40 eV. The equilibrium internuclear distance in KF(g) is 0.217 nm.

SOLUTION:

The work required to separate $K^+ F^-$ from the equilibrium internuclear distance to infinity is

$$\phi = \frac{Q_1Q_2}{4\pi\varepsilon_o R_e} = \frac{(1.602 \times 10^{-19}\ C)^2(0.8988 \times 10^{10}\ N\ m^2\ C^{-2})}{217 \times 10^{-12}\ m}$$

$$= 1.063 \times 10^{-18}\ J = \frac{1.063 \times 10^{-18}\ J}{1.602 \times 10^{-19}\ J\ eV^{-1}} = 6.64\ eV$$

Since KF actually dissociates into neutral atoms, the ionization potential of the metal has to be subtracted and the electron affinity of the nonmetal has to be added to calculate D_e:

$$D_e = 6.64 - 4.34 + 3.40 = 5.70\ eV$$

This is higher than the experimental value (5.18 eV) because there is some repulsion of K^+ and F^- due to inner electron clouds. The dipole moment expected for the oversimplified structure $K^+ F^-$ is

$$(1.602 \times 10^{-19}\ C)(217 \times 10^{-12}\ m) = 34.7 \times 10^{-30}\ C\ m$$

12.18 The equilibrium internuclear distance for NaCl(g) is 236.1 pm. What dipole moment is expected? The actual value is 3.003×10^{-29} C m. How do you explain the difference?

SOLUTION:

$$\mu = (1.602 \times 10^{-19}\ \mathrm{C})\ (0.2361 \times 10^{-9}\ \mathrm{m})$$

$$= 3.783 \times 10^{-29}\ \mathrm{C\ m}$$

The actual value is less because the ions polarize each other. The positive ion attracts the electrons of the negative ion toward itself, and the negative ion repels the electrons of the positive ion. The more polarizable the ions the larger the effect is.

12.19 Calculate the dipole moment HCl would have if it consisted of a proton and a chloride ion (considered to be a point charge separated by 127 pm (the internuclear distance obtained from the infrared spectrum). The experimental value is 3.44×10^{-30} C m. How do you explain the difference?

SOLUTION:

$$\mu = QR = (1.602 \times 10^{-19}\ \mathrm{C})(0.127 \times 10^{-9}\ \mathrm{m})$$

$$= 20.4 \times 10^{-30}\ \mathrm{C\ m}$$

$$\mu_{expt} = 3.44 \times 10^{-30}\ \mathrm{C\ m}$$

The charges are not completely separated in HCl.

12.20 Show that the SI units are the same on the two sides of equation 12.84.

SOLUTION:

$$\frac{\kappa-1}{\kappa+2}\frac{M}{\rho} = \frac{N_A}{3\varepsilon_o}\left(\frac{\mu^2}{3kT} + \alpha_e + \alpha_v\right) \qquad \mu = C\ m$$

$$\alpha = \frac{\text{dipole moment}}{\text{electric field strength}} = \frac{C\ m}{V\ m^{-1}} = \frac{C\ m^2}{J\ C^{-1}} = \frac{C^2\ m^2}{kg\ m^2\ s^{-2}}$$

$\varepsilon_o = C^2\ N^{-1}\ m^{-2}$

Left side: $$\frac{kg\ mol^{-1}}{kg\ m^{-3}} = m^3\ mol^{-1}$$

Right side: $$\frac{mol^{-1}}{C^2\ N^{-1}\ m^{-2}}\left(\frac{C^2\ m^2}{kg\ m^2\ s^{-2}} + \frac{C^2\ m^2}{kg\ m^2\ s^{-2}}\right) = m^3\ mol^{-1}$$

12.21 The equilibrium distances in HCl, HBr, and HI are 127 pm, 141 pm, and 161 pm, respectively. Given the dipole moments in Table 12.4, find the effective fractional charges on the H and X ions. Is this in accord with the order of the electronegativities of the halogens?

SOLUTION:

q = fractional charge

$\mu = qeR_e$

$$q_{HCl} = \frac{3.44 \times 10^{-30}\ C\text{-}m}{1.60 \times 10^{-19}\ C \times 127 \times 10^{-12}\ m} = 0.17$$

$$q_{HBr} = 0.12$$

$$q_{HI} = 0.039$$

In accord with the electronegativities.

12.22 $1\sigma_g^2 1\sigma_u 2\sigma_g$

12.23

$$\begin{vmatrix} \alpha-E & \beta & 0 & \beta \\ \beta & \alpha-E & \beta & 0 \\ 0 & \beta & \alpha-E & \beta \\ \beta & 0 & \beta & \alpha-E \end{vmatrix} = 0$$

$E = \alpha + 2\beta, \alpha, \alpha, \alpha-2\beta$

$E_\pi = 2(\alpha+2\beta) + 2\alpha = 4\alpha + 4\beta$ No extra stabilization

12.24 (a) Ground state is $\psi_1^2\psi_2^2$.

First excited state is $\psi_1^2\psi_2\psi_3$.

$\Delta E = 1.49 \times 10^{-18}$ J, $\lambda = 133$ nm.

(b) Ground state is $\psi_1^2\psi_2^2\psi_3^2$.

First excited state is $\psi_1^2\psi_2^2\psi_3\psi_4$.

$\Delta E = 7.50 \times 10^{-19}$ J, $\lambda = 265$ nm.

12.26 0.666×10^{-6} pm^{-3}, 0.312×10^{-6} pm^{-3}

12.27 Li_2^+ $1\sigma_g^2 1\sigma_u^2 2\sigma_g$ B.O. = 1/2

Be_2^+ $1\sigma_g^2 1\sigma_u^2 2\sigma_g^2 2\sigma_u$ B.O. = 1/2

B_2^+ $1\sigma_g^2 1\sigma_u^2 2\sigma_g^2 2\sigma_u^2 3\sigma_g$ B.O. = 1/2

N_2^+ $1\sigma_g^2 1\sigma_u^2 2\sigma_g^2 2\sigma_u^2 3\sigma_g^2 1\pi_u^3$ B.O. = 5/2

12.28 14.9%

12.29

$$\begin{vmatrix} \alpha-E & \beta_1 & 0 & \beta_2 \\ \beta_1 & \alpha-E & \beta_2 & 0 \\ 0 & \beta_2 & \alpha-E & \beta_1 \\ \beta_2 & 0 & \beta_1 & \alpha-E \end{vmatrix} = 0$$

$E_1 \approx \alpha + (1 + \gamma)\beta_1$

$E_2 \approx \alpha + (1 - \gamma)\beta_1$

$E_3 \approx \alpha - (1 - \gamma)\beta_1$

$E_4 \approx \alpha - (1 + \gamma)\beta_1$

12.30 + ion: $\psi_1^2\psi_2$

- ion: $\psi_1^2\psi_2^2\psi_3$

+ ion $|\psi_2|^2$: $(0.602)^2$, $(0.372)^2$, $(-0.372)^2$, $(-0.602)^2$

The unpaired electron densities are the same for the negative ion.

12.31 935 pm

12.32 31.4 eV

12.33 493.4 kJ mol^{-1} (Table C.2 gives 493.6 kJ mol^{-1})

12.36 787.38 kJ mol^{-1}

12.37

	Cl	Br	I
Electroneg.	3.0	2.8	2.5
μ	1.85	1.45	1.35

Because of the higher electronegativity of Cl, the Cl atom is more negative in HCl than the Br atom in HBr, causing the dipole moment of HCl to be larger than that of HBr.

12.38 5.29×10^{-30} C m

12.40 429 pm

CHAPTER 13: Symmetry

13.1 H_2S

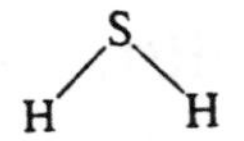

SOLUTION: C_{2v}: E, $C_2(z)$, $\sigma_v(xz)$, $\sigma_v'(yz)$

13.2 PCl_3

SOLUTION: C_{3v}: E, C_3, $3\sigma_v$

13.3 trans-$[CrBr_2(H_2O)_4]^+$ (ignoring the H's)

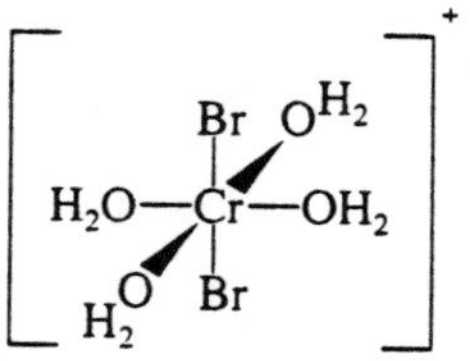

SOLUTION: E, $C_4(z)$, $S_4(z)$, $C_2'(x)$, $C_2'(y)$, $2C_2''$, i, σ_h, $\sigma_v(xz)$, $\sigma_v(yz)$, $2\sigma_d$

13.4 Gauche CH_2ClCH_2Cl

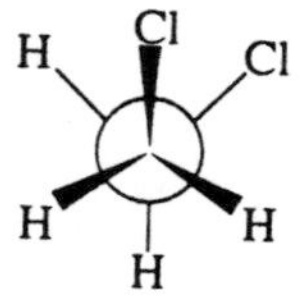

SOLUTION: C_2: E, C_2

13.5 1,3,5-tribromobenzene

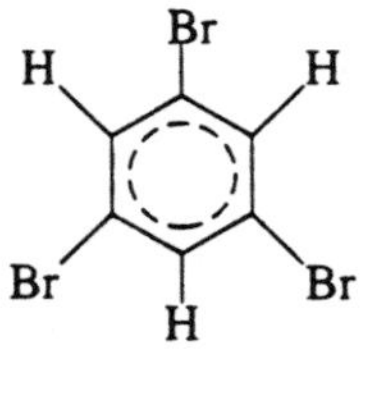

SOLUTION: D_{3h}:

E, $C_3(z)$, $S_3(z)$, $3C_2$, σ_h, $3\sigma_v$

13.6 CHClBr(CH_3)

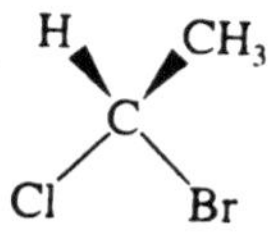

SOLUTION: C_1: E

13.7 IF_5

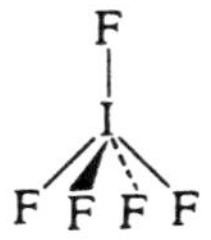

SOLUTION: C_{4v}: E, C_4, C_2, $2\sigma_v$, $2\sigma_d$

13.8 C_6H_{12}(cyclohexane)

SOLUTION: D_{3d}: E, C_3, $3C_2$, i, $3\sigma_d$, $2S_6$

13.9 B_2H_6

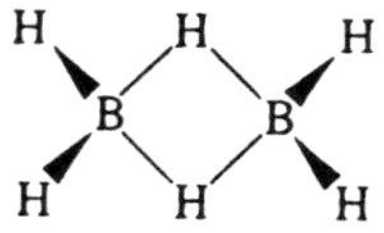

SOLUTION: D_{2h}: E $C_2(z)$, $C_2(y)$, $C_2(x)$, i,$\sigma(xy)$, $\sigma(xz)$, $\sigma(yz)$

13.10 $C_{10}H_8$ (naphthalene)

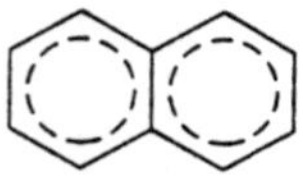

(planar)

SOLUTION: D_{2h}: E, $C_2(z)$, $C_2(y)$, $C_2(x)$, i, $\sigma(xy)$, $\sigma(xz)$, $\sigma(yz)$

13.11 C_5H_8 (spiropentane)

(triangles ⊥ to each other)

SOLUTION: D_{2d}: E, $S_4(z)$, $C_2(z)$, $C_2'(x)$, $C_2'(y)$, $2\sigma_d$

13.12 C_4H_4S (thiophene)

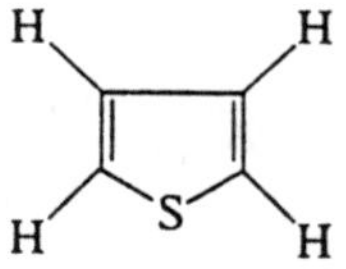

(planar)

SOLUTION: C_{2v}: E, $C_2(z)$, $\sigma_v(xz)$, $\sigma_v'(yz)$

13.13 p-dichlorobenzene $C_6H_4Cl_2$

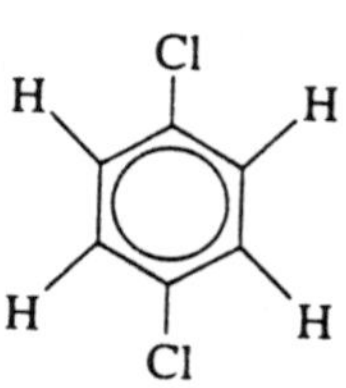

SOLUTION: D_{2h}: E $C_2(z)$, $C_2(y)$, $C_2(x)$, i, $\sigma(xy)$, $\sigma(xz)$, $\sigma(yz)$

13.14 trans CFClBrCFClBr

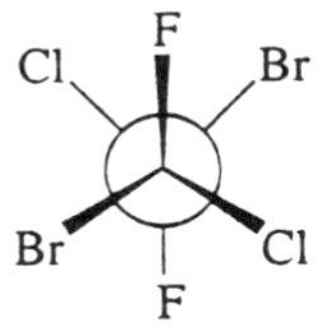

<u>SOLUTION</u>: C_i: E, i

13.15 Construct the operator multiplication table for the point gorup C_{2h}.

<u>SOLUTION</u>:

	E	C_2^1	σ_h	i
E	E	C_2^1	σ_h	i
C_2^1	C_2^1	E	i	σ_h
σ_h	σ_h	i	E	C_2^1
i	i	σ_h	C_2^1	E

13.16 List the operators associated with the S_6 elements and their equivalents, if any. How many distinct operations are produced?

SOLUTION:

$$S_6^1 \quad S_6^2 \quad S_6^3 \quad S_6^4 \quad S_6^5 \quad S_6^6$$

$$C_3^1 \quad i \quad C_3^2 \quad E$$

Thus S_6 axis implies the existence of both a C_3 axis coincident with the S_6 axis and a center of symmetry (i). Only the S_6^1 and S_6^5 operations are distinct operations characteristic of only the S_6 axis.

13.17 Consider the three distinct isomers of dichloroethylene, $C_2H_2Cl_2$. To which symmetry group does each belong? Which can have a permanent dipole moment?

SOLUTION:

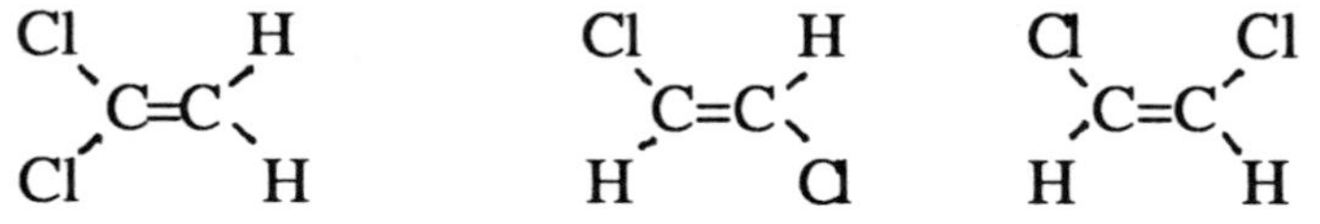

C_{2v}	C_{2h}	C_{2v}
dipole	no dipole	dipole

13.18 The first excited singlet state of ethylene is twisted so that the two hydrogens and carbon on one side are in a plane perpendicular to the plane containing the other three atoms. To which symmetry gorup does it belong? Does it have a dipole moment?

SOLUTION:

D_{2d}, no dipole moment

13.19 Which of the molecules in problems 13.1 - 13.14 can have a permanent dipole moment?

SOLUTION:

H_2S, PCl_3, $C_2H_4Cl_2$, CMeClBrH, IF_5, thiophene

13.20 Which of the molecules in problems 13.1 - 13.14 can be optically active?

SOLUTION: C(Me)ClBrH

13.21 Consider the three distinct isomers of dichlorobenzene. To which symmetry group does each belong? Which can have a dipole moment?

SOLUTION:

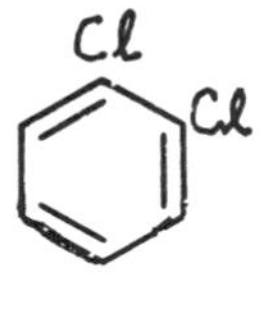

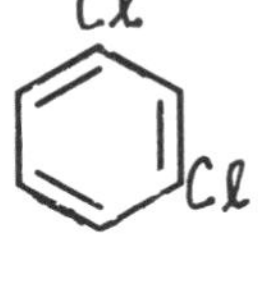

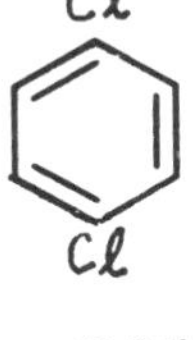

ortho	meta	para
C_{2v}	C_{2v}	D_{2h}
dipole	dipole	no dipole

13.22 There are 10 distinct isomers of dichloronaphthalene, $C_{10}H_6Cl_2$. Two of them do not have a dipole moment. List these two and find the symmetry group to which each belongs.

<u>SOLUTION</u>:

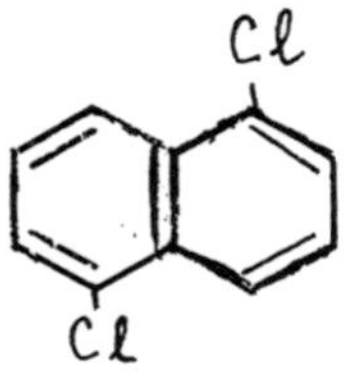

C_{2h}

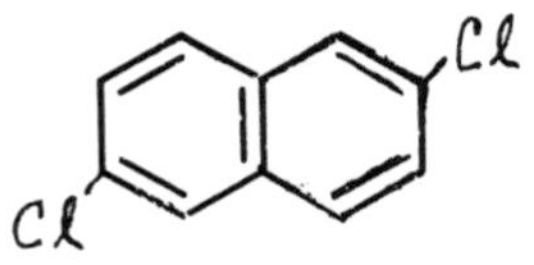

C_{2h}

13.23 Some of the excited electronic states of acetylene are cis-bent and some are trans-bent. What is the symmetry group of each?

<u>SOLUTION</u>:

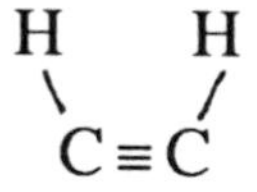

cis-bent

C_{2v}

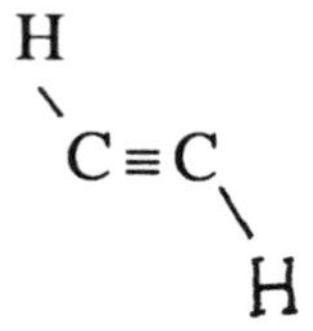

trans-bent

C_{2h}

13.24 D_{2d}: $E, 2S_4, C_2, 2\sigma_d, 2C'_2$

13.25 C_s: E, σ

13.26 D_{5h}: $E, 5C_5, 5S_5, \sigma_h, 5\sigma_2, 5\sigma_v$

13.27 C_{2v}: $E, C_2, \sigma_v, \sigma_{v'}$

13.28 O_h: $E, 4C_3, 4S_6, 3C_4, 3S_4, 3C_2, 6C_2'$

13.29 D_{3h}: $E, 2C_3, 2S_3, \sigma_h, 3\sigma_v, 3C_2$

13.30 T_d: $E, 4C_3, 3S_4, 3C_2', 6\sigma_d$

13.31 C_{2v}: $E, C_2, \sigma_v, \sigma_{v'}$

13.32 C_s: E, σ

13.33 D_{3h}: $E, C_3, S_3, 3C_2, \sigma_h, 3\sigma_v$

13.34 C_3: E, C_3

13.35 C_{3v}: $E, C_3, 3\sigma_v$

13.36 T_d: $E, 4C_3, 3S_4, 3C_2, 6\sigma_d$

13.37 C_s: E, σ

13.38 D_{4h}: $E, C_4, C_2, S_4, 2C_2', 2C_2'', i, \sigma_h, 2\sigma_{v'}, 2\sigma_d$

13.39

E	C_3	C_3^2	$\sigma_v^{(1)}$	$\sigma_v^{(2)}$	$\sigma_v^{(3)}$
C_3	C_3^2	E	$\sigma_v^{(3)}$	$\sigma_v^{(1)}$	$\sigma_v^{(2)}$
C_3^2	E	C_3	$\sigma_v^{(2)}$	$\sigma_v^{(3)}$	$\sigma_v^{(1)}$
$\sigma_v^{(1)}$	$\sigma_v^{(2)}$	$\sigma_v^{(3)}$	E	C_3	C_3^2
$\sigma_v^{(2)}$	$\sigma_v^{(3)}$	$\sigma_v^{(1)}$	C_3^2	E	C_3
$\sigma_v^{(3)}$	$\sigma_v^{(1)}$	$\sigma_v^{(2)}$	C_3	C_3^2	E

13.40 D_{2h}

13.41 $C_{\infty v}$, yes

CHAPTER 14: Rotational and Vibrational Spectroscopy

14.1 Since the energy of a molecular quantum state is divided by kT in the Boltzmann distribution, it is of interest to calculate the temperature at which kT is equal to the energy of photons of different wavelength. Calculate the temperature at which kT is equal to the energy of photons of wavelength 10^3 cm, 10^{-1} cm, 10^{-3} cm, 10^{-5} cm.

SOLUTION:

$$kT = \frac{hc}{\lambda}$$

$$T = \frac{hc}{k\lambda} = \frac{(6.626 \times 10^{-34}\ \text{J s})(2.998 \times 10^{8}\ \text{m s}^{-1})}{(1.381 \times 10^{-23}\ \text{J K}^{-1})(10\ \text{m})}$$

$$= 1.44 \times 10^{-4}\ \text{K for } \lambda = 10^3\ \text{cm}$$

For $\lambda = 10^{-1}$ cm, T = 14.4 K

For $\lambda = 10^{-3}$ cm, T = 1440 K

For $\lambda = 10^{-5}$ cm, T = 144,000 K

Note that conversion factors between temperature and various forms of energy are given in the back cover.

14.2 Most chemical reactions require activation energies ranging between 40 a 400 kJ mol^{-1}. What are the equivalents of 40 and 400 kJ mol^{-1} in terms (a) nm, (b) wave numbers, and (c) electron volts?

SOLUTION:

For 40 kJ mol^{-1}

(a) $\lambda = \frac{hc}{E}$

$$= \frac{(6.626 \times 10^{-34}\ J\ s)(2.998 \times 10^{8}\ m\ s^{-1})(6.022 \times 120^{23}\ mol^{-1})}{(4 \times 10^{4}\ J\ mol^{-1})}$$

$= 2.991 \times 10^{-6}\ m = 2991\ nm$

(b) $\tilde{\nu} = \frac{1}{\lambda} = \frac{1}{2.991 \times 10^{-4}\ cm} = 3343\ cm^{-1}$

(c) $\frac{(4 \times 10^{4}\ J\ mol^{-1})}{(6.022 \times 10^{23}\ mol^{-1})(1.602 \times 10^{-19}\ J\ eV^{-1})} = 0.415\ eV$

For 400 kJ mol^{-1}

(a) $\lambda = (2991\ nm)\frac{4 \times 10^{4}\ J\ mol^{-1}}{4 \times 10^{5}\ J\ mol^{-1}} = 299.1\ nm$

(b) $\tilde{\nu} = \frac{1}{\lambda} = \frac{1}{2.991 \times 10^{-5}\ cm} = 33{,}430\ cm^{-1}$

(c) $(0.415\ eV)\frac{4 \times 10^{5}\ J\ mol^{-1}}{4 \times 10^{4}\ J\ mol^{-1}} = 4.15\ eV$

14.3 (a) What vibrational frequency in wave number corresponds to a thermal energy of kT at 25 °C? (b) What is the wavelength of this radiation?

<u>SOLUTION:</u>

$hc\tilde{\nu} = kT$

$$\tilde{\nu} = \frac{kT}{hc} = \frac{(1.381 \times 10^{-23}\ J\ K^{-1})(298.15\ K)(10^{-2}\ m\ cm^{-1})}{(6.626 \times 10^{-34}\ J\ K^{-1})(2.998 \times 10^{8}\ ms^{-1})}$$

$= 207\ cm^{-1}$

$$\lambda = \frac{1}{\tilde{\nu}} = 4.83 \times 10^{-3} \text{ cm}$$

14.4 Calculate the reduced mass and the moment of inertia of $D^{35}Cl$, given that $R_e = 127.5$ pm.

SOLUTION:

$$\mu = \frac{m_D m_{Cl}}{m_D + m_{Cl}}$$

$$= \frac{(2.014\ 10 \times 10^{-3} \text{kg mol}^{-1})(34.968\ 85 \times 10^{-3} \text{ kg mol}^{-1})}{[(2.014\ 10 + 34.968\ 85) \times 10^{-3} \text{kg mol}^{-1}](6.022 \times 10^{23} \text{ mol}^{-1})}$$

$$= 3.162 \times 10^{-27} \text{ kg}$$

$$I = \mu r^2 = (3.162 \times 10^{-27} \text{ kg})(1.275 \times 10^{-10} \text{ m})^2$$

$$= 5.141 \times 10^{-47} \text{ kg m}^2$$

14.5 Calculate the frequency in wave numbers and the wavelength in cm of th first rotational transition ($J = 0 \rightarrow 1$) for $D^{35}Cl$.

SOLUTION:

$$\tilde{\nu} = 2BJ = \frac{h}{4\pi^2 cI} \quad \text{since } J = 1$$

The moment of inertia was calculated in the preceding problem.

$$\tilde{\nu} = \frac{(6.626 \times 10^{-34} \text{ J s})}{4\pi^2(2.998 \text{ X } 10^8 \text{ m s}^{-1})(5.141 \times 10^{-47} \text{ kg m}^2)} = 1089 \text{ m}^{-1}$$

$$= (1089\ m^{-1})(10^{-2}\ m\ cm^{-1}) = 10.89\ cm^{-1}$$

$$\lambda \quad = \frac{1}{\tilde{\nu}} = \frac{1}{(10.89\ cm^{-1})} = 0.09183\ cm$$

14.6 The pure rotational spectrum of $^{12}C^{16}O$ has transitions at 3.863 cm^{-1} and 7.725 cm^{-1}. Calculate the internuclear distance in $^{12}C^{16}O$. Predict the positions, in cm^{-1}, of the next two lines.

SOLUTION:

$$\tilde{\nu} = \frac{h}{8\pi^2 Ic}[J(J+2) - J(J+1)] = \frac{2(J+1)h}{8\pi^2 Ic}$$

the only two transitions that are related 2:1 are $0 \rightarrow 1$ amd $1 \rightarrow 2$.

$$J = 0 \quad \tilde{\nu} = 3.863\ cm^{-1} = \frac{h}{4\pi^2 Ic}$$

$$J = 1 \quad \tilde{\nu} = 7.725\ cm^{-1} = \frac{h}{2\pi^2 Ic}$$

$$\therefore \quad I \quad = \frac{6.626 \times 10^{-34}\ J\ s}{4(3.13159)^2(3.863\ cm^{-1})(10^2\ cm\ m^{-1})(2.998 \times 10^8\ m\ s^{-1})}$$

$$= 1.449 \times 10^{-46}\ kg\ m^2$$

$$I \quad = \mu R_e^2 \qquad \mu = \frac{m_o m_c}{m_o + m_c} = 1.139 \times 10^{-26}\ kg$$

$$\therefore \quad R_e = 1.128 \times 10^{-10}\ m = 112.8\ pm$$

Next two lines are J = 2→3 and J = 3→4

$$\tilde{\nu}_2 = \frac{6\,h}{8\pi^2 Ic} \qquad = 11.589 \text{ cm}^{-1}$$

$$\tilde{\nu}_3 = 15.452 \text{ cm}^{-1}$$

14.7 Assume the bond distances in $^{13}C^{16}O$, $^{13}C^{17}O$, and $^{12}C^{17}O$ are the same as in $^{12}C^{16}O$. Calculate the position, in cm^{-1}, of the first rotational transitions in these four molecules. (Use the information in problem 14.6).

SOLUTION:

$$I = \mu R_e^2 \qquad \text{and} \quad \tilde{\nu}\ \frac{h}{4\pi^2 Ic} = \frac{h}{4\pi^2 \mu R_e^2 c}$$

$$\tilde{\nu}(^{12}C^{16}O) = 3.863 \text{ cm}^{-1} \text{ from problem 14.6.}$$

$$\therefore \quad \frac{\tilde{\nu}(^{12}C^{16}O)}{\tilde{\nu}(^{13}C^{16}O)} = \frac{\mu(^{13}C^{16}O)}{\mu(^{12}C^{16}O)} \text{ etc.}$$

$$\tilde{\nu}(^{13}C^{16}O) = \frac{\mu(^{12}C^{16}O)}{\mu(^{13}C^{16}O)}\tilde{\nu}(^{12}C^{16}O)$$

$$= \frac{m(^{12}C)\,[m(^{13}C) + m(^{16}O)]}{m(^{13}C)\,[m(^{12}C) + m(^{16}O)]} \bullet \tilde{\nu}(^{12}C^{16}O)$$

$$= 3.693 \text{ cm}^{-1}$$

Similarly

$$\tilde{\nu}(^{13}C^{17}O) = 3.596 \text{ cm}^{-1}$$

$$\tilde{\nu}{}^{12}C^{17}O) = 3.766 \text{ cm}^{-1}$$

14.8 The far infrared spectrum of HI consists of a series of equally spaced lines with $\Delta\tilde{\nu} = 12.8\ cm^{-1}$. What is (a) the moment of inertia, and (b) the internuclear distance?

SOLUTION:

(a) $$\Delta\tilde{\nu} = 2B = \frac{2h}{8\pi^2 cI}$$

$$I = \frac{h}{8\pi^2 cB} = \frac{(6.63 \times 10^{-34}\ J\ s)}{8\pi^2(2.998 \times 10^8\ m\ s^{-1})(6.4\ cm^{-1})(10\ cm\ m^{-1}}$$

$$= 4.37 \times 10^{-47}\ kg\ m^2$$

(b) $$I = \frac{m_A m_B}{m_A + m_B} R^2$$

$$R_e = \left[\frac{(4.37 \times 10^{-47}\ kg\ m^2)(128 \times 10^{-3}\ kg)(6.02 \times 10^{23}\ mol^{-1})}{(127 \times 10^{-3}\ kg)(1 \times 10^{13}\ kg)}\right]^{1/2}$$

$$= 163\ pm$$

14.9 Using equation 14.35, show that J for the maximally populated level is given by

$$J_{max} = \sqrt{\frac{kT}{2hcB}} - \frac{1}{2}$$

SOLUTION:

$$N_J = C(2J+1)\ e^{-BhcJ(J+1)/kT}$$

$$\frac{\partial N_J}{\partial J} = 0 = 2e^{-[]} - (2J+1)^2 \frac{Bhc}{kT}\ e^{-[]}$$

$$\therefore \quad J_{max} = \left(\frac{kT}{2hcB}\right)^{1/2} - \frac{1}{2}$$

14.10 Using the result of problem 14.9, find the J nearest J_{max} at room temperature for $H^{35}Cl$ and $^{12}C^{16}O$. What is the ratio of the population at that J to the population of J = 0? What is the energy of that J relative to J = 0 in units of kT?

SOLUTION:

(a) for $H^{35}Cl$, $B = 10.59\ cm^{-1} = 10.59 \times 10^2\ m^{-1}$

$$J_{max} = \left[\frac{1.381 \times 10^{-23} \times 198}{2 \times 6.626 \times 10^{-34} \times 1.998 \times 10^8 \times 10.59 \times 10^2}\right]^{1/2} - \frac{1}{2}$$

$$= 2.64$$

$$\therefore \quad J = 3$$

$$\frac{E(3)-E(0)}{kT} = \frac{12B}{kT}$$

$$= \frac{12 \times 10.59 \times 10^2\ m^{-1} \times 6.626 \times 10^{-34} \times 2.998 \times 10^8}{1.381 \times 10^{-23} \times 298})$$

$$= 0.6134$$

$$\frac{N(3)}{N(0)} = 7\ e^{-0.6134} = 3.79$$

(b) for $^{12}C^{16}O$, $B = 1.931\ cm^{-1} = 193.1\ m^{-1}$

$J_{max} = 6.85 \quad \therefore J = 7$

$$\frac{E(7)-E(0)}{kT} = \frac{56B}{kT} = 0.5220$$

$$\frac{N(7)}{N(0)} = \frac{15}{1}\ e^{-0.5220} = 8.90$$

14.11 The moment of inertia of $^{16}O^{12}C^{16}O$ is 7.167×10^{-46} kg m^2. (a) Calculate the CO bond length, R_{CO}, in CO_2. (b) Assuming that isotopic substitution does not alter R_{CO}, calculate the moments of inertia of (1) $^{18}O^{12}C^{18}O$ and (2) $^{16}O^{13}C^{16}O$.

SOLUTION:

(a) Since the CO_2 molecule is symmetrical, the carbon atom is on the axis rotation and does not contribute to the moment of inertia.

$$I = 2mR^2_{CO}$$

$$R_{CO} = (I/2m)^{1/2}$$

$$= \left[\frac{(71.67 \times 10^{-47}\ kg\ m^2)(6.022 \times 10^{23}\ mol^{-1})}{2(15.994\ 91 \times 10^{-3}\ kg\ mol^{-1})}\right]^{1/2}$$

$= 0.1162$ nm

(b) For $^{18}O^{12}C^{18}O$

$I \quad = 2mR^2_{CO}$

$$= \left[\frac{2(17.999\ 126 \times 10^{-3}\ kg\ mol^{-1})(0.1162 \times 10^{-9}\ m)^2}{6.022 \times 10^{23}\ mol^{-1}}\right]$$

$= 8.071 \times 10^{-46}$ kg m^2

For $^{16}O^{13}C^{16}O$ the moment of inertia is the same as for $^{16}O^{12}C^{16}O$.

14.12 Derive the expression for the moment of inertia of a symmetrical tetrahedral molecule like CH_4 in terms of the bond length R and the masses of the four tetrahedral atoms. The easiest way to derive the expression is to consider an axis along one CH bond. Show that the same result is obtained if the axis is taken perpendicular to the plane defined by one group of three atoms HCH.

SOLUTION:

109.471°

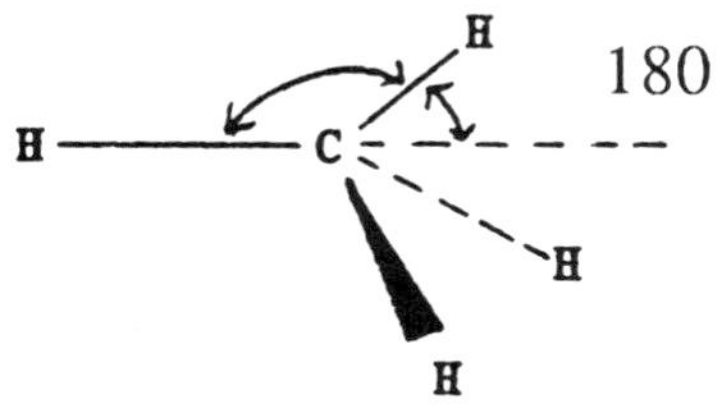

$180° - 109.471° = 70.529°$

$\sin 70.529° = h/R$ where h is the perpendicular distance to the axis

$$I = \sum m_i h_i^2 = 3mR^2 \sin^2(70.529°) = \frac{8}{3} mR^2$$

If the axis is taken perpendicular to the page through C

$$I = 2mR^2 + 2[R^2 \sin^2(90° - \frac{109.471}{2}] = \frac{8}{3} mR^2$$

14.13 What are the values of A and B (from equation 14.51) for the symmetric top NH_3 if $I_{||} = 4.41 \times 10^{-47}$ kg m^2 and $I_{\perp} = 2.81 \times 10^{-47}$ kg m^2? What is the wavelength of the J = 0 to J = 1 transition? What are the wavelengths of the J = 1 to J = 2 transitions (remember the selection rules: $\Delta J = \pm 1$, $\Delta K = 0$, and find all allowed transitions).

SOLUTION:

$$B = \frac{\hbar}{4\pi c I_{\perp}} = \frac{1.054 \times 10^{-34} \text{J s}}{4(3.14159)(2.998 \times 10^{8} \text{ ms}^{-1})(2.81 \times 10^{-47} \text{ kg m}^2)}$$

$$= 9.95 \times 10^{2} \text{ m}^{-1} = 9.995 \text{ cm}^{-1}$$

$$= \frac{\hbar}{4\pi c I_{||}} = 6.35 \text{ cm}^{-1}$$

$$\tilde{\nu}_{0\to 1} = 2B = 19.90 \text{ cm}^{-1}$$

$$\lambda = 1.01 \times 10^{-3} \text{ m}$$

$$\tilde{\nu}^{K=0}_{0\to2} = 4B = 39.84 \text{ cm}^{-1}$$

$$\lambda = 2.51 \times 10^{-4} \text{ m}$$

$$\tilde{\nu}^{K=\pm1}_{1\to2} = 4B = 29.84 \text{ cm}^{-1}$$

$$\lambda = 2.51 \times 10^{-4} \text{ m}$$

14.14 Consider a linear triatomic molecule, ABC. Find the center of mass (which by symmetry lies on the molecular axis). Show that the moment of inertia is given by

$I = \frac{1}{M}\{R_{AB}^2 m_A m_B + R_{BC}^2 m_B m_C + (R_{AB} + R_{BC})^2 m_A m_C\}$ where R_{AB} is the AB bond distance, R_{BC} is the BC bond distance, m_i are the masses of the atoms, and $M = m_A+m_B+m_C$. Show that if $R_{AB} = R_{BC}$ and $m_A = m_C$, $I = 2m_A R_{AB}^2$.

SOLUTION:

(a) Center of mass: choose the origin at atom A, then

$$R_{CM} = \frac{m_B R_{AB}+m_C(R_{AB}+R_{BC})}{m_A+m_B+m_C} = \frac{m_B R_{AB}+m_C(R_{AB}+R_{BC})}{M})$$

(b) moment of inertia

$$I = m_A(0-R_{CM})^2 + m_B(R_{AB}-R_{CM})^2 + m_C(R_{AB}+R_{BC}-R_{CM})^2$$

$$= \frac{1}{M^2}[m_A(m_BR_{AB}+m_C(R_{AB}+R_{BC}))^2 + m_BM^2(R_{AB}-R_{CM})^2$$

$$+ m_CM^2(R_{AB}+R_{BC}-R_{CM})^2]$$

$$= \frac{1}{M^2}[R_{AB}^2(m_A(m_B+m_C)M)+R_{BC}^2Mm_C(m_A+m_B)+2R_{AB}R_{BC}Mm_Am_C]$$

$$=(\frac{1}{M})[R_{AB}^2m_A(m_B+m_C)+R_{BC}^2(m_A+m_B)m_C+2R_{AB}R_{BC}m_Am_C]$$

$$I = \frac{1}{M}\{m_Am_BR_{AB}^2+m_Bm_CR_{BC}^2+m_Am_C(R_{AB}+R_{BC})^2\}$$

If $R_{AB} = R_{BC}$ and $m_A = m_C$

$$I = \frac{1}{M}\{2m_Bm_AR_{AB}^2+4m_A^2R_{AB}^2\}$$

$$= \frac{(2m_A+m_B)}{M}(2m_AR_{AB}^2) = 2m_AR_{AB}^2$$

14.15 The fundamental vibration frequency of $H^{35}Cl$ is $8.967 \times 10^{13} s^{-1}$ and of $D^{35}Cl$ is $6.428 \times 10^{13} s^{-1}$. (a) What is the difference in wavelengths between infrared absorption lines of $H^{35}Cl$ and $H^{37}Cl$? (b) What is the difference in wavelengths between infrared absorption lines of $D^{35}Cl$ and $D^{37}Cl$? Assume that the force constants of the bonds are the same in each pair.

SOLUTION:

(a) $\mu = \dfrac{m_H m_{Cl}}{m_H + m_{Cl}}$

for $H^{35}Cl$

$$\mu_{35} = \frac{(1.008 \times 10^{-3}\ kg\ mol^{-1})(34.97 \times 10^{-3}\ kg\ mol^{-1})}{(35.98 \times 10^{-3} kg\ mol^{-1})(6.022 \times 10^{23}\ mol^{-1})}$$

$$= 1.627 \times 10^{-27}\ kg$$

for $H^{37}Cl$

$$\mu_{37} = 1.629 \times 10^{-27}\ kg$$

$$\nu = \frac{1}{2\pi}\left(\frac{k}{\mu}\right)^{1/2}$$

$$\therefore \frac{\nu_{35}}{\nu_{37}} = \left(\frac{\mu_{37}}{\mu_{35}}\right)^{1/2} = \frac{\lambda_{37}}{\lambda_{35}}$$

$$\lambda_{37} = \lambda_{35}\left(\frac{\mu_{37}}{\mu_{35}}\right)^{1/2} \quad \therefore \quad \lambda_{37} - \lambda_{35} = \lambda_{35}\left[\left(\frac{\mu_{37}}{\mu_{35}}\right)^{1/2} - 1\right]$$

$$\lambda_{35} = \frac{2.998 \times 10^{8}\ ms^{-1}}{8.967 \times 10^{13} s^{-1}} = 3343 \times 10^{-9}\ m$$

$$\lambda_{37} - \lambda_{35} = 3343\left[\left(\frac{1.629}{1.627}\right)^{1/2} - 1\right] = 2.5\ nm$$

(b) DCl

$$\therefore \frac{\nu_{35}}{\nu_{37}} = (\frac{\mu_{37}}{\mu_{35}})^{1/2} = 1.0015$$

$$\nu_{35} = 6.428 \times 10^{13} \text{ s}^{-1}$$

$$\lambda_{35} = 4664 \text{ nm}$$

$$\lambda_{37} - \lambda_{35} = 4664 \ ((1.0015)^{1/2}-1) = 3.5 \text{ nm}$$

14.16 Find the force constants of the halogens, $^{127}I_2$, $^{79}Br_2$, and $^{35}Cl_2$ using the data of Table 14.4. Is the order of these the same as the order of the bond energies?

SOLUTION:

$$\tilde{\nu} = \frac{1}{2\pi c}\sqrt{\frac{k}{\mu}}$$

$$k = 4\pi^2 c^2 \tilde{\nu}^2 \mu$$

$\mu_i = \frac{m_i}{2} \times 10^{-3}$, if m_i is the atomic mass of the atom

$$k_{I_2} = 4(3.13159)^2 \frac{m_I 10^{-3}}{2} \frac{(214.5 \times 10^2)^2 \text{m}^{-2}}{6.022 \times 10^{23}} (2.998 \times 10^8)^2 \text{m}^2\text{s}^{-2}$$

$$= 172.2 \text{ N/m}^2$$

$$k_{Br_2} = 246.3 \text{ N/m}^2$$

$$k_{Cl_2} = 323.0 \text{ N/m}^2$$

∴

	I_2	Br_2	Cl_2
D_o	1.54 eV	1.97 eV	2.48 eV
k	172.2N/m^2	246.3 N/m^2	323.0 N/m^2

∴ same order

14.17 Given the following fundamental frequencies of vibration, calculate ΔH^o for the reaction

$H^{35}Cl\ (v=0) + {}^2D_2\ (v=0) = {}^2D^{35}Cl\ (v=0) + H^2D\ (v=0)$.

$H^{35}Cl$:2989 cm^{-1} $\quad$ H^2D:3817 cm^{-1}

${}^2D^{35}Cl$:2144 cm^{-1} $\quad$ ${}^2D^2D$:3119 cm^{-1}

<u>SOLUTION</u>:

Since the electronic part of the bond energies D_o are the same on left and right-hand side, only the zero point energy contributes to $\Delta H°$.

N.B. $D_o = D_o - \frac{1}{2} hc\tilde{v}$.

$$\Delta H° = E_{DCl} + E_{HD} - E_{D_2} - E_{HCl}$$

$$= +\frac{1}{2} hc(\tilde{v}_{DCl} + \tilde{v}_{HD} - \tilde{v}_{D_2} - \tilde{v}_{HCl})$$

$$= +\frac{1}{2} (6.626 \times 10^{-34} \times 1.998 \times 10^8)(2144 + 3817 - 2989 - 3119)$$

$$= (+147\ cm^{-1})\frac{1}{2} (6.626 \times 10^{-34} \times 2.998 \times 10^8) \times 10^2\ m^{-1}\ cm$$

$$= -\ 1.46 \times 10^{-21}\ J$$

$$= -\ 1.46 \times 10^{-21} \times 6.022 \times 10^{23} = -\ 879\ J\ mol^{-1}$$

14.18 If the fundamental vibration frequency of $^1H_2$ is 4401.21 cm^{-1}, compute the fundamental vibration frequency of $^2D_2$ and $^1H^2D$ assuming the same force constants. If D_o for $^1H_2$ is 4.4781 eV, what is D_o for $^2D_2$ and ^1HD? Neglect anharmonicities.

<u>SOLUTION</u>:

$$\frac{\tilde{\nu}_{^2D_2}}{\tilde{\nu}_{^1H_2}} = \left(\frac{\mu_{^1H_2}}{\mu_{^2D_2}}\right)^{1/2} = \left(\frac{m_{^1H}}{m_{^2D}}\right)^{1/2} = 0.707$$

$$\frac{\tilde{\nu}_{^1H_2D}}{\tilde{\nu}_{^1H_2}} = \left(\frac{\mu_{^1H_2}}{\mu_{^1H^2D}}\right)^{1/2} = \left[\frac{\frac{1}{2}m_{^1H}(m_{^1H}+m_{^2D})}{(m_{^1H}m_{^2D})}\right]^{1/2} = \left(\frac{m_{^1H}+m_{^2D}}{2m_{^2D}}\right)^{1/2}$$

$$= 0.866$$

$\therefore$ $\tilde{\nu}_{^2D_2} = 3112.1\ cm^{-1}$ $\quad \tilde{\nu}_{^1H^2D} = 3811.6\ cm^{-1}$

$$D_o = D_e - \frac{1}{2} h\tilde{\nu}c$$

$\therefore$ $D_o(D_2) = D_o(H_2) + \frac{hc}{2}(\tilde{\nu}_{H_2} - \tilde{\nu}_{D_2}) = 4.4781\ eV +$

$$\frac{6.626 \times 10^{-34} \times 2.998 \times 10^{8}}{2(1.602 \times 10^{-19})} [4401.2 - 3112.1] \times 10^{2}$$

$$= 4.4781 + 0.080\ eV = 4.558\ eV$$

$$D_o(HD) = 4.4781 + 0.037\ eV = 4.515\ eV$$

14.19 Using the values for ω_e and $\omega_e x_e$ in Table 14.4 for $^1H^{35}Cl$ estimate the dissociation energy assuming the Morse potential is applicable.

SOLUTION:

$\omega_e = 2990.95\ cm^{-1}$ $\quad \omega_e x_e = 52.819\ cm^{-1}$ $\quad x_e = 0.01766$

The zero point energy lies at

$$G(0) = \frac{1}{2}\omega_e - \frac{1}{4}\omega_e x_e = \frac{1}{2}(2991\ cm^{-1}) - \frac{1}{4}(52.8\ cm^{-1})$$

$$= 1482\ cm^{-1}$$

$$D_e = \frac{\omega_e}{4x_e} = \frac{2990.95\ cm^{-1}}{4(0.017\ 66)} = 42{,}341\ cm^{-1}$$

$$D_0 = D_e - G(0) = 42{,}341 - 1482 = 40{,}859\ cm^{-1}$$

$$= (40{,}859\ cm^{-1})(1.2399 \times 10^{-4}\ eV\ cm) = 5.07\ eV$$

The actual value is 4.434 eV

14.20 (a) What fraction of $H_2(g)$ molecules are in the v = 1 states at room temperature? (b) What fractions of $Br_2(g)$ molecules are in the v = 1, 2, and 3 states at room temperature?

SOLUTION:

(a) Using equation $f_1 = (1 - e^{-h\nu/kT})\, e^{-h\nu/kT}$

$$\frac{h}{kT} = \frac{hc\bar{\nu}}{kT} = \frac{(6.626 \times 10^{-34}\ J\ s)(2.998 \times 10^{10}\ cm\ s^{-1})(4401\ cm^{-1})}{(1.381 \times 10^{-23}\ J\ K^{-1})(298\ K)}$$

$$= 21.2$$

$$f_1 = (1 - e^{-21.2})e^{-21.2} = 6.2 \times 10^{-10}$$

(b) For Br_2, $\omega_e = 325.321\ cm^{-1}$

$$\frac{h\nu}{kT} = \frac{hc\bar{\nu}}{kT} = \frac{(6.626 \times 10^{-34}\ \mathrm{J\ s})(2.998 \times 10^{10}\ \mathrm{cm\ s^{-1}})(325.3\ \mathrm{cm^{-1}})}{(1.381 \times 10^{-23}\ \mathrm{J\ K^{-1}})(298\ \mathrm{K})}$$

$$= 1.570$$

$$f_1 = (1 - e^{-1.570})\, e^{-1.570} = 0.165$$

$$f_2 = (1 - e^{-1.570})\, e^{-(2)(1.570)} = 0.034$$

$$f_3 = (1 - e^{-1.570})\, e^{-(3)(1.570)} = 0.007$$

14.21 The first three lines in the R branch of the fundamental vibration-rotation band of $H^{35}Cl$ have the following frequencies in cm^{-1}: 2906.25(0), 2925.78(1), 2944.89(2), where the numbers in parentheses are the J values for the initial level. What are the values of ω_0, B'_v, B''_v, B_e, and α?

SOLUTION:

According to equation 14.76

$$\tilde{\nu} = \omega_o + (B'_v + B''_v)m + (B'_v - B''_v)m^2$$

$$= a + bm + cm^2 \quad \text{where } m = J'' + 1$$

$$2906.25 = a + b + c$$

$$2925.78 = a + 2b + 4c$$

$$2944.89 = a + 3b + 9c$$

$$a = 2886.30\ \mathrm{cm^{-1}} = \omega_o$$

$$b = 20.16 = B'_v + B''_v$$

$c \qquad = -0.21 \qquad = B'_v - B''_v$

$B'_v \qquad = 9.98 \text{ cm}^{-1} = B_e - \frac{3}{2}\alpha$

$B''_v \qquad = 10.19 \text{ cm}^{-1} = B_e - \frac{1}{2}\alpha$

$\alpha \qquad = 0.21 \text{ cm}^{-1}$

14.22 In Table 12.1, D_e for H_2 is given as 4.7483 eV or 458.135 kJ mol^{-1}. Given the vibrational parameters for H_2 in Table 14.4 calculate the value you would expect for $\Delta_f H^o$ for H(g) at 0 K.

SOLUTION:

$$D_o = D_e - \frac{1}{2}\omega_e + \frac{1}{4}\omega_e x_e$$

$$= (4.7483 \text{ eV})(8065.478 \text{ cm}^{-1} \text{ eV}^{-1}) - \frac{1}{2}(4400.39) + \frac{1}{4}(120.815)$$

$$= 36.127 \text{ cm}^{-1} \text{ or } 432.175 \text{ kJ mol}^{-1}$$

$H_2(g) \qquad = 2H(g)$

$\Delta H^o_o \qquad = 2\Delta_f H^o(H) = 432.175 \text{ kJ mol}^{-1}$

$\Delta_f H^o(H) \qquad = 216.088 \text{ kJ mol}^{-1}$

Table C.2 gives 216.037 kJ mol^{-1}

14.23 Calculate the wavelengths in (a) wave numbers, and (b) micrometers of the center two lines in the vibration-rotation spectrum of HBr for the fundamental vibration. The necessary data are to be found in Table 14.4.

SOLUTION:

The reduced mass of $H^{80}Br$ is given in Table 14.4 as

$$\mu = \frac{0.995\ 58 \times 10^{-3}\ \text{kg mol}^{-1}}{6.022\ 137 \times 10^{-23}\ \text{mol}^{-1}} = 1.653\ 20 \times 10^{-27}\ \text{kg}$$

$$I = \mu R^2 = (1.653\ 20 \times 10^{-27}\ \text{kg})(1.4138 \times 10^{-10}\ \text{m})^2$$

$$= 3.304\ 47 \times 10^{-47}\ \text{kg m}^2$$

$$B = \frac{h}{8\pi^2 c I}$$

$$= \frac{(6.626 \times 10^{-34}\ \text{J s})(10^{-2}\ \text{m cm}^{-1})}{8\pi^2(2.998 \times 10^8\ \text{m s}^{-1})(3.3045 \times 10^{-47}\ \text{kg m}^2)} = 8.47\ \text{cm}^{-1}$$

(a) $\tilde{\nu}_P = \tilde{\nu}_o - 2BJ''$ where $J'' = 1,2,3,\cdots$.

$$= 2649.67\ \text{cm}^{-1} - 2(8.47\ \text{cm}^{-1}) = 2632.72\ \text{cm}^{-1}$$

$\tilde{\nu}_R = \tilde{\nu}_o + 2B + 2BJ''$ where $J'' = 0,1,2,\cdots$.

$$= 2649.67\ \text{cm}^{-1} + 2(8.47\ \text{cm}^{-1}) = 2666.61\ \text{cm}^{-1}$$

(b) $$\lambda_P = \frac{1}{\tilde{\nu}_P} = 3.798\ 34 \times 10^{-4}\ \text{cm} = 3.798\ 34\ \mu\text{m}$$

$$\lambda_R = \frac{1}{\tilde{\nu}_R} = 3.750\ 08 \times 10^{-4}\ \text{cm} = 3.750\ 08\ \mu\text{m}$$

14.24 How many normal modes of vibration are there for (a) SO_2(bent), (b) H_2O_2(bent), (c) HC≡CH(linear), and (d) C_6H_6?

SOLUTION:

The number of normal modes of vibration is 3N-6 for a nonlinear molecule and 3N-5 for a linear molecule.

(a) 3N-6 = 9-6 = 3 (c) 3N-5 = 12-5 = 7

(b) 3N-6 = 12-6 = 6 (d) 3N-6 = 36-6 = 30

14.25 List the numbers of translational, rotational, and vibrational degrees of freedom for (a) Ne, (b) N_2, (c) CO_2, and (d) CH_2O.

SOLUTION:

	Molecule	Trans	Rot	Vib	Total = 3N
(a)	Ne	3	0	0	3
(b)	N_2	3	2	1	6
(c)	CO_2	3	2	4*	9
(d)	CH_2O	3	3	6**	12

*For linear molecules 3N-5 **For nonlinear molecules 3N-6

14.26 Acetylene is a symmetrical linear molecule. It has seven normal modes of vibration, two of which are doubly degenerate. These normal modes may be represented as follows:

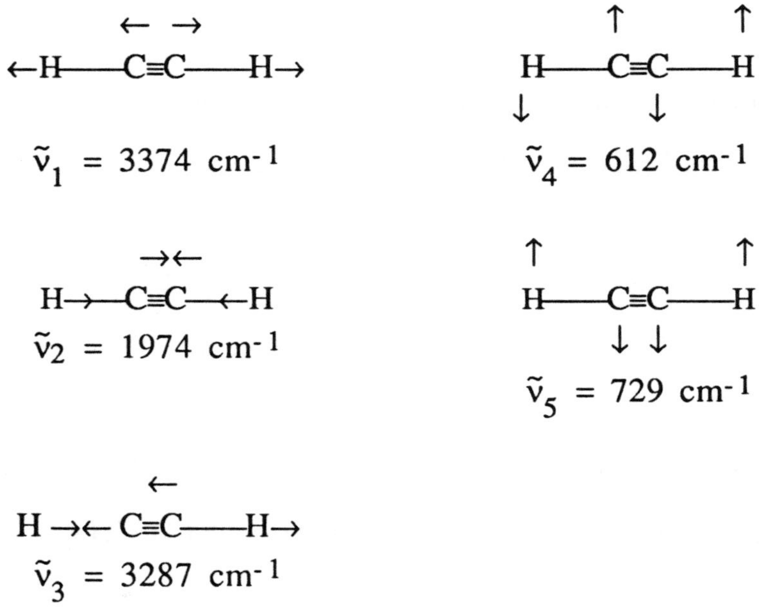

(a) Which are the doubly degenerate vibrations?

(b) Which vibrations are infrared active?

(c) Which vibrations are Raman active?

<u>SOLUTION</u>:

(a) $\tilde{\nu}_4$ and $\tilde{\nu}_5$ since they can also take place in a plane perpendicular to the plane of the paper

(b) $\tilde{\nu}_3$ and $\tilde{\nu}_5$ because the dipole moment changes during the vibration

(c) $\tilde{\nu}_1$, $\tilde{\nu}_2$, and $\tilde{\nu}_4$ because the polarizabiility of the molecule changes during the vibrations.

14.27 When CCl_4 is irradiated with the 435.8 nm mercury line, Raman lines are obtained at 439.9, 444.6, and 450.7 nm. Calculate the Raman frequencies of CCl_4 (expressed in wave numbers). Also calculate the wavelengths (expressed in μm) in the infrared at which absorption might be expected.

SOLUTION:

$$\Delta\tilde{\nu}_R = \frac{1}{\lambda_{incid}} - \frac{1}{\lambda_{scatt}}$$

$$= \frac{1}{435.8 \times 10^{-7}\ cm} - \frac{1}{439.9 \times 10^{-7}\ cm} = 214\ cm^{-1}$$

$$\lambda = \frac{1}{\Delta\tilde{\nu}} = \frac{1}{214\ cm^{-1}} = 46.7 \times 10^{-6}\ m$$

For the remaining lines

$\Delta\tilde{\nu}_R\ /cm^{-1}$	312	454	759
$\lambda/\mu m$	32.0	22.0	13.2

14.28 The first several Raman frequencies of $^{14}N_2$ are 19.908, 27.857, 35.812, 43.762, 51.721, and 59.662 cm^{-1}. These lines are due to pure rotational transitions with J = 1,2,3,4,5, and 6. The spacing between the lines is $4B_e$. What is the internuclear distance?

SOLUTION:

$$\mu = \frac{(14.003\ 07 \times 10^{-3}\ kg)^2}{(28.006\ 14 \times 10^{-3}\ kg)(6.022\ 1367 \times 10^{23}\ mol^{-1})}$$

$$= 1.162\ 633 \times 10^{-26}\ kg$$

The average spacing between lines is 7.951 cm^{-1}, and so

$$B_e = (7.951\ cm^{-1})(10^2\ cm\ m^{-1})/4 = 198.78\ m^{-1} \text{ since } \Delta J = 2.$$

$$B_e = \frac{h}{8\pi^2 c\mu R_e^2}$$

$$R_e = \left(\frac{h}{8\pi^2 c\mu B_e^2}\right)^{1/2}$$

$$= \left[\frac{6.626\ 0755 \times 10^{-34} J\ s}{8\pi^2(2.997\ 924\ 58 \times 10^8\ m\ s^{-1})(1.162\ 651 \times 10^{-26}\ kg)(198.78\ m^{-1})^2}\right]^{1/}$$

$$= 110\ .06\ pm$$

14.29 What Raman shifts are expected for the first four Stokes lines for CO_2?

<u>SOLUTION</u>:

$$I = 7.167 \times 10^{-46}\ kg\ m^2 \qquad \text{(problem 14.11)}$$

$$B_e = \frac{h}{8\pi^2 Ic} = \frac{(6.626 \times 10^{-34}\ J\ s)(10^{-2}\ m\ cm^{-1})}{8\pi^2(7.167 \times 10^{-46}\ kg\ m^2)(2.9979 \times 10^8\ m\ s^{-1})}$$

$$= 0.3906\ cm^{-1} \qquad \Delta\tilde{\nu}_R = 2B_e(2J + 3)$$

J"	$\Delta\tilde{\nu}_R$
0	2.3436 cm-1
1	3.9060
2	5.4684
3	7.0308

14.30	11,600, 1.16×10^6, 1.16×10^7 K; 0.024 eV
14.31	(a) 1.1385×10^{-26} kg, (b) 1.4491×10^{-46} kg m^2
14.32	21.2, 42.36, 63.54, 190.6 cm^{-1}
	472, 236, 157, 52.5 μm
14.33	(a) 4.607×10^{-48}, (b) 6.139×10^{-48}, (c) 6.904×10^{-48},
	(d) 9.205×10^{-48} kg m^2
14.34	37.822 cm^{-1}, 0.452 kJ mol^{-1} for OH.
	20.02 cm^{-1}, 0.239 kJ mol^{-1} for OD
14.35	1.938×10^{-47}, 1.950×10^{-47} kg m^2, J = 2, 86.64 cm^{-1} for ^{12}CH
	86.10 cm^{-1} for ^{13}CH
14.36	1.129×10^{-10} m
14.38	6B, 10B, 14B; -6B, -10B, -14B.
14.40	121.69×10^8, 243.38×10^8, 365.07×10^8 s^{-1}
14.41	1.188×10^{12}, 2.376×10^{12}, 3.564×10^{12} s^{-1}
14.42	9.10, 14.39 K
14.43	38,812 cm^{-1}; 42,358 cm^{-1}; 33,658 cm^{-1}
14.44	3.908, 4.618, 3.196 eV
14.45	516.4, 411.6, 314.1 N m^2
	427.78, 362.56, 294.66 kJ mol^{-1}
14.46	(a) 5.94×10^{-7}, (b) 0.360
14.47	3.60, 3.03, 2.31 eV
14.48	5.29, 5.42, 5.49, 0.13 cm^{-1}; 127.0 pm ($^1H^{35}Cl$, R_e = 127.455 pm)
14.49	1882, 8.56 cm^{-1}

14.50

	trans	rot	vib
Cl_2	3	2	1
H_2D	3	3	3
C_2H_2	3	2	7

14.51

		trans	rot	vib
	NNO	3	2	4
	NH_3	3	3	6

14.52 75.19 pm

14.53 112.6 pm

14.54 655 J mol^{-1}

CHAPTER 15: Electronic Spectroscopy of Molecules

15.1 The spectroscopic dissociation energy of $H_2(g)$ into ground state hydrogen atoms is 4.4763 eV. What is the spectroscopic dissociation energy of $H_2(g)$ into one ground state H and one H atom in the 2p state? If H_2 is dissociated with photons of energy 15 eV, what is the velocity of the H atoms coming off in the 1s and 2p states?

SOLUTION:

$$H_2(g) = 2H(g;1s) \quad \Delta E = 4.4763 \text{ eV}$$

$$H(g;1s) = H(g;2p) \quad \Delta E = 13.6(\frac{3}{4}) = 9.45 \text{ eV}$$

$$H_2(g) = H(g;1s) + H(g;2p) \quad \Delta E = 13.93 \text{ eV}$$

The excess energy is 15 - 13.93 = 1.07 eV, all of which goes into kinetic energy of the two atoms:

$$2(\frac{1}{2} mv^2) = 1.07 \text{ eV}$$

$$v = \sqrt{\frac{1.07 \text{ eV} \times 1.602 \times 10^{-19} \text{ J/eV}}{(1.008 \times 10^{-3}/6.022 \times 10^{23})\text{kgm}}}$$

$$= 1.012 \times 10^4 \text{ m/s}$$

15.2 According to the hypothesis of Franck, the molecules of the halogens dissociate into one normal atom and one excited atom. The wavelength of the convergence limit in the spectrum of iodine is 499.5 nm. (a) What is the energy of dissociation in kJ mol^{-1} of iodine into one normal and one excited atom? (b) The thermochemical value of the heat of dissociation of I_2 into ground state atoms can be found in Table C-2. Calculate the energy of the excited state of I that is formed from the spectroscopic dissociation in kJ mole^{-1} and eV.

SOLUTION:

(a) $E = hcN_A/\lambda$

$$= \frac{(6.626 \times 10^{-3}\ \text{Js})(2.998 \times 10^{8}\ \text{m s}^{-1})(6.022 \times 10^{23}\ \text{mol}^{-1})}{(4995 \times 10^{-10}\ \text{m})(10^{3}\ \text{J kJ}^{-1})}$$

$= 239.5$ kJ mol^{-1}

(b) $I_2 = I + I^*$ $\Delta E = 2139.5$ kJ/mol

$I_2 = 2I$ $\Delta E = 2(107.25) - 65.52 = 148.98$ kJ/mol

$I = I^*$ $\Delta E = 90.53$ kJ/mol

$= 0.938$ eV

15.3 The ultraviolet absorption of O_2 includes a series of lines (the Schumann-Runge bands) due to transitions from the $^3\Sigma_g^-$ ground state to the excited electronic state $^3\Sigma_u^-$, which are shown in Fig. 15.1. These lines converge to 175.9 nm, which corresponds to dissociation to one O atom in its ground state ^3P and one O atom in an excited state ^1D. What is D_0 for O_2? How does this compare with the enthalpy of formation at 0 K? Given: The ^1D state of O is 1.970 eV above the ground state ^3P.

SOLUTION:

The dissociation energy D_0 for $O_2(^3\Sigma_g^-) = O(^3P) + O(^1D)$ is

$$\Delta E = \frac{hcN_A}{\lambda} =$$

$$= \frac{(6.626 \times 10^{-34} \text{ J s})(2.998 \times 10^{8} \text{ m s}^{-1})(6.022 \times 10^{23} \text{ mol}^{-1})}{(175 \times 10^{-9} \text{ m})}$$

$$= 680.2 \text{ kJ mol}^{-1} = \frac{(680.2 \text{ kJ mol}^{-1})}{(96.485 \text{ kJ eV}^{-1})}$$

$= 7.050$ eV

Since $O(^3P) = O('D)$ $\Delta E = 1.970$ eV

The dissociation energy of $O_2(^3\Sigma_g^-)$ into $2O(^3P)$ is

$D_o = 7.0\text{-}50 - 1.970 = 5{,}080$ eV or 490.14 kJ mol^{-1}.

From Table C.2 $\Delta_f H^o$ (0 K) = 2(246.785) = 493.57 kJ mol^{-1}.

15.4 The spectroscopic dissociation energy of $^{127}I_2$ is 1.542 38 eV according to Table 14.4. What wavelength of light would you use to dissociate ground state molecules to ground state atoms if you wanted the atoms to fly away with velocities of
(a) 0 m s^{-1}, (b) 10^3 m s^{-1}, (c) 10^5 m s^{-1}?

<u>SOLUTION:</u>

(a) $E = 1.542\ 38\ (96{,}485)$ J/mol $= 1.48\ 82 \times 10^5$ J/mol

$\lambda = \frac{h}{c} = 804.35$ nm

(b) $E = (1.542\ 38 \text{ eV})(96{,}485 \text{ J mol}^{-1} \text{ eV}^{-1}) +$

$$2(\frac{1}{2}\ 126.904 \times 10^{-3}\ \text{kg mol}^{-1})(10^3\ \text{m s}^{-1})^2$$

$$= (1.488 \times 10^{-5} + 1.269 \times 10^5)\ \text{J mol}^{-1}$$

$$= 2.757 \times 10^5\ \text{J mol}^{-1} = N_A hc/\lambda$$

$$\lambda = \frac{(6.022 \times 10^{23}\ \text{mol}^{-1})(6.626 \times 10^{-34}\ \text{J s})(2.998 \times 10^8\ \text{m s}^{-1})}{2.757 \times 10^5\ \text{J mol}^{-1}}$$

$$= 4.339 \times 10^{-7}\ \text{m} = 433.9\ \text{nm}$$

(c) $$E = 2.47\ 13 \times 10^{-19}\ \text{J} + 2\left[\frac{1}{2}\frac{(126.904 \times 10^{-3})(10^{10})}{6.022 \times 10^{23}}\right]$$

$$= 2.1076 \times 10^{-15}\ \text{J}$$

$$\lambda = 94.32\text{pm}$$

15.5 A solution of a dye containing 0.1 mol L^{-1} transmits 80% of the light at 435.6 nm in a glass cell 1 cm thick. (a) What percent of light will be absorbed by a solution containing 2 mol L^{-1} in a cell 1 cm thick? (b) What concentration will be required to absorb 50% of the light? (c) What percent of the light will be transmitted by a solution of the dye containing 0.1 mol L^{-1} in a cell 5 cm thick? (d) What thickness should the cell be in order to absorb 90% of the light with solution of this concentration?

<u>SOLUTION</u>:

$\log (I_o/I) = \varepsilon c l$

$\log (100/80) = \varepsilon(0.1\ \text{mol L}^{-1})\ (1\ \text{cm})$

$\varepsilon = 0.969\ \text{L mol}^{-1}\ \text{cm}^{-1}$

(a) $I = I_o e^{-2.303\ \varepsilon c l}$

$= (100)\ \exp\ [-(2.303)(0.969\ \text{L mol}^{-1}\ \text{cm}^{-1})(2\ \text{mol L}^{-1})(1\ \text{cm})]$

$= 1.2\%$

(b) $\log\ (100/50) = (0.969\ \text{L mol}^{-1}\ \text{cm}^{-1})(1\ \text{cm})c$

$c = 0.311\ \text{mol L}^{-1}$

(c) $I = (100)\exp[-(2.303)(0.969\ \text{L mol}^{-1}\text{cm}^{-1})(0.1\ \text{mol L}^{-1})(5\ \text{cm})]$

$= 32.8\%$

(d) $\log(100/10) = (0.969\ \text{L mol}^{-1}\ \text{cm}^{-1})(0.1\ \text{mol L}^{-1})1$

$1 = 10.3\ \text{cm}$

15.6 Derive equation 15.14 for the integrated intensity of a Gaussian absorption line. A Gaussian line has the form $\varepsilon_m\ e^{-\sigma(\tilde{\nu} - \tilde{\nu}_m)^2}$, where $\tilde{\nu}$ is the frequency at the intensity maximum, ε_m.

[Hint: relate σ to the width of the line at half maximum intensity and use the integral

$$\int_{-\infty}^{+\infty} e^{-\sigma x^2}\ dx = (\pi/\sigma)^{1/2}]$$

SOLUTION:

$$\varepsilon(\tilde{\nu}) = \varepsilon_m\ e^{-\sigma(\tilde{\nu}-\tilde{\nu}_m)^2}$$

$$\int_{-\infty}^{+\infty} \varepsilon(\tilde{\nu})d\tilde{\nu} = \varepsilon_m \int_{-\infty}^{+\infty} d\tilde{\nu} e^{-\sigma(\tilde{\nu}-\tilde{\nu}_m)^2}$$

$$= \varepsilon_m \int_{-\infty}^{+\infty} dx\ e^{-\sigma x^2}$$

$$= \varepsilon_m \left(\frac{\pi}{\sigma}\right)^{1/2}$$

At $\tilde{v} = \tilde{v}_{1/2}$, $\varepsilon(\tilde{v}) = \varepsilon_m/2$

$$\therefore \tilde{v}_{1/2} = \tilde{v}_m \pm \frac{1}{\sigma^{1/2}}(\ell n2)^{1/2}$$

Full width at half maximum,

$$\therefore \Delta\tilde{v}_{1/2} = 2\sigma^{-1/2}\ (\ell n2)^{1/2}$$

$$\therefore \frac{1}{\sigma^{1/2}} = \frac{\Delta\tilde{v}_{1/2}}{2(\ell n2)^{1/2}}$$

$$\therefore \int_{-\infty}^{+\infty} \varepsilon\ (\tilde{v})d\tilde{v} = \varepsilon_m \cdot \frac{\pi^{1/2}}{2(\ln 2)^{1/2}} \cdot \Delta\tilde{v}_{1/2} = (1.064)\ \varepsilon_m \Delta\tilde{v}_{1/2}$$

15.7 The following absorption data are obtained for solutions of oxyhemoglobin in pH 7 buffer at 575 nm in a 1-cm cell:

g/cm^3	Transmission, %
3×10^{-4}	53.5
5×10^{-4}	35.1
10×10^{-4}	12.3

The molar mass of hemoglobin in 64.0 kg mol^{-1}. (a) Is Beer's law obeyed? What is the molar absorption coefficient? (b) Calculate the percent transmission for a solution containing 10^{-4} g/cm^3.

SOLUTION:

(a)

grams per cm^3	mol L^{-1}	I/I_o	$\log(I/I_o)$	ε/L $mol^{-1}cm^{-1}$
3×10^{-4}	4.69×10-6	0.535	-0.272	5.80×10^4
5×10^{-4}	7.82×10^{-6}	0.351	-0.455	5.82×10^4
10×10^{-4}	15.64×10^{-6}	0.123	-0.910	5.82×10^4

Beer's law is obeyed and the molar absorption coefficient is 5.81×10^4 L mol^{-1} cm^{-1}.

(b) $\log(I/I_o) = -\varepsilon c l$

$= -[5.81 \times 10^4 \text{ L mol}^{-1} \text{ cm}^{-1}](1.564 \times 10^{-6} \text{ mol L}^{-1})(1 \text{ cm})$

$= -0.091$

$I/I_o = 0.81$ or 81% transmission

15.8 The protein metmyoglobin and imidazole form a complex in solution. The molar absorption coefficients in L mol^{-1} cm^{-1} of the metmyoglobin (Mb) and the complex (C) are as follows:

λ/nm	ε_{Mb}/10^3 L mol^{-1} cm^{-1}	ε_C/10^3 L mol^{-1} cm^{-1}
500	9.42	6.88
630	3.58	1.30

An equilibrium mixture in a cell of 1-cm path length has an absorbance of 0.435 at 500 nm and 0.121 at 630 nm. What are the concentrations of metmyoglobin and complex?

<u>SOLUTION</u>:

$\log(I_o/I) = A = (\varepsilon_1 c_1 + \varepsilon_2 c_2)1$

At 500 nm, $0.435 = 9.42 \times 10^3 c_1 + 6.88 \times 10^3 c_2$

At 630 nm, $0.121 = 3.58 \times 10^3 c_1 + 1.30 \times 10^3 c_2$

Solving these equations simultaneously

$c_1 = 2.17 \times 10^{-5}$ mol L^{-1} metmyoglobin

$c_2 = 3.37 \times 10^{-5}$ mol L^{-1} complex

15.9 The absorption spectrum for benzene in Fig. 15.6 shows maxima at about 180, 200, and 250 nm. Estimate the integrated absorption coefficients using ε_{max} and $\Delta\tilde{v}_{1/2}$ and assuming that the width at half maximum is 5000 cm^{-1} in each case. What are the three oscillator strengths? (See Example 15.6)

SOLUTION:

At 180 nm

$$\varepsilon_{max}\Delta\tilde{v}_{1/2} = (50.1 \times 10^3 \text{ L mol}^{-1} \text{ cm}^{-1})(5000 \text{ cm}^{-1})$$

$$= 2.50 \times 10^8 \text{ L mol}^{-1} \text{ cm}^{-2}$$

$$f = (4.33 \times 10^{-9} \text{ L mol}^{-1} \text{ cm}^2)(2.50 \times 10^8 \text{ L mol}^{-1} \text{ cm}^{-2}) = 1.08$$

At 200 nm

$$\varepsilon_{max}\Delta\tilde{v}_{1/2} = (7000 \text{ L mol}^{-1} \text{ cm}^{-1})(5000 \text{ cm}^{-1})$$

$$= 3.5 \times 10^7 \text{ L mol}^{-1} \text{ cm}^{-2} = 0.152$$

At 250 nm

$$\varepsilon_{max}\Delta\tilde{v}_{1/2} = (100 \text{ L mol}^{-1} \text{ cm}^{-1})(5000 \text{ cm}^{-1})$$

$$= 5 \times 10^5 \text{ L mol}^{-1} \text{ cm}^{-2}$$

$$f = (4.33 \times 10^{-9} \text{ mol L}^{-1} \text{cm}^2)(5 \times 10^5 \text{ L mol}^{-1} \text{ cm}^{-2}) = 0.0022$$

15.10 Relatively strong absorption bands have ε_{max} = 10^4 to 10^5 L mol^{-1} cm^{-1} and $\Delta\tilde{\nu}_{1/2}$ of the order 1000 to 5000 cm^{-1}, while weak absorption bands have ε_{max} = 10 L mol^{-1} c m^{-1} and $\Delta\tilde{\nu}_{1/2}$ of the order 100 cm^{-1}. Assuming the absorption lines are Gaussian, compute the integrated absorption coefficient and the oscillator strengths for these bands.

SOLUTION:

$$\int \varepsilon\, d\tilde{\nu} = 1.06\ \varepsilon_{max}\ \Delta\tilde{\nu}_{1/2}\ ,\ 4.32 \times 10^{-9} \int \varepsilon\, d\tilde{\nu} = f$$

Strong

$$\int \varepsilon\, d\tilde{\nu} = 1.06 \times \begin{matrix} 10^4 \times 1000 \\ 10^5 \times 5000 \end{matrix} = \begin{matrix} 1.06 \times 10^7 \text{ L mol}^{-1} \text{ cm}^{-2} \\ 5.30 \times 10^8 \text{ L mol}^{-1} \text{ cm}^{-2} \end{matrix}$$

$$f = 4.58 \times 10^{-2} - 2.29$$

Weak

$$\int \varepsilon\, d\tilde{\nu} = 1.06 \times 10 \times 100 = 1.06 \times 10^3 \text{ L mol}^{-1} \text{ cm}^{-2}$$

$$f = 4.58 \times 10^{-6}$$

15.11 The measured oscillator strength of a transition can be used to compute the transition moment, $|\mu_{12}|^2$ by combining equation 15.16 and 15.17a to find

$|\mu_{12}|^2 = (f3he^2/8\pi^2 m_e \nu)$

For strong transitions (for which $f \cong 1$), moderately weak transitions ($f \simeq 10^{-3}$) and weak transitions ($f \simeq 10^{-6}$), calculate $|\mu_{12}|$ and $|R_{12}| = |\mu_{12}|/e$, assuming a transition energy of 25000 cm^{-1}.

SOLUTION:

$$|\mu_{12}|^2 = \left\{\frac{f3he^2}{8\pi^2 m_e(25000) \times 3 \times 10^{10}\ s^{-1}}\right\}^{1/2}$$

$$= e\, f^{1/2}\ (60.7\ pm)$$

	$\lvert\mu_{12}\rvert$	$\lvert R_{12}\rvert$
∴ strong	9.72×10^{-30} C m	60.7 pm
moderate	3.08×10^{-31} C m	1.92 pm
weak	9.72×10^{-33} C m	0.061 pm

1 Debye = 3.34×10^{-30} C m

15.12 In Chapter 12, the Hückel molecular orbital model was introduced to describe the electronic states of conjugated molecules. In this chapter, the free electron model (FEMO) was introduced for the same systems. Consider the butadiene molecule in both descriptions. The Hückel model (equation 12.64) gives the energies and wave functions for four orbitals, while the FEMO model gives an infinite number of orbitals. Consider the lowest four in the FEMO model. Do they have the same number of nodes the Hückel orbitals? Is there any way of choosing α and β in the Hückel model or a in the FEMO model to make the predictions for the energies of all four orbitals agree? Suppose we are content to make the lowest electronic absorption energy agree in both models, what is the formula for β in terms of a?

SOLUTION:

FEMO $\quad E = \dfrac{n^2h^2}{8ma^2} \quad n = 1,2,3,4$

$\phi_n = N \sin \dfrac{n\pi x}{a} \quad 0 \le x \le a$ so # nodes = n - 1 same as Hückel

Hückel: Eq 12.64

$$E_1 = \alpha + 1.618\ \beta$$

$$E_2 = \alpha + 0.618\ \beta$$

$$E_3 = \alpha - 0.618\ \beta$$

$$E_4 = \alpha - 1.618\ \beta$$

Since $(E_2\text{-}E_1)_{FEMO} = \dfrac{3h^2}{8ma^2}$ $\qquad (E_2\text{-}E_1)_{HU} = -\beta$

$$(E_3\text{-}E_2)_{FEMO} = \frac{5h^2}{8ma^2} \qquad (E_3\text{-}E_2)_{HU} = -1.236\beta$$

$$(E_4\text{-}E_3)_{FEMO} = \frac{7h^2}{8ma^2} \qquad (E_4\text{-}E_3)_{HU} = -\beta$$

OR: 3:5:7 1:1.236:1

There is *no* way to make the predictions agree for all four orbitals.

First excitation energy:

	FEMO	HÜCKEL
$(E_3\text{-}E_2)$	$\dfrac{5h_2}{8ma^2}$	-1.236β

$$\boxed{\therefore\ \beta = -\frac{5h^2}{8(1.236)ma^2}}$$

15.13 The lifetimes of vibrationally excited states of molecules of a liquid are limited by the collision rates in the liquid. If one in ten collisions deactivates a vibrationally excited state, what is the broadening of vibrational lines if a molecule undergoes 10^{13} collisions per second?

SOLUTION:

$$\Delta E = hc\Delta\tilde{v} \geq \frac{h}{2\pi\Delta t}$$

$$\Delta\tilde{v} \geq (2\pi c\Delta t)^{-1} = [2\pi(3 \times 10^{10}\ cm\ s^{-1})(10^{-12}\ s)]^{-1}$$

$$\Delta\tilde{v} \geq 5.3\ cm^{-1}$$

15.14 The first ionization potentials of Ar, Kr, and Xe are 15.755 eV, 13.966 eV, and 12.130 eV, respectively. Calculate the velocity of the emitted electrons when photons from a He discharge lamp with λ = 58.43 nm are used to record the photoelectron spectrum of these gases.

SOLUTION:

$$\lambda = 58.43\ nm;\ E_{photon} = \frac{hc}{\lambda} = \frac{6.626 \times 10^{34} \times 2.998 \times 10^{9}}{58.43 \times 10^{-9}}\ J$$

$$= 3.3998 \times 10^{-18}\ J$$

$$= 21.22\ eV$$

(a) Ar

$$\Delta E = 21.22 - 15.755\ eV = 5.47\ eV$$

$$v = \sqrt{\frac{2}{m_{Ar}} \bullet \Delta E} = \sqrt{\frac{2 \times 5.47 \times 1.602 \times 10^{-19}\ J}{(39.948/6.022 \times 12023) \times 10^{-3}\ kg}}$$

$$= 5.14\ km/s$$

(b) Kr

$\Delta E = 21.22 - 13.966 = 7.26$ eV

$$v = \left(\frac{2 \times 7.26 \times 1.602 \times 10^{-19}}{(83.80/(6.022 \times 10^{23})) \times 10^{-3}}\right)^{1/2} = 4.09 \text{ km/s}$$

(c) Xe

$\Delta E = 21.22 - 12.13 = 9.09$ eV

$$v \quad \left(\frac{2 \times 7.26 \times 1.602 \times 10^{-19}}{(131.29/(6.022 \times 10^{23})) \times 10^{-3}}\right)^{1/2} = 3.27 \text{ km/s}$$

15.15 A sample of oxygen gas is irradiated with $MgK_{\alpha_1\alpha_2}$ radiation of 0.99 nm (1253.6 eV). A strong emission of electrons with velocities of 1.57×10^7 m $s^{-1}$is found. What is the binding energy of these electrons?

<u>SOLUTION</u>:

$$\frac{1}{2} mc^2 = h\nu - I$$

$$I = h\nu - \frac{1}{2} mv^2$$

$$= 1253.6 \text{ eV} - \frac{1}{2}\frac{(9.109 \times 10^{-31} \text{ kg})(1.57 \times 10^7 \text{ m s}^{-1})^2}{(1.602 \times 10^{-19} \text{ J eV}^{-1})}$$

15.16 The photoelectron spectrum of molecules shows that similar atoms in different chemical environments have slightly different core orbital binding energies. For example, the 1s binding energy of carbon in CH_4 is 290 eV, while it is 293 eV in CH_3F. (a) Explain this shift based on the electronegativity difference between carbon and fluorine. (b) In the molecule $F_3CCOOCH_2CH_3$, predict the order of the carbon 1s binding energies in the 4 carbon atoms.

SOLUTION:

(a) Since F is more electronegative than C and H, it pulls electrons towards itself, thereby decreasing the shielding at the carbon nucleus, and increasing the binding energy of the 1s electrons

(b) The binding energy should be greatest for the CF_3 carbon, next largest for the COO carbon, next for the OCH_2 carbon, and smallest for the CH_3 carbon.

15.17 When α-D-mannose ($[\alpha]_D^{20}$ = +29.3°) is dissolved in water, the optical rotation decreases as β-D-mannose is formed until at equilibrium $[\alpha]_D^{20}$ = +14.2°). This process is referred to as mutarotation. As expected, when β-D-mannose ($[\alpha]_D^{20}$ =-17.2°) is dissolved in water, the optical rotation increases until $[\alpha]_D^{20}$ = +14.2° is obtained. Calculate the percentage of α form in the equilibrium mixture.

SOLUTION:

$[\alpha]_\alpha = 29.3°$ $\qquad\qquad$ $[\alpha]_\beta = -17.0°$

$$[\alpha]_{mixture} = +14.2° \quad = 29.3° f_\alpha - 17.0° f_\beta$$

$$= 29.3° f_\alpha - 17.0° (1 - f_\alpha)$$

$$= - 17.0° + 36.3 f_\alpha$$

$$f_\alpha = \frac{31.2°}{46.3°} = 0.67 \quad = \text{fraction of } \alpha \text{ form}$$

15.18 -92.2 kJ mol^{-1}

15.19 (a) 16,065, (b) 1.992 eV

15.20 (a)

	$E/10^{-19}$ J	E/kJ mol^{-1}
n = 1	2.410	145.14
n = 2	9.640	580.57

(b) a275 nm

15.21 210 nm

15.22 196 L mol^{-1} cm^{-1}, 24.7 %

15.23 8.48 x 10^7 L mol^{-1} cm^{-2}, 0.366

15.24 0.091, 88.2%

15.25 7.7 x 10^{-5} g cm^{-3}

15.26 2.6 μg/μL

15.27 (a) 0.324 x 10^{-4}, 0.512 x 10^{-4} mol L^{-1}

(b) 2.95 x 10^4 L mol^{-1}

15.28 $\varepsilon_{app} = \varepsilon_{base}/[1 + 10^{-(pH-pK)}]$

15.29 7.3 x 10^{-4}

15.30 14.04 eV

15.31 5.20 eV

15.32 63.6%

CHAPTER 16: Magnetic Resonance Spectroscopy

16.1 NMR spectrometers usually have fixed frequencies between 60 and 500 MHz. Calculate the magnetic flux densities needed to give a NMR transition frequency for hydrogen equal to these frequencies.

SOLUTION:

a) 60 MHz

$$B = \frac{h\nu}{|g_N|\mu_N} = \frac{(6.626 \times 10^{-34}\ Js)(6 \times 10^{7} s^{-1})}{(5.585)(5.0508 \times 10^{-27}\ JT^{-1})}$$

$$= 1.409\ T$$

b) 500 MHz

$$B = 11.74\ T$$

16.2 For the frequencies of problem 16.1, calculate the corresponding energies in kJ mol^{-1} and compare these with RT at 300 K.

SOLUTION:

a) 60 MHz

$$E = h\nu = \frac{6.626\times10^{-34} \times 6\times10^{7}\ s^{-1} \times 6.022\times10^{23}\ mol^{-1}}{10^{3} J\ kJ^{-1}}$$

$$= 2.394 \times 10^{-5}\ kJ\ mol^{-1}$$

$$RT = 8.314 \times 10^{-3} \times 300\ kJ\ mol^{-1}$$

$$= 2.494\ kJ\ mol^{-1}$$

$$\therefore \quad \frac{E(60\ MHz)}{RT} = 0.960 \times 10^{-5}$$

b) 500 MHz

$$E = 1.995 \times 10^{-2} \text{ kJ mol}^{-1}$$

$$\frac{E(500 \text{ mHz})}{RT} = 8.00 \times 10^{-3} = 0.800 \times 10^{-2}$$

16.3 (a) What are the energy levels for a ^{23}Na nucleus in a magnetic field of 2 T? (b) What is the absorption frequency?

SOLUTION:

$$E = |g_N|\mu_N m_I B \quad = (1.478)(5.0508 \times 10^{-27} \text{ J T}^{-1})(2 \text{ T})\, m_I$$
$$= (1.493 \times 10^{-26} \text{ J})m_I$$

Since $I = \frac{3}{2}$, $\quad m_I = -\frac{3}{2}, -\frac{1}{2}, +\frac{1}{2}, \frac{3}{2}$ and

$$E = -2.240 \times 10^{-26}, \quad -0.7465 \times 10^{-26}, \quad 0.7465 \times 10^{-26}, \text{ and } 2.240 \times 10^{-26}$$

$$\nu = \frac{\Delta E}{h} = \frac{1.493 \times 10^{-26} \text{ J}}{6.6262 \times 10^{-34} \text{ J s}} = 22.53 \text{ MHz}$$

16.4 The gyromagnetic ratio γ_N for a nucleus is defined by $\mu_N = \gamma_N \hbar I$. What is the value of γ_N for H?

SOLUTION:

$\mu_N = g_N \mu_N I$ Therefore, $\gamma_N = g_N \mu_N / h$

$$= \frac{2\pi(5.585)(5.0508 \times 10^{-17}\ J\ T^{-1})}{(6.6262 \times 10^{-34}\ J\ s)} = 2.675 \times 10^{8}\ s\ T^{-1}$$

16.5 What is the difference in fractional populations of ^{13}C spins between the upper and lower states in a magnetic field of 2 T at room temperature?

SOLUTION:

$$\frac{N_\alpha - N_\beta}{N_\alpha + N_\beta} = \frac{g_N \mu_N\ B/kT}{1 + g_N \mu_N\ B/kT}$$

$g_N = 1.405$

$$\frac{N_\alpha - N_\beta}{N_\alpha + N_\beta} = \frac{(1.405)(5.052 \times 10^{-27})2/(1.3807 \times 10^{-23} \times 298.15)}{1 + (3.5 \times 10^{-6})}$$

$$= 3.4485 \times 10^{-6}$$

16.6 In a magnetic field of 2 T, what fraction of the protons have their spin lined up with the field at room temperature?

SOLUTION:

$$\frac{N_l}{N_h} = 1 + \frac{g_N \mu_N B}{kT}$$

$$= 1 + \frac{(5.585)(5.05 \times 10^{-27}\ \text{J T}^{-1})(2\text{T})}{(1.38 \times 10^{-23}\ \text{J K}^{-1})(298\ \text{K})} = 1.000\ 013\ 72$$

$$\frac{N_1}{N_1 + N_h} = \frac{1}{1 + \frac{N_h}{N_1}} = \frac{1}{1 + \frac{1}{(N_h/N_1)}}$$

$$= \frac{1}{1 + \frac{1}{1.000\ 013\ 72}} = 0.500\ 003\ 43$$

16.7 It is now possible to do NMR experiments at very low temperatures. Calculate the ratio of the number of protons in the upper spin state to that in the lower spin state in a magnetic field of 2 T at 1 mK and 10 mK.

SOLUTION:

$$\frac{N_u}{N_d} = e^{-\Delta E/kT} = e^{-g_N \mu_N B/kT}$$

a) 1 mK

$$\frac{g_N \mu_N B}{kT} = \frac{(5.585)(5.05 \times 10^{-27}\ \text{J T}^{-1})(2\ \text{T})}{(1.38 \times 10^{-23}\ \text{J K}^{-1})(0.001\text{K})}$$

$$= 4.08$$

$$\frac{N_u}{N_d} = e^{-4.08} = 1.69 \times 10^{-2}$$

b)10 mK

$$\frac{N_u}{N_d} = e^{-0.408} = 0.665$$

16.8 Using information from Tables 16.2 and 16.3, sketch the spectrum you would expect for ethyl acetate ($CH_3CO_2CH_3$).

<u>SOLUTION</u>:

```
            O
            ‖
CH3—C—O—CH2—CH3
 ↑             ↑     ↑
δ ≈2        δ≈3.6  δ≈1
```

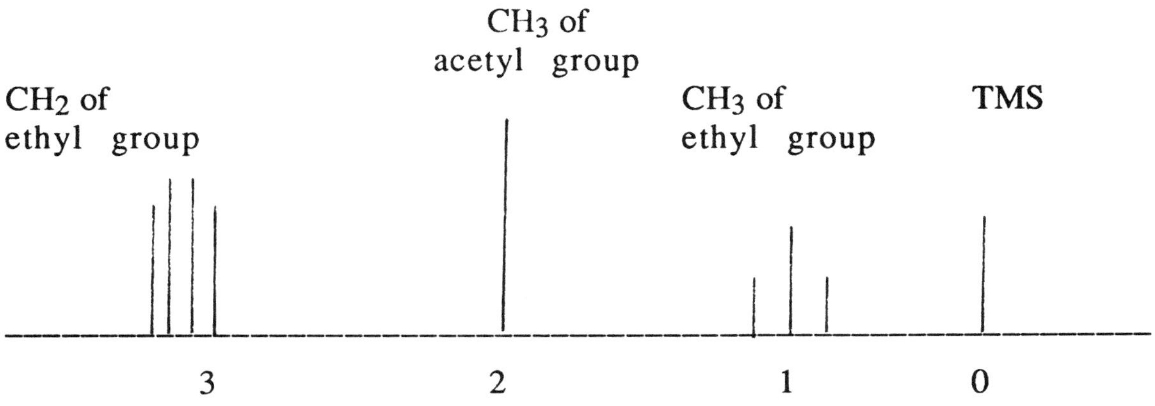

16.9 Chemical shifts δ are expressed in ppm, but can also be expressed in Hz. What magnetic fields are necessary to produce frequency shifts of 100 Hz and 500 Hz for protons with a δ = 1?

<u>SOLUTION</u>:

$$\nu_i = |g_i|\, \mu_N (1-\sigma_i)\, \frac{B}{h} \qquad \Delta\nu = |g_i|\, \mu_N\, \sigma_i\, \frac{B}{h}$$

(a) $\therefore\ B = \dfrac{h\Delta\nu}{|g_i|\, \mu_N\, \sigma_i} = \dfrac{(6.626 \times 10^{-34}\text{ J s})(100\text{ s}^{-1})}{(5.585)(5.051 \times 10^{-27}\text{ JT}^{-1})(10^{-6})}$

$= 2.35$ T

(b) 500 Hz $B = 5 \times 2.35$ T $= 11.7$ T

16.10 (a) What is the separation of the methyl and methylene proton resonances in ethanol at 60 MHz? (b) At 500 MHz? (See Table 16.2.)

SOLUTION:

$\delta_{CH_3} = 1.17 \times 10^{-6}$ $\delta_{CH_2} = 3.59 \times 10^{-6}$

(a) $\nu_0\delta = (60 \times 10^6 \text{ Hz})(2.42 \times 10^{-6}) = 14.52$ Hz

(b) $\nu_0\delta = (500 \times 10^6 \text{ Hz})(2.42 \times 10^{-6}) = 1210$ Hz

16.11 At a magnetic flux density of 1.41 T, the frequency separation between protons in benzene and protons in tetramethylsilane is 436.2 Hz. What is the chemical shift?

SOLUTION:

$$\nu_0 = \frac{Bg_N\mu_N}{h} = \frac{(1.41 \text{ T})(5.585)(5.0508 \times 10^{-27} \text{ J T}^{-1})}{6.6262 \times 10^{-34} \text{ J s}}$$

$= 60.0$ MHz

$\delta = 436.2 \text{ Hz}/60 \text{ MHz} = 7.270 \times 10^{-6}$

which is usually given as 7.270

16.12 Sketch the proton resonance spectrum of $D_2CHCOCD_3$ (deuteroacetone containing a little hydrogen). Indicate the relative intensities of the lines.

SOLUTION:

The deuteron has a spin of I = 1. Some of the methyl groups will be CD_2H-groups, so that the proton sees two equivalent deuterons. The first deuteron will split the proton resonance into a triplet, and the second deuteron will further split these lines with the same spin-spin coupling constant.

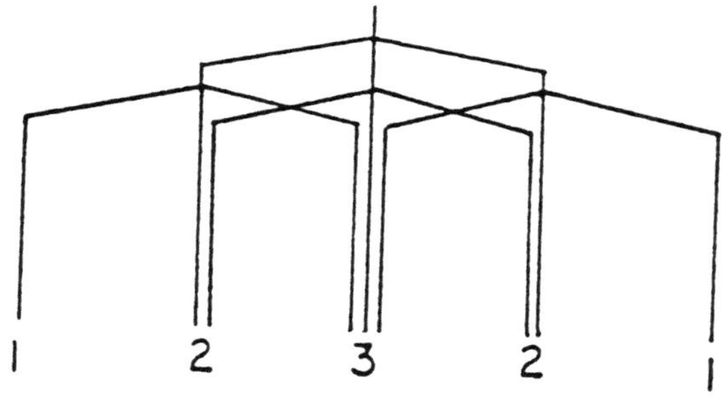

16.13 The proton resonance pattern of 2,3-dibromothiophene shows an AB-type spectrum with lines at 405.22, 410.85, 425.07, and 430.84 Hz measured from tetramethylsilane at 1.41 T [K.F. Kuhlmann and C. L. Braun, *J. Chem. Ed.*, **56**, 750 (1969)]. (a) What is the coupling constant J?: (b) What is the difference in the chemical shifts of the A and B hydrogens? (c) At what frequencies would the lines be found at 2 T? Use the results of Sec. 16.7.

SOLUTION:

(a) The average spacing of the two doublets is J = 5.70 ± 0.07 Hz.

(b) $\nu_0\delta = \left[(a-d)(b-c)\right]^{1/2} = \left[(25.62)(14.22)\right]^{1/2}$

$= 19.09$ Hz

$\delta = (19.09 \text{ Hz})/(60 \times 10^6 \text{ Hz}) = 0.318 \times 10^{-6} = 0.318$ ppm

(c) At 2 T $\nu'_0 = \frac{2}{1.41} 60 \times 10^6 \text{ Hz} = 85.1 \times 10^6$ Hz

The center of the spectrum shifts by ν'_0/ν_0

$$418.00 \frac{85.1 \times 10^6}{60 \times 10^6} = 592.86$$

$$\text{Distance of b and c from center} = \frac{[(\nu_0\delta)^2 + J^2]^{1/2} - J}{2}$$

$$= \frac{1}{2}\left\{[(85.1 \times 0.318)^2 + 5.70^2]^{1/2} - 5.40\right\} = 10.98$$

a = 592.86	- 10.908	- 5.70	= 576.18 Hz
b = 592.86	- 10.908		= 581.88 Hz
c = 592.86	+10.908		= 603.84 Hz
d = = 592.86	+10.908	+ 5.70	= 609.54 Hz

16.14 At room temperature the chemical shift of cyclohexane protons is an average of the chemical shifts of the axial and equatorial protons. Explain.

SOLUTION:

At room temperature the rate of conversion of cyclohexane from boat to chair forms is so fast that the protons are at the average local magnetic field.

16.15 As shown in Fig. 16.20, the two lines in the proton magnetic resonance spectrum for the two methyl groups connected to nitrogen in N,N-dimethylacetamide coalesce when the temperature is raised. What is the rate constant for the cis-trans isomerization when the multiplet structure is just lost at 331 K? The difference in chemical shifts between the two peaks is 10.85 Hz.

SOLUTION:

$$k = \frac{\pi \Delta \nu}{\sqrt{2}} = \frac{\pi (10.85 \text{ Hz})}{\sqrt{2}} = 24.1 \text{ s}^{-1}$$

16.16 Calculate the transition (Larmor) frequency of a free electron in a 3 T field. What energy in cm^{-1} does this correspond to?

SOLUTION:

$$\nu = \frac{B g_e \mu_B}{h} = \frac{(3\text{T})(2.0023)(9.2742 \times 10^{-24} \text{ J T}^{-1})}{6.6262 \times 10^{-34} \text{ J s}}$$

$$= 8.4074 \times 10^{10} \text{ s}^{-1} = 84074 \text{ MHz}$$

$$\frac{1}{\lambda} = \frac{\nu}{c} = \frac{84074 \times 10^{10} \text{ s}^{-1}}{3 \times 10^{10} \text{ cm s}^{-1}} = 2.80 \text{ cm}^{-1}$$

16.17 Line separations in ESR may be expressed in G or MHz. Show how the conversion factor $1 \text{ T} = 2.80 \times 10^4$ MHz is obtained.

SOLUTION:

The resonance frequency for electrons in a 1 tesla field is given by

$$\nu = \frac{g_e B \mu_B}{h} = \frac{(2.00)(9.274 \times 10^{-24}\ J\ T^{-1})(1\ T)}{6.626 \times 10^{-34}\ J\ s}$$

$$= 2.80 \times 10^{10}\ s^{-1} = 2.80 \times 10^{4}\ MHz$$

16.18 Sketch the ESR spectrum expected for p-benzosemiquinone radical ion.

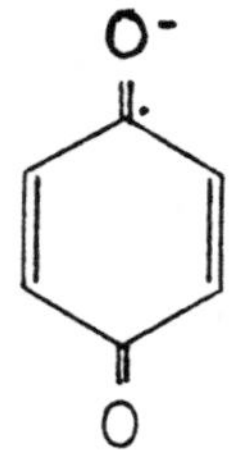

The four hydrogens are magnetically equivalent.

SOLUTION:

			1		1				1H
		1		2		1			2H
	1		3		3		1		3H
1		4		6		4		1	4H

Thus there will be five equally spaced lines with relative intensities of 1,4,6,4,1.

16.19 Sketch the ESR spectrum for an unpaired electron in the presence of three protons for the following cases: (a) the protons are not equivalent, (b) the protons are equivalent, and (c) two protons are equivalent and the third is different.

SOLUTION:

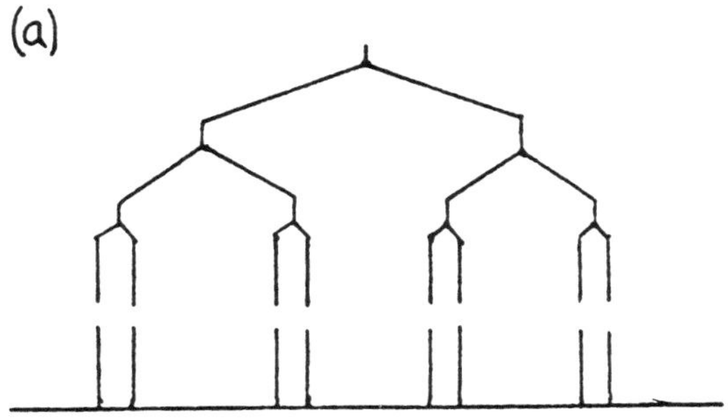

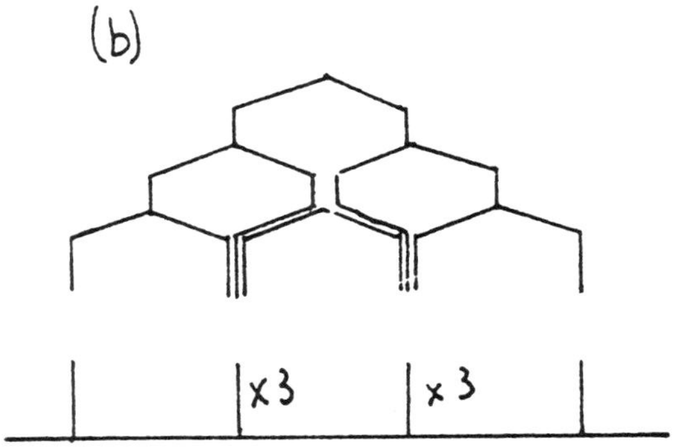

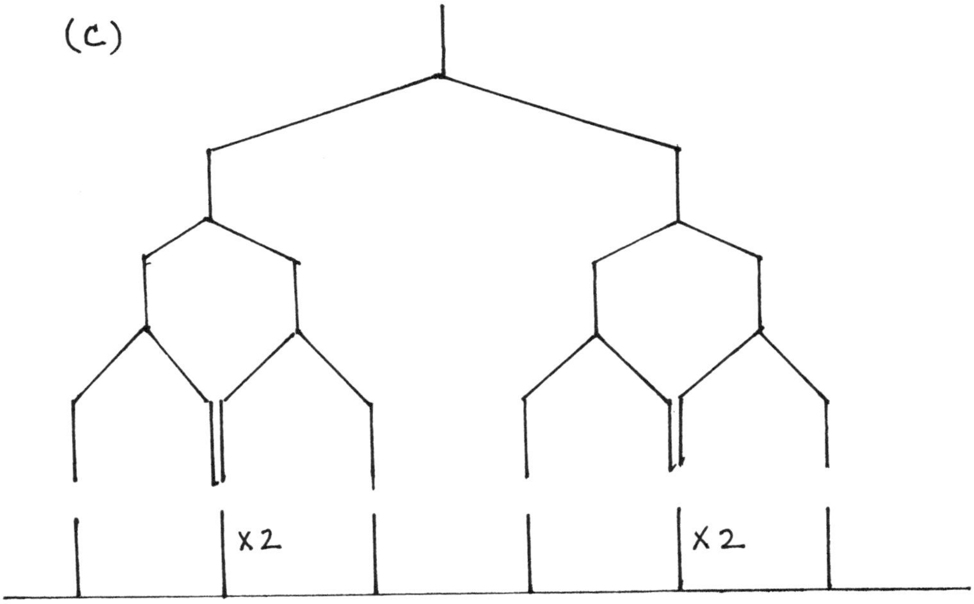

16.20 An unpaired electron in the presence of two protons gives the following four-line ESR spectrum: $\Delta B/10^{-4}$ T = 0, 0.5, 3.5, 4. What are the two coupling constants in T and in MHz?

SOLUTION:

$a_1 = 3.5 \times 10^{-4}$ T

$a_2 = 0.5 \times 10^{-4}$ T

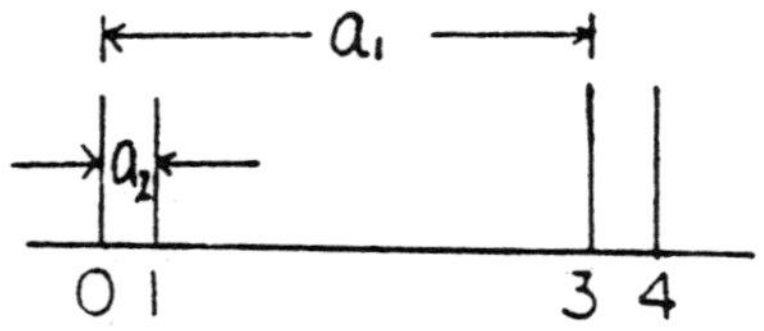

Multiplying by the factor in problem 16.17

$a_1 = (3.5 \times 10^{-4} \text{ T})(2.80 \times 10^4 \text{ MHz T}^{-1}) = 9.60$ MHz

$a_2 = (0.5 \times 10^{-4} \text{ T})(2.80 \times 10^4 \text{ MHz T}^{-1}) = 1.40$ MHz

16.21 Sketch the ESR spectrum of CH_3. radical

SOLUTION:

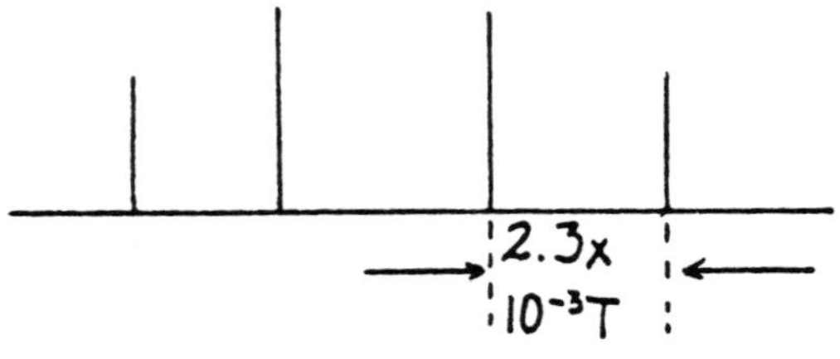

16.22 1.410×10^{-28} J T^{-1}

16.23 (a) 0.4669, (b) 0.1248 T

16.24 511 MHz, 0.017 cm^{-1}

16.25 4.99, 18.67 T

16.26 For protons, 6.856×10^{-6}

For fluorines, 6.453×10^{-6}

16.27

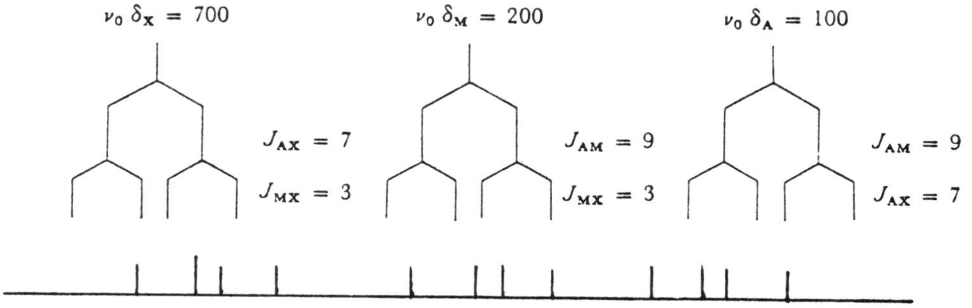

16.28 The proton resonance would be split into three lines by the deuteron, and the deuteron resonance would be split into two lines by the proton.

16.29 $J_{24} \sim 0.5\text{-}4$, $J_{25} \sim 0$, $J_{45} \sim 6\text{-}9$

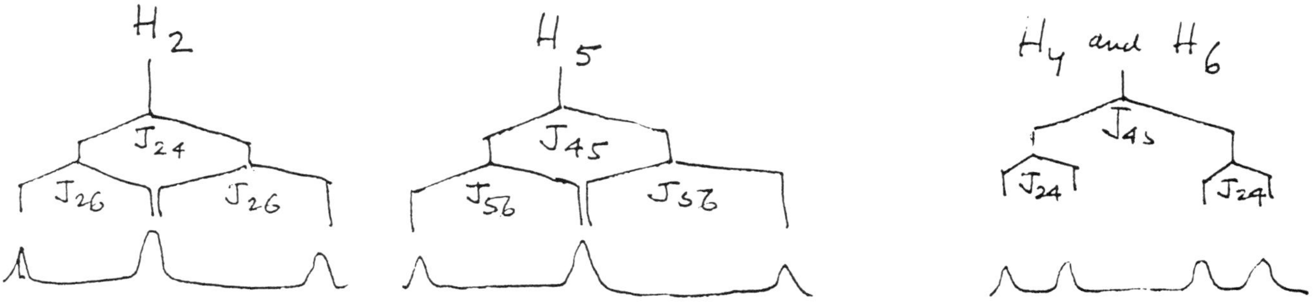

16.30 0.31 cm^{-1}

16.31

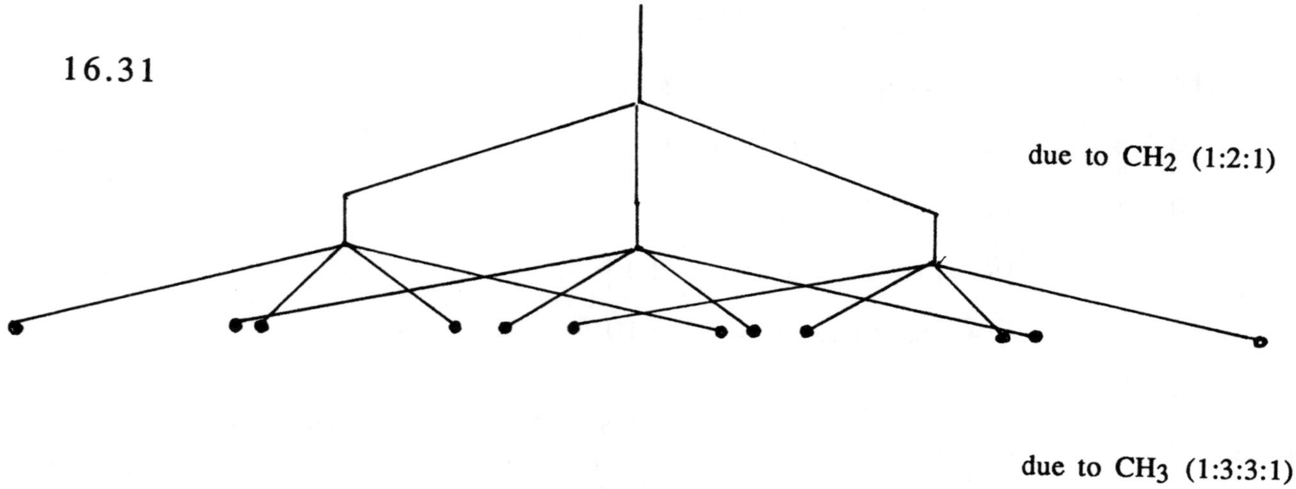

16.32

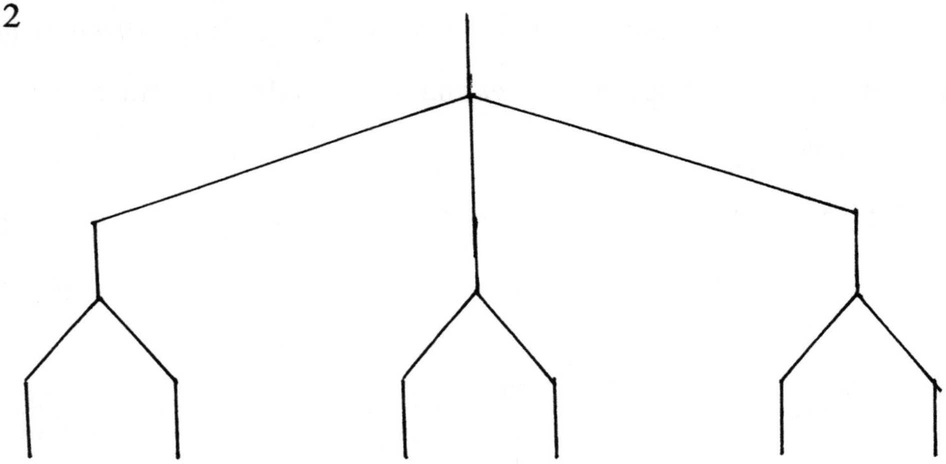

CHAPTER 17: Statistical Mechanics

17.1 Show that the differential of $ln\Omega$ (equation 17.6) is given by equation 17.12. Start with $dln\Omega=\sum_i(\partial ln\Omega(\partial N_i)dN_i$.

SOLUTION:

$$ln\Omega = NlnN + \sum_i N_i\, ln(g_i/N_i)$$

$$dln\Omega = \sum_i(\partial ln\Omega/dN_i)dN_i$$

$$= \sum_i ln(g_i/N_i)dN_i + \sum_i N_i\,[\partial ln(g_i/N_i)/\partial N_i]dN_i$$

$$= \sum_i ln(g_i/N_i)dN_i + \sum_i \left(\frac{N_i}{g_i}\right)^2 \left[\frac{\partial(g_i/N_i)}{\partial N_i}\right]dN_i$$

$$= \sum_i ln(g_i/N_i)dN_i - \sum_i dN_i$$

$$= \sum_i ln(g_i/N_i)dN_i$$

since $\sum_i dN_i = 0$.

17.2 Using the Boltzmann distribution calculate the ratio of populations at 25 °C of energy levels separated by (a) 100 cm^{-1}, and (b) 10 kJ mol^{-1}.

SOLUTION:

$$\text{(a) } \exp\left[-\frac{hc\tilde{\nu}}{kT}\right] = \exp\left[-\frac{(6.626\times10^{-34}\text{ J s})(2.998\times10^{8}\text{ m s}^{-1})(10^{5}\text{ m}^{-1})}{(1.380\times10^{-23}\text{ J K}^{-1})(298.15\text{ K})}\right]$$

$$\text{(b) } \exp\left[-\frac{10^{4}\text{ J mol}^{-1}}{(8.314\text{ J K}^{-1}\text{ mol}^{-1})(298\text{ K})}\right] = 0.0177$$

17.3 Calculate the ratio of populations at 25 °C of energy levels separated by (a) 1 eV and (b) 10 eV. (c) Calculate the ratios at 1000 °C.

SOLUTION:

(a) $$\frac{N_A}{N_B} = \frac{e^{-E_A/RT}}{e^{-E_B/RT}} = e^{-(E_A - E_B)/RT}$$

$$1 \text{ eV} = (1.602 \times 10^{-19} \text{ J})(6.022 \times 10^{23} \text{ mol}^{-1})$$

$$= 9.647 \times 10^{4} \text{ J mol}^{-1}$$

$$\frac{N_A}{N_B} = \exp\left(\frac{-9.647 \times 10^{4} \text{ J mol}^{-1}}{8.314 \text{ J K}^{-1} \text{ mol}^{-1} \times 298.15 \text{ K}}\right)$$

$$= 1.253 \times 10^{-17}$$

(b) $$\frac{N_A}{N_B} = \exp\left(\frac{-9.647 \times 10^{4} \text{ J mol}^{-1}}{8.315 \text{ J K}^{-1} \text{ mol}^{-1} \times 298.15 \text{ K}}\right)$$

$$= 1.147 \times 10^{-169}$$

(c) At 1000 K: $[\frac{N_A}{N_B}]_{1eV} = 1.100 \times 10^{-4}$; $[\frac{N_A}{N_B}]_{10eV} = 2.596 \times 10^{-40}$

17.4 Calculate the fractional occupations (N_i/N) of the energy levels of a particle in a one-dimensional box at a temperature at which the population of the n = 2 level is 0.421 of the population of the n = 1 level. The energy levels of a particle in a one-dimensional box are nondegenerate. The relative energies of the levels can simply be taken as 1, 4, 9, 16, 25, ... (See Fig. 17.2a.)

SOLUTION

Since $E = h^2n^2/8ma^2$, the relative energies of the levels are given by

n	= 1	2	3	4	5	6	...
E	= 1	4	9	16	25	36	...

The fractional occupancies of the levels are given by

$$N_i/N = \frac{\exp(-\varepsilon_i/kT)}{\Sigma\exp(-\varepsilon_i/kT)}$$

The rate of the population of the n = 2 level to the n = 1 level is

$$\frac{N_2}{N_i} = \frac{\exp(-4/k'T)}{\exp(-1/k'T)} = \exp(-3/k'T)$$

where the prime reminds us that relative energies are being used. We do not have to calculate T because only k'T is needed.

$$\ln(N_2/N_i) = -3/k'T$$

$$k'T = 3/\ln 0.421 = 3.47$$

The molecular partition function is

$$q = \exp(-1/3.47) + \exp(-4/3.47) + \exp(-9/3.47) + \exp(-16/3.47) + \exp(-25/3.47) + \cdots$$

$$= 0.7496 + 0.3158 + 0.0747 + 0.0099 + 0.0007 +$$

$$= 1.1507$$

$N_1/N = 0.7496/1.1507 = 0.651$

$N_2/N = 0.274$, $N_3N = 0.0649$, $N_4/N = 0.0086$, $N_5/N = 0.0006$

17.5 The temperature of the system in the preceding problem is doubled by heating it. Calculate the fractional occupations (N_i/N) of the energy levels at the higher temperature. (See Fig. 17.2a.)

SOLUTION

Since the relative temperature was 3.47 in the preceding problem, we recalculate the molecular partition function with

$k'T = 2(3.47) = 6.94$

$$q = \exp(-1/6.94) + \exp(-4/6.94) + \exp(-9/6.94) + \exp(-16/6.94) + \exp(-25/694) + \exp(-36/694$$

$$= 0.8658 + 0.5619 + 0.2734 + 0.0997 + 0.0273 + 0.056 + \cdots$$

$$= 1.8337$$

The fractional occupations of the energy levels are given by

$N_i/N = 0.472, 0.306, 0.149, 0.055, 0.015, 0.003$

17.6 Work is done on the system described in problem 17.4 adiabatically and reversibly so that the length of the box is reduced to 0.707 of its original length. Calculate the fractional occupations (N_i/N) of the energy levels of the compressed system. (See Fig. 17.2b.)

SOLUTION

Since the length of the box is reduced by a factor of 0.707, the energy levels are all raised by a factor of 2. Thus the relative energies of the successive levels are 2, 8, 18, 32, ... Since the temperature is twice as big as in problem 17.4, the partition function calculated in that problem is unchanged. Therefore, the fractional occupancies are unchanged, as shown in Fig. 17.2b.

17.7 Starting with the definition of the molecular partition function (equation 17.17) and $U = \sum_i N_i \varepsilon_i$, derive equation 17.39 for U.

SOLUTION:

$$U = (\frac{N}{q}) \sum_i g_i \, \varepsilon_i \, \exp(-\varepsilon_i/kT)$$

$$q = \sum_i g_i \, \exp(-\varepsilon_i/kT)$$

$$(\frac{\partial q}{\partial T})_V = \frac{1}{kT^2} \sum_i g_i\varepsilon_i \, \exp(-\varepsilon_i/kT) = q(\frac{\partial lnq}{\partial T})_V$$

Eliminating the summation between the first and third equations,

$$\frac{1}{kT^2} (\frac{qU}{N}) = q(\frac{\partial lnq}{\partial T})_V$$

$$U = NkT^2 (\frac{\partial lnq}{\partial \tau})_V$$

17.8 Show that the same expression is obtained for the chemical potential from A_{indis} (equation 17.51) as from G_{indis} (equation 17.53) for an ideal gas.

SOLUTION:

$$A_{indis} = -NkT[ln(\frac{q}{N}) + 1]$$

$$= -kT(Nlnq - NlnN + N)$$

$$\mu = (\partial A_{indis}/\partial N)_{T,V}$$

$$\mu = -kT(lnq - lnN)$$

$$= -kTln(\frac{q}{N})$$

$$G_{indis} = -NkT\,ln(\frac{q}{N})$$

$$= -kT(N\,lnq - N\,lnN)$$

$$\mu = (\frac{\partial G_{indis}}{\partial N})_{T,P}$$

Since we have to hold P constant for the differentiation, G_{indis} has to be written as a function of P. The molecular partition function for our ideal gas is directly proportional to the volume; q = q' V. For an ideal gas, $V = \frac{NkT}{P}$ so that

$$G_{indis} = -NkT\,ln(q'\frac{kT}{P})$$

$$= -kT[N\,ln(q'kT) - N\,lnP]$$

$$\mu = (\frac{\partial G_{indis}}{\partial N})_{T,P}$$

$$= -kT[ln(q'kT) - lnP]$$

$$= -kT\,ln(\frac{q'kT}{P})$$

$$= -kT\,ln(\frac{qkT}{PV})$$

$$= -kT\,ln(\frac{q}{N})$$

17.9 What is the ratio of the thermal de Broglie wavelength to the length of one side of the container for (a) a hydrogen atom in a cube 1 nm on a side at 2 K and (b) an oxygen molecule in 0.25 m^3 at 300 K?

SOLUTION:

(a) $\Lambda = (\frac{h^2}{2\pi mkT})^{1/2}$

For H(g) at 2 K

$$\Lambda = \frac{6.626 \times 10^{-34} \text{ J s}}{[2\pi(\frac{1.008 \times 10^{-3}\text{kg mol}^{-1}}{6.022 \times 10^{23} \text{ mol}^{-1}})(1.381\text{x}10^{-23} \text{ J K})(2\text{K})]^{1/2}}$$

$$= 1.23 \times 10^{-9} \text{ m}$$

$$\frac{\Lambda}{10^{-9} \text{ m}} = 1.23$$

(b) For O_2(g) at 300 K

$$\Lambda = \frac{6.626 \times 10^{-34} \text{ J s}}{[2\pi(\frac{32 \times 10^{-3}}{6.022 \times 10^{23}})(1.381\text{x}10^{-23} \text{ J K})(300 \text{ K})]^{1/2}}$$

$$= 1.782 \times 10^{-11} \text{ m}$$

$$\frac{\Lambda}{(0.25 \text{ m}^3)^{1/3}} = \frac{1.782\text{x}10^{-11} \text{ m}}{(0.25 \text{ m}^3)^{1/3}} = 2.83 \times 10^{-11}$$

17.10 Calculate the translational partition function for H_2(g) at 1000 K and 1 bar.

SOLUTION:

$$V = \frac{nRT}{P} = \frac{(1 \text{ mol})(8.314 \text{ J K}^{-1})(1000 \text{ K})}{10^5 \text{ N m}^{-2}} = 0.08314 \text{ m}^3$$

$$m = \frac{2(1.0078 \times 10^{-3} \text{ kg mol}^{-1})}{(6.022 \times 10^{23} \text{ mol}^{-1})}$$

$= 3.347 \times 10^{-27}$ kg

$$q = \left(\frac{2\pi mkT}{h^2}\right)^{3/2} V$$

$$= \left[\frac{2\pi(3.347 \times 10^{-27} \text{ kg})(1.3806 \times 10^{-23} \text{ J K}^{-1})(1000 \text{ K})}{(6.626 \times 10^{-34} \text{ J s})^2}\right]^{3/2}$$

$\times (0.083\ 14 \text{ m}^3)$

$= 1.414 \times 10^{30}$

17.11 What are the translational partition functions of hydrogen atoms and hydrogen molecules at 500 K in a volume of 4.157×10^{-2} m^3? (This is the molar volume of an ideal gas at this temperature and a pressure of 1 bar.)x (0.083 14 m^3)

<u>SOLUTION</u>:

$$q_t = \left(\frac{2\pi mkT}{h^2}\right)^{3/2} V$$

For H

$$q_t = \left[\frac{2\pi(1.008 \times 10^{-3} \text{ kg mol}^{-1})(1.3806 \times 10^{-23} \text{ J K}^{-1})(500 \text{ K})}{(6.022 \times 10^{23} \text{ mol}^{-1})(6.626 \times 10^{-34} \text{ J s})^2 \times (4.157 \times 10^{-2} \text{ m}^3)}\right]$$

$= 8.84 \times 10^{28}$

For H_2, $m_{H_2} = 2m_{H_2}$

$$q_{tH_2} = 2^{3/2}\, q_{tH} = 2.50 \times 10^{29}$$

17.12 Calculate the molar entropy of one mole of H-atom gas at 1000 K and (a) 1 bar, and (b) 1000 bar.

SOLUTION:

(a) Since there are only translational contributions and electronic contributions,

$$\bar{S}° = R\{ln[(\frac{2\pi mkT}{k2})^{3/2}(\frac{kT}{P})] + 5/2\} + Rln2$$

$$(\frac{2\pi mkT}{k^2})^{3/2} = [\frac{2\pi(1.673 \times 10^{-27}kg)(1.381 \times 10^{-23} JK^{-1})(1000 K)}{(6.626 \times 10^{-34} J s)^2}]$$

$$= 6.012 \times 10^{30} m^{-3}$$

$$\bar{S}° = Rln[\frac{(6.012 \times 10^{30} m^{-3})(1.381 \times 10^{-23} J K^{-1})(1000 K)}{(10^5 Pa)}]$$

$$+ Rln2 + Rln\frac{5}{2}$$

$$= 139.86 J K^{-1} mol^{-1}$$

where $R = 8.314 J K^{-1} mol^{-1}$

(b) At 1000 K and 1000 bar,

$$\bar{S}° = 82.4 J K^{-1} mol^{-1}$$

17.13 Calculate the molar entropy of neon at 25 °C and 1 bar.

SOLUTION:

$$m = \frac{20.179 \times 10^{-3}\ \mathrm{kg\ mol^{-1}}}{6.022 \times 10^{23}\ \mathrm{mol^{-1}}} = 3.350 \times 10^{-26}\ \mathrm{kg}$$

$$\bar{V} = \frac{RT}{P} = \frac{(8.314\ \mathrm{J\ K^{-1}\ mol^{-1}})(298.15\ \mathrm{K})}{10^5\ \mathrm{N\ m^{-2}}}$$

$$= 2.479 \times 10^{-2}\ \mathrm{m^3\ mol^{-1}}$$

$$\bar{S}^\circ = R\{ln[(\frac{2\pi mkT}{h^2})^{3/2}\frac{\bar{V}}{N_A}] + \frac{5}{2}\} = 146.3\ \mathrm{J\ K^{-1}\ mol^{-1}}$$

17.14 What are the most probable populations of the successive vibrational levels of Cl_2 at 298 K and 1000 K? Given: $\tilde{\nu} = 559.7\ \mathrm{cm^{-1}}$.

SOLUTION:

Putting together equations 17.32 (the Maxwell-Boltzmann equation) and equation 17.81 for the vibrational partition function yields

$$\frac{N_i}{N} = (1 - e^{-h\nu/kT})e^{-ih\nu/kT} = P_i$$

since the vibrational levels are not degenerate.

At 298 K,

$$\frac{h\nu}{kT} = \frac{hc\tilde{\nu}}{kT} = \frac{(6.626 \times 10^{-34}\ \mathrm{J\ s})(2.998 \times 10^8\ \mathrm{ms^{-1}})(55970\ \mathrm{m^{-1}})}{(1.381 \times 10^{-23}\ \mathrm{J\ K^{-1}})(298\mathrm{K})}$$

$= 2.702$

$P_i = (1 - e^{-2.702})\, e^{-1.702i}$

$P_o = 0.0329$

$P_1 = 0.0626$

$P_2 = 0.0042$

At 1000 K

$\frac{h\nu}{kT} = 0.8051$

$P_i = (1 - e^{-0.8051})e^{-0.8051i}$

$P_o = 0.553$

$P_1 = 0.247$

$P_2 = 0.111$

$P_3 = 0.049$

$P_4 = 0.022$

17.15 Derive the expression for the vibrational contribution to the internal energy

$U_v = \frac{RTx}{e^x - 1}$ where $x = \frac{h\nu}{kT}$.

SOLUTION:

$$q_v = \frac{1}{1 - e^{-h\nu/kT}}$$

$$U_v = RT^2 \frac{\partial ln\, q_v}{\partial T} \qquad \ln q_v = -ln(1 - e^{-h\nu/kT})$$

$$\frac{\partial ln\ q_v}{\partial T} = \frac{e^{-h\nu/kT}}{1-e^{-h\nu/kT}}\left(\frac{h\nu}{kT^2}\right) = \frac{x}{T(e^x - 1)}$$ where $x = h\nu/kT$.

$$U_v = \frac{RTx}{e^x - 1}$$

17.16 By use of series expansions show the vibrational contribution to $\bar{C}_V^o$ for a diatomic molecule approaches R as $T\to\infty$.

SOLUTION:

$$(\bar{C}_V^o)_v = R\left(\frac{\theta_v}{T}\right)^2 \frac{e^{\theta_v/T}}{(e^{\theta_v/T} - 1)^2}$$

$$e^x = 1 + x + \frac{x^2}{2!} + \cdots \qquad e^{\theta_v/T} = 1 + \frac{\theta_v}{T} + \cdots$$

$$(\bar{C}_V^o)_v = R\left(\frac{\theta_v}{T}\right)^2 \frac{(1 + e^{\theta_v/T} + \cdots)}{\left(\frac{\theta_v}{T}\right)^2}$$

As $T\to\infty$, $(\bar{C}_V^o)_v \to R$.

17.17 What are the rotational contributions to $\bar{C}_P^o$, $\bar{S}^o$, $\bar{H}^o$, and $\bar{G}^o$ for $NH_3(g)$ at25°C?

SOLUTION:

$$(\bar{C}_P^o)_r = \frac{3}{2} R = 12.471 \text{ J K}^{-1}\text{mol}^{-1}$$

$(\bar{H}_r^o) \quad = \frac{3}{2} R = 3.718 \text{ kJ mol}^{-1}$

$$\bar{S}_r^o = R\left\{ ln\left[\frac{\left(\frac{T^3\pi}{\theta_a\theta_b\theta_c}\right)^{1/2}}{\sigma}\right] + \frac{3}{2}\right\}$$

$= 47.822 \text{ J K}^{-1} \text{ mol}^{-1}$

since $\theta_a\theta_b\theta_c = 1876.0$ K and $\sigma = 3$.

$$\bar{G}_r^o = -RT\, ln\left[\frac{\left(\frac{T^3\pi}{\theta_a\theta_b\theta_c}\right)^{1/2}}{\sigma}\right]$$

$= -10.540 \text{ kJ mol}^{-1}$

17.18 What fraction of HCl molecules is in the state v = 2, J = 7 at 500 °C? The characteristic vibrational and rotational temperatures are given in Table 17.2.

SOLUTION:

$$\frac{N_i}{N} = \frac{g_i \exp(-\varepsilon_i/kT)}{q} = \frac{g_v \exp(-\varepsilon_v/kT) g_r \exp(-\varepsilon_r/kT)}{q_v q_r}$$

$g_v = 1 \quad g_r = 2J + 1 = 15$

$$\frac{\varepsilon_v}{kT} = \frac{vh\nu}{kT} = \frac{v\theta_v}{T} = \frac{2(4301K)}{773K}$$

$$\frac{\varepsilon_r}{kT} = \frac{J(J+1)h^2}{8\pi^2 IkT} = \frac{J(J+1)\theta_r}{T} = \frac{(7)(8)(15.2K)}{773K}$$

$$q_v = (1 - e^{-\theta_v/T})^{-1} \quad \text{(equation 17.81)}$$

$$q_r = \frac{T}{\sigma\theta_r} \text{ (equation 17.87)}$$

$$q_v = (1 - e^{-4301/773})^{-1} = 1.0039$$

$$q_r = \frac{773}{15.2} = 50.85$$

$$\frac{N_i}{N} = \frac{(\exp[-\frac{(2)(4301)}{773}] \times 15 \times \exp[-\frac{(56)(15.2)}{773}])}{(1.0039)(50.85)}$$

$= 1.43 \times 10^{-6}$

17.19 Calculate the translational partition functions for H, H_2, and H_3 at 1000 K and 1 bar. What are the rotational partition functions of H_2 and H_3 (linear) at 1000 K? The internuclear distances in H_3 are 94 pm.

SOLUTION:

$$V = \frac{nRT}{P} = (1 \text{ mol}) \frac{(8.314\ 51 \text{ J K}^{-1}\text{mol}^{-1})(1000 \text{ K})}{10^5 \text{ N m}^{-2}}$$

$$= 0.083\ 1451 \text{ m}^3$$

$$q_t = \frac{(2\pi mkT)^{3/2}\ V}{h^3}$$

$$= \frac{[2\pi(\frac{1.0079 \times 10^{-3}\ kg}{6.022 \times 10^{23}\ mol^{-1}})(1.38 \times 10^{-23}\ J\ K^{-1})(10^3 K)]^{3/2} V}{(6.62 \times 10^{-34}\ J\ s)^3}$$

$$= (6.026 \times 10^{30}\ m^{-3})\ V = 5.00 \times 10^{29}$$

For H_2 $\quad q_t = 6.026 \times 10^{30}\ V\ 2^{3/2} = 1.42 \times 10^{30}$

For H_3 $\quad q_t = 6.026 \times 10^{30}\ V\ 2^{3/2} = 2.60 \times 10^{30}$

$$q_r = \frac{8\pi^2 IkT}{2h^2}$$

For H_2

$$q_r = \frac{8\pi^2(4.6054 \times 10^{-48}\ kg\ m^2)(1.3806 \times 10^{-23}\ J\ K^{-1})(1000K)}{2(6.626 \times 10^{-34}\ J\ s)^2} = 5.72$$

For H_3

$$I = \frac{m_1 m_3}{m_1 + m_3} R^2$$

$$= \frac{(1.0079 \times 10^{-3}\ kg\ mol^{-1})^2(1.88 \times 10^{-10}\ m)^2}{2(1.00789 \times 10^{-3}\ kg\ mol^{-1})(6.022 \times 10^{23}\ mol^{-1})}$$

$$= 2.96 \times 10^{-47}\ kg\ m^2$$

(Note that m_2 is on the axis of rotation, and does not contribute to the moment of inertia.)

$$q_t = 5.72\ \frac{29.6 \times 10^{-48}\ kg\ m^2}{4.60 \times 120^{-48}\ kg\ m^2} = 36.8$$

17.20 What are the symmetry numbers of the following organic molecules, assuming free rotation of methyl groups? (a) ethane, (b) propane, (c) 2-methylpropane, (d) 2,2-dimethylpropane.

SOLUTION:

(a) CH_3CH_3 $\sigma = 3^2 \times 2 = 18$ (b) $CH_3CH_2CH_3$ $\sigma = 3^2 = 9$

(c)

```
        H
        |
CH3—C—CH3
        |
       CH3
```

$\sigma = 3^2 \times 3 = 81$

(d)

```
       CH3
        |
CH3—C—CH3
        |
       CH3
```

$\sigma = 12 \times 3^4 = 972$

17.21 Calculate the symmetry numbers of methane (CH_4) as ethylene (C_2H_4) by adding up the number of distinct proper rotational operations in Table 13.2 plus the identity operation.

SOLUTION

CH_4 is T_d and has $8C_3$, $3C_2$, and E. Thus its symmetry number is 12. C_2H_4 is D_{2h} and has $C_2(x)$, $C_2(y)$, $C_2(z)$, and E. Thus its symmetry number is 4.

17.22 (a) Calculate the symmetry number for the ethane structures shown in Table 13.2. This is the symmetry number of the rigid structure.

(b) Since there is essentially free rotation about the C—C bond, thre are 3 equivalent positions of the second CH_3 group with respect to the first. What is the symmetry number of a freely rotating ethane molecule?

SOLUTION

(a) C_2H_6 is D_{3d} and had $3C_3$, $3C_2$, and E. Thus the symmetry number is 6.

(b) $\sigma = 3 \times 6 = 18$.

17.23 What are the electronic contributions to $\overline{S}^o$ and $\overline{G}^o$ for I(g) at 298.15 K and 3000 K.

SOLUTION:

For I(g) $g_o = 4$, $g_1 = 2$, θ_e /K = 10939.3(2)

$$\overline{S}^o_e = R\ln\,(g_o + g_1\, e^{-\varepsilon_i/kT} + \cdots)$$

$$\overline{S}^o_e = 8.314\,\ln(4 + 2\, e^{-10939.3/298})$$

$$= 11.53\ \text{J K}^{-1}\text{mol}^{-1}$$

$$\overline{G}^o_e = -RT\ln(g_o + g_1\, e^{-\varepsilon_i/kT} + \cdots)$$

$$\overline{G}^o_e = -8.314(298.15)\ln(4 + 2e^{-0939.3/248})$$

$$= -3.44\ \text{kJ mol}^{-1}$$

17.24 Calculate the molar entropies of H(g) and N(g) at 25 °C and 1 bar. The degeneracies of the ground states are 2 and 4, respectively. Compare these values with those in Table C.1.

SOLUTION:

$\overline{S} = R\{ln[(2\pi mkT/h^2)^{3/2}(kT/P)] + 5/2 + lng_i\}$

$= R[-1.151\ 693 + \frac{3}{2} ln\ A_r - ln(P/P^o) + \frac{5}{2} ln(T/K) + lng_i\}$

For H(g),

$\overline{S} = 8.31451[-1.151\ 693 + \frac{3}{2} ln\ 1.00794 + \frac{5}{2} ln 298.15 + ln2]$

$= 114.718\ J\ K^{-1}\ mol^{-1}$ (JANAF, 114.716)

For N(g),

$\overline{S} = 8.31451[-1.151\ 693 + \frac{3}{2} ln\ 4.0067 + \frac{5}{2} ln 298.15 + ln4]$

$= 153.301\ J\ K^{-1}\ mol^{-1}$ (JANAF, 153.300)

17.25 A molecule has a ground state and two excited electronic energy levels, all of which are nondegenerate; $\varepsilon_o = 0$, $\varepsilon_1 = 1x10^{-20}$ J, $\varepsilon_2 = 3x10^{-20}$ J. What fraction of each level is occupied at 298 K and 1000 K?

SOLUTION:

$N_i/N = (1/q)\ e^{-\varepsilon_i/kT}$

$q = 1 + exp(-1x10^{-20}/kT) + exp\ (-3x10^{-20}/kT)$

At 298 K

$q = 1 + 0.0879 + 0.00068 = 1.0886$

$N_o = \frac{1}{1.0886} = 0.919$

$N_1/N = 0.0879/1.0886 = 0.081$

$N_2N = 0.00068/1.0886 = 0.006$

At 1000 K

$q \quad = 1 + 0.485 + 0.114 \ = 1.599$

$N_o \quad = \dfrac{1}{1.599} = 0.625$

$N_1/N = \ 0.485/1.599 \quad = 0.303$

$N_2/N = \ 0.114/1.599 \quad = 0.072$

17.26 The ground state of Cl(g) is fourfold degenerate. The first excited state is 875.4 cm^{-1} higher in energy and is twofold degenerate. What is the value of the electronic partition function at 25 °C? At 1000 K?

<u>SOLUTION:</u>

$q \quad = g_0 e^{-0} + s_1 e^{-\varepsilon_1/kT} = 4 + 2e^{-hc\tilde{\nu}/kT}$

$$= 4 + 2\ \exp\left[\frac{-(6.626 \times 10^{-34}\ J\ s)(2.9979 \times 10^8\ m\ s^{-1})(8.754 \times 10^4\ m^{-1})}{(1.3806 \times 10^{-23}\ J\ K^{-1})(298.15\ K)}\right]$$

$= 4 + 2e^{-4.224} = 4.029$

At 1000 K $q = 4 + 2e^{-1.259} = 4.568$

17.27 What is the partition function for oxygen atoms at 1000 K according to the data in Table 17.2? What are the relative populations of these levels at equilibrium?

<u>SOLUTION:</u>

$q_e \quad = 5 + 3e^{-228.1/1000} + e^{-325.9/1000} = 8.11$

$P_D \quad = 5/q_e = 0.617$

$$P_1 = \frac{3e^{-228.1/1000}}{q_e} = 0.294$$

$$P_2 = \frac{e^{-325.9/1000}}{q_e} = 0.089$$

17.28 Derive the expression for the electronic internal energy of an atom or molecule. What is the electronic energy per mole for a chlorine atom at 298 K and 1000 K? (See Problem 17.21).

SOLUTION:

$$U = N\frac{\sum_i \varepsilon_i e^{-\varepsilon_i/kT}}{\sum_i e^{-\varepsilon_i/kT}}$$

$$= N\frac{2\varepsilon_i e^{-\varepsilon_i/kT}}{4 + 2e^{-\varepsilon_i/kT}}$$

where ε_0 is taken as zero and $\varepsilon_1 = 875.4\ \text{cm}^{-1}$

$\varepsilon_1 = (875.4\ \text{cm}^{-1})(1.439\ \text{J cm}^{-1}) = 1{,}260\ \text{K}$

$$N\varepsilon_1 = Nhc\tilde{\nu}_1 = (6.022 \times 10^{23}\ \text{mol}^{-1})(6.626\text{x}10^{-34}\ \text{J s})$$

$$(2.998 \times 10^{8}\ \text{m s}^{-1})(875.4\ \text{cm}^{-1})(100\ \text{cm m}^{-1})$$

$$= 10.47\ \text{kJ mol}^{-1}$$

$$U_{298} = \frac{(10.47\ \text{kJ mol}^{-1})e^{-1260/298}}{2 + e^{-1260/298}} = 0.076\ \text{kJ mol}^{-1}$$

$$U_{1000} = \frac{(10.47\ \text{kJ mol}^{-1})e^{-1260/1000}}{2 + e^{-1260/298}} = 1.301\ \text{kJ mol}^{-1}$$

17.29 Calculate the fraction of hydrogen atoms that are at equilibrium at 1000 °C would have n = 2.

SOLUTION:

Since the fraction will be very small, it is given by the ratio of the number with n = 2 to the number n = 1.

$$\text{Fraction} = \frac{e^{-E_2/kT}}{e^{-E_1/kT}} = e^{-(E_2-E_1)/kT}$$

From Example 11.1 $\quad E = \dfrac{-2.179 \times 10^{-18}\ \text{J}}{n^2}$

$$\text{Fraction} = \exp\left[\frac{(-2.179 \times 10^{-18}\ \text{J})(-0.75)}{(1.380 \times 10^{-23}\ \text{J K}^{-1})(1273\ \text{K})}\right]$$

$$= 4 \times 10^{-41}$$

17.30 A quantum mechanical system has two energy levels ε_1 and ε_2. Derive equations for the probability p_1 that the system will be in state 1 and the probability that the system will be in state 2. What are the probabilities at T/K = 0 and ∞? What are the values of p_1 and p_2 at $\Delta\varepsilon = kT$?

SOLUTION:

$$\frac{p_2}{p_1} = \exp(-\Delta\varepsilon/kT)$$

$p_1 + p_2 = 1$

$p_1 + p_1 \exp(-\Delta\varepsilon/kT) = 1$

$$p_1 = \frac{1}{1+\exp(-\Delta\varepsilon/kT)}$$

$$p_2 = \frac{\exp(-\Delta\varepsilon/kT)}{1+\exp(-\Delta\varepsilon/kT)}$$

T	p_1	p_2
0	1	0
$\Delta\varepsilon$	0.73	0.27
∞	1/2	1/2

17.31 Calculate $\bar{C}_p^o$ for $NH_3(g)$ at 1000 K. The characteristic vibrational temperatures for the six normal modes are given in Table 17.2.

SOLUTION:

The vibrational contribution of a normal mode is calculated from

$$\frac{\bar{C}_p}{R} = \frac{(\theta_{vi}/T)^2 \exp(\theta_{vi}/T)}{[\exp(\theta_{vi}/T) - 1]^2}$$

For θ_{vi} = 1367, 2341, 4800, and 4955 K, these contributions are 0.858, 0.646, 0.193, and 0.176, respectively. The degeneracies of these normal modes have to be taken into account in adding up the contributions.

$$\frac{\bar{C}_p}{R} = 5/2 + 0.858 + 2(0.646) + 0.193 + 2(0.176) + 3/2$$

$$= 6.695$$

$$\bar{C}_p^o = 55.67 \text{ J K}^{-1} \text{ mol}^{-1}$$ (Table C.2 gives 56.491 J $K^{-1}mol^{-1}$.)

17.32 Calculate the entropy of nitrogen gas at 25 °C and 1 bar pressure. The equilibrium separation of atoms is 109.5 pm and the vibrational wave number is 2330.7 cm^{-1}.

<u>SOLUTION:</u>

$$m = \frac{1(14.0067 \times 10^{-3} \text{ kg mol}^{-1})}{6.022 \times 10^{23} \text{ mol}^{-1}} = 4.651 \times 10^{-26} \text{ kg}$$

$$\frac{kT}{P^o} = \frac{(1.380\ 6 \times 10^{-23} \text{ J K}^{-1})(298.15 \text{ K})}{10^5 \text{N m}^{-2}} = 4.1164 \times 10^{-26} \text{ m}^3$$

$$\bar{S}_t^o = R\{ln[(\frac{2\pi mkT}{h^2})^{3/2} \frac{kT}{P^o}] + \frac{5}{2}\}$$

$$ln\ [(\frac{2\pi(4.651 \times 10^{-26} \text{ kg})(1.3806 \times 10^{-23} \text{ J K}^{-1})(298.15 \text{ K})}{(6.626 \times 10^{-34} \text{ J s})^2})^{3/2}$$

$$\times (4.116\ 4 \times 10^{-26} \text{ m}^3)] = 15.591$$

$$\bar{S}_t^o = (8.314 \text{ J K}^{-1} \text{ mol}^{-1})(15.591 + 5/2)$$

$$= 150.42 \text{ J K}^{-1} \text{ mol}^{-1}$$

$$\theta_r = \frac{h^2}{8\pi^2 IK}$$

$$\mu = \frac{m_N^2}{2} = \frac{m_N}{2} = \frac{14.0067 \times 10^{-13} \text{ kg mol}^{-1}}{2(6.022 \times 10^{23} \text{ mol}^{-1})} = 1.163 \times 10^{-26} \text{ kg}$$

$$I = \mu R^2 = (1.163 \times 10^{-26} \text{ kg})(1.095 \times 10^{-10} \text{ m})^2$$

$$= 1.394 \times 10^{-46} \text{ kg m}^2$$

$$\theta_r = \frac{(6.626 \times 10^{-34} \text{ J s})^2}{8\pi^2(1.394 \times 10^{-46} \text{ kg m}^2)(1.380 \times 10^{-23} \text{ J K}^{-1})}$$

$$= 2.888 \text{ K}$$

$$\overline{S_r^o} = R[ln(\frac{T}{\sigma\theta_r}) + 1]$$

$$= (8.314 \text{ J K}^{-1} \text{ mol}^{-1}) \{ln\ [(\frac{(298.15 \text{ K})}{(2)(2.888 \text{ K})}] + 1\}$$

$$= 41.10 \text{ J K}^{-1} \text{ mol}^{-1}$$

$$x = \frac{h\tilde{\nu}}{kT} = \frac{hc\tilde{\nu}}{kT}$$

$$= \frac{(6.626 \times 10^{-34} \text{ J s})(2.9979 \times 10^8 \text{ m s}^{-1})(2.3307 \times 10^5 \text{ m}^-}{(1.380 \times 10^{-23} \text{ J K}^{-1})(298.15 \text{ K})}$$

$$= 11.25$$

$$\overline{S}_v^o = R[\frac{x}{e^x - 1} - ln(1 - e^{-x})]$$

$$= (8.314 \text{ J K}^{-1} \text{ mol}^{-1})[\frac{11.25}{e^{11.25} - 1} - ln(1 - e^{-11.25})]$$

$$= 1.21 \times 10^{-3} \text{ J K}^{-1} \text{ mol}^{-1}$$

$$\overline{S}^o = \overline{S}_t^o + \overline{S}_r^o + \overline{S}_v^o = 150.42 + 41.10 + 0 = 191.52 \text{ J K}^{-1} \text{ mol}^{-1}$$

17.33 Calculate $\bar{C}_P$ for CO_2 at 1000 K. compare the actual contributions to $\bar{C}_P$ from the various normal modes with the classical expectations.

SOLUTION

$$\bar{C}_P^o = \frac{5}{2}R + R + \sum_{i=1}^{4} \frac{Rx_i^2 e^{x_i}}{(e^{x_i} - 1)^2}$$

$$x_i = \frac{hc\tilde{\nu}}{kT} = (1.438 \times 10^{-5} \text{ m})\tilde{\nu}$$

$\frac{\nu_i}{m^{-1}}$	1.3512×10^5	6.722×10^4	6.722×10^4	2.3964×10^5
x_i	1.943	0.967	0.967	3.446

$$\bar{C}_p^o = \frac{7}{2}(8.314) + 6.127 + 2(7.695) + 3.357$$

$$= 53.97 \text{ J K}^{-1}\text{mol}^{-1}$$

Classically $\bar{C}^o_P = \bar{C}^o_{Pt} + \bar{C}^o_{Pv} = \frac{5}{2}R + R + 4R$

$$= \frac{15}{2}R \quad = 62.36 \text{ J K}^{-1} \text{ mol}^{-1}$$

17.34 Calculate the equilibrium constant for the isotope exchange reaction $D + D_2 = H + DH$ at 25 °C. Assume that the equilibrium distance and force constants of H_2 and DH are the same.

SOLUTION:

Since the force constants are the same

$$\nu_{HD} = \nu_{H_2}\left(\frac{\mu_{H_2}}{\mu_{HD}}\right)^{1/2} = \nu_{H_2}\left[\frac{(m_H + m_D)}{2m_D}\right]^{1/2}$$

The change in energy for the reaction is due to the difference in zero poi energies for H_2 and HD.

$$\Delta\varepsilon_o = \frac{h}{2}(\nu_{HD} - \nu_{H_2}) = \frac{h}{2}\nu_{H_2}\left[\frac{(m_H + m_D)}{2m_D}\right]^{1/2} - 1]$$

$$= \frac{h}{2}(1.319 \times 10^{-14} \text{ s}^{-1})\left[\left(\frac{3}{4}\right)^{1/2} - 1\right] = -5.85 \times 10^{-21} \text{ J}$$

$$K = \frac{\left(\frac{f_{HD}}{P^o}\right)\left(\frac{f_H}{P^o}\right)}{\left(\frac{f_{H_2}}{P^o}\right)\left(\frac{f_H}{P^o}\right)} = \frac{q_H q_{HD}}{q_D q_{H_2}} e^{-\Delta\varepsilon_o/kT}$$

$$\frac{q_{tH}q_{tHD}}{q_{tD}q_{tH_2}} = \left(\frac{m_H\, m_{HD}}{m_D m_{H_2}}\right)^{3/2} = \left(\frac{3}{4}\right)^{3/2} = 0.650$$

$$\frac{q_{rHD}}{q_{rH_2}} = \frac{2(\frac{1}{m_H}+\frac{1}{m_H})}{(\frac{1}{m_H}+\frac{1}{m_D})} = \frac{2(2/m_H)}{(3/2m_H)} = \frac{8}{3} = 2.66$$

$$\frac{q_{vHD}}{q_{vH_2}} = 1$$

$$K = (0.650)(2.66)\ \exp[(5.85 \times 10^{-21} J)/(1.38 \times 10^{-23}\ J\ K^{-1})(298\ K)]$$

$$= 7.18$$

17.35 Calculate the equilibrium constant at 25 °C for the reaction $H_2 + D_2 = 2HD$. It may be assumed that the equilibrium distance and forace constant k are the same for all three molecular species, so that the additional vibrational frequencies required may be calculated from $2\pi\nu = (\frac{k}{\mu})^{1/2}$. Because of the zero point vibration, $\Delta\varepsilon_o$ for this reaction is given by

$$\Delta\varepsilon_o = \frac{1}{2} N_A h(2\nu_{HD} - \nu_{H_2} - \nu_{D_2})$$

SOLUTION:

Since the force constants are the same

$$\nu_H = \nu_{H_2} (\frac{\mu_{H_2}}{\mu_{HD}})^{1/2} = \nu_{H_2} [(\frac{(m_H + m_D)}{2m_D}]^{1/2}$$

$$\nu_{D_2} = \nu_{H_2} (\frac{m_H}{m_D})^{1/2}$$

$$\Delta\varepsilon_o = \frac{1}{2} h\nu_{H_2} [\sqrt{2}(\frac{m_H+m_D}{m_D})^{1/2} - (\frac{m_H}{m_D})^{1/2} - 1]$$

$$= \frac{h\nu_{H2}}{2} [\sqrt{2} (\frac{3}{2})^{1/2} - (\frac{1}{2})^{1/2} - 1]$$

$$= \frac{(6.626 \text{ x } 10^{-34} \text{ J s})(1.319 \text{ x } 10^{14} \text{ s}^{-1})(0.0249)}{2}$$

$$= 1.09 \text{ x } 10^{-21} \text{ J}$$

$$K = \frac{(\frac{f_{HD}}{p^o})^2}{(\frac{f_{H_2}}{p^o})(\frac{f_{D_2}}{p^o})} = \frac{q^2_{HD}}{q_{H_2}q_{D_2}} e^{-\Delta U_o/RT}$$

$$= \frac{q^2_{t\ HD}}{q_{tH}q_{tD_2}} \ \frac{q^2_r\ HD}{q_{rH_2}q_{rD_2}} e^{-\Delta U_o/RT}$$

The vibrational partition functions are essentially equal to unity.

$$\frac{q^2_{t\ HD}}{q_{tH_2}q_{tD_2}} = \frac{m^3_{HD}}{m_{H_2}{}^{3/2} m_{D_2}{}^{3/2}} = \frac{3^3}{2^{3/2}4^{3/2}} = 1.19$$

$$\frac{q^2_r\ HD}{q_{rH_2}q_{rD_2}} = \frac{2^2(\frac{1}{m_H} + \frac{1}{m_H})}{(\frac{1}{m_H} + \frac{1}{m_D})^2} = \frac{16\ m_H\ m_D}{(m_D + m_D)^2}$$

$$= \frac{16(1)(2)}{3^2} = 3.55$$

$$K = (1.19)(3.55)\ e^{-(1.09 \times 10^{-21} J)/kT} = 4.22\ e^{-79/T}$$

$$= 4.11\ e^{-79/298} = 3.23$$

17.36 Express the equilibrium constant for the reaction $H_2 + I_2 = 2HI$ in terms of molecular properties.

SOLUTION:

$$K = \frac{(q_{HI}/V)^2}{(q_{H_2}/V)(q_{I_2}/V)}$$

$$= \left(\frac{m_{HI}^2}{m_{H_2} m_{I_2}}\right)^{3/2} \frac{4\theta_{rH_2}\theta_{rI_2}}{\theta_{r_{HI}}^2}$$

$$\frac{(1 - e^{-\theta v_{H_2}/kT})(1 - e^{-\theta v_{I_2}/kT})}{(1 - e^{-\theta v_{HI}/kT})}$$

$$\times \exp\left[\frac{-(2D_{oHI} - D_{oH_2} - D_{oI_2})}{RT}\right]$$

17.37 The classical limits of heat capacities of molecules of ideal gases are readily calculated using the principle of equipartition. Calculate $\bar{C}_p^o/R$ and $\bar{C}_v^o/R$ for Ar, O_2, CO_2, and CH_4 and compare $\bar{C}_p^o/R$ with values in Table C.2 at 3000 K.

SOLUTION:

	trans	rot.	vib	$\bar{C}_v^o/R$	$\bar{C}_p^o/R$	Table C.2
Ar	1.5	0	0	1.5	2.5	2.500
O_2	1.5	1	1	3.5	4.5	4.806
CO_2	1.5	1	4	6.5	7.5	7.404
CH_4	1.5	1.5	9	12	13	12.195

The $\bar{C}_p^o$ of oxygen at 3000 K is high because of electronic excitation.

17.38 Considering H_2O to be a rigid nonlinear molecule, what value of $\bar{C}_p^o$ for the gas would be expected classically? If vibration is taken into account, what value is expected? Compare these values of $\bar{C}_p^o$ with the actual values of 298 and 300 K in Table C.1.

SOLUTION:

A rigid molecule has translational and rotational energy. The translational contribution to $\bar{C}_V$ is $\frac{3}{2}$ R = 12.47 J K^{-1} mol^{-1}. Since H_2O is a nonlinear molecule, it has three rotational degrees of freedom, and so the rotational contribution to

$\bar{C}_V$ is $\frac{3}{2}R$ = 12.47 J $K^{-1}mol^{-1}$. Thus $\bar{C}_P$ for the rigid molecule is $\bar{C}_P = \bar{C}_V$ + R = 33.26 J K^{-1} mol^{-1}.

Since H_2O is a nonlinear molecule, the number of vibrational degrees of freedom is 3N - 6 = 3. Since each vibrational degree of freedom contributes R to the heat capacity, classical theory predicts

$$\bar{C}_P = 33.26 \text{ J K}^{-1} \text{ mol}^{-1} + 3R$$
$$= 58.20 \text{ J K}^{-1} \text{ mol}^{-1}$$

The experimental value of $\bar{C}_P$ at 298 K is 33.58 J K^{-1} mol^{-1}, which is only lightly higher than the value expected for a rigid molecule. The experimental value of $\bar{C}_P$ at 3000 K is 55.66 J K^{-1} mol^{-1}, which is only slightly less than the classical expectations for a vibrating water molecule.

17.39 Use the law of corresponding states to estimate the Lennard-Jones parameters for methane from T_c = 191 K and $\bar{V}_c$ = 100 cm^3 mol^{-1}. The values obtained from second virial coefficient data are σ = 0.378 nm, and ε/k = 148.9 K.

SOLUTION:

$$\sigma = \left(\frac{V_c}{2.7N_A}\right) = \left[\frac{(100 \text{ cm}^3 \text{ mol}^{-1})(10^{-2} \text{ m cm}^{-1})^3}{(2.7)(6.022 \times 10^{23} \text{ mol}^{-1})}\right]^{1/3}$$

$$= 0.395 \text{ nm}$$

$$\varepsilon/k = T_c/1.3 = \frac{191 \text{ K}}{1.3} = 146 \text{ K}$$

17.40 Estimate the second virial coefficient for methane at 300 K and 600 K from its Lennard-Jones parameters in Table 12.5. Compare these values with Fig. 1.4.

SOLUTION:

$\varepsilon/k = 151$ K $\quad \sigma = 380 \times 10^{-12}$m

At 300 K, $T^* = kT/\varepsilon = 300 \text{ K}/151 \text{ K} = 2.0$

From Fig. 17.2, $B^* = -0.9$

$$B(300 \text{ K}) = \frac{-0.9(2\pi)(6.02 \times 10^{23} \text{ mol}^{-1})(3.80 \times 10^{-12} \text{ m})^3}{3}$$

$$= (-6.1 \times 10^{-5} \text{ m}^3\text{mol}^{-1})(10^2 \text{ cm m}^{-1})^3$$

$$= -61 \text{ cm}^3 \text{ mol}^{-1}$$

From Fig. 1.4, $B(300 \text{ K}) = -45 \text{ cm}^3\text{mol}^{-1}$

At 600 K, $T^* = 600 \text{ K}/151 \text{ K} = 4.0$

From Fig. 17.2, $B^* = 0.1$

$$B(600 \text{ K}) = \frac{0.1(2\pi)(6.02\times10^{23} \text{ mol}^{-1})(380\times10^{-12} \text{ m})^3}{3}$$

$$= (6.9 \times 20^{-6} \text{ m}^3\text{mol}^{-1})(10^2\text{cm m}^{-1})^3$$

$$= 6.9 \text{ cm}^3\text{mol}^{-1}$$

From Fig. 1.4, $B(600 \text{ K}) = 10 \text{ cm}^3 \text{ mol}^{-1}$

17.41 Show how $P = kT (\partial ln\, Q/\partial V)_T$ leads to $PV = nRT$.

SOLUTION:

$$Q = \frac{q^N}{N!} = \frac{q^N}{N^N e^{-N}}$$

$$\ln Q = N\ln q - N\ln N + N$$

The only contribution to the molecular partition function that depends on the volume is the translational contribution. $q_t = VV/\Lambda^3$, where Λ is the thermal de Broglie wavelength

$$\frac{\partial \ln Q}{\partial V} = N\frac{\partial \ln Q}{\partial V} = \frac{N}{V}$$

$$P = \frac{NkT}{V} = \frac{NRT}{N_A V}$$

$PV =$ nRT, where $n = N/N_A$

17.42 (a) 3.85×10^{-173} (b) 2.47×10^{-23}

17.44 $\mu = -kT\,ln(q/N)$

17.45 7.74×10^{21}, 4.76×10^{22}, 5.32×10^{23}

17.46 (a) 4.596×10^{-11} m (b) The de Broglie wavelength is a little larger than the thermal de Broglie wavelength. (c) Both the de Broglie wavelength and the thermal de Broglie wavelength are small compared with the mean distance between atoms.

17.47 154.8, 20.8 J K^{-1} mol^{-1}

17.48 126.154 J K^{-1} mol^{-1}

17.45 $q_t(I)/q(H) = (126.90/1.0079)^{3/2}$

17.50 (a) 3.08×10^{-10} , 1.79×10^{-11} m

(b) 5.18×10^{-10}, 3.45×10^{-11} m

(c) Applicable at 298 K, but not at 1 K.

17.51 0.8971, 0.0923, 0.0095, 0.00098

17.52 $S = R[x/(\exp x-1) - ln(1-\exp(-x))]$

17.53 0.048, 48, 48 000, 48×10^{6} K

17.54 12.47, 42.37 J K^{-1} mol^{-1}

17.55 4, 3, 18, 27, 324

17.56 (a) 2, (b) 12

17.57 0.0407

17.58 22,000 K

17.59 (a) 0.0828×10^{-20} J, 0.519×10^{-20} J

(b) same as (a)

17.60 (a) 3.369, (b) 8.144

17.61 $<10^{-99}$, 1.49×10^{-23}

17.62 No substance remains a gas all the way down to 0 K. The Debye theory shows that $S \rightarrow 0$ for a solid as $T \rightarrow 0$.

17.63 -1.151 693

17.64 29.10, 32.93 J K^{-1} mol^{-1}

17.65 222.86 J K^{-1} mol^{-1}

17.66 20.79 J K^{-1} mol^{-1}, 162.71 J K^{-1} mol^{-1}, 62.36 kJ^{-1} mol^{-1}, -425.76 kJ mol^{-1}

17.67 34.91 J K^{-1} mol^{-1}, -337.52 kJ mol^{-1}, 201.01 J K^{-1} mol^{-1}, -940.56 kJ mol^{-1}.

17.68 6.67 J K^{-1} mol^{-1}, 462.24 kJ mol^{-1}, 124.39 J K^{-1} mol^{-1}, 89.03 kJ mol^{-1}, 2.82×10^{-2} ,0.084

17.69 432.070, 493.580, 239.242, 427.783, 1071.780 kJ mol^{-1}

17.65 2.5, 4.5, 7, 10

17.66 0, -50 $cm^3 mol^{-1}$. Yes.

CHAPTER 18: Kinetic Theory of Gases

18.1 If the diameter of a gas molecule is 0.4 nm and each is imagined to be in a separate cube, what is the length of the side of the cube in molecular diameters at 0°C and pressure of (a) 1 bar, and (b) 1 Pa.

SOLUTION

(a) $$\left[\frac{(22.7\ \text{L mol}^{-1})(1000\ \text{cm}^3\ \text{L}^{-1})(10^7\ \text{nm cm}^{-1})^3}{6.02 \times 10^{23}\ \text{mol}^{-1}}\right]^{1/3}$$

$$= 3.43\ \text{nm} = \frac{3.43\ \text{nm}}{0.4\ \text{nm}} = 8.6\ \text{molecular diameters}$$

(b) $$\left[\frac{(2271\ \text{m}^3\ \text{mol}^{-1})(10^9\ \text{nm m}^{-1})^3}{6.022 \times 10^{23}\ \text{mol}^{-1}}\right]^{1/3}$$

$$= 156\ \text{nm} = \frac{156\ \text{nm}}{0.4\ \text{nm}} = 390\ \text{molecular diameters}$$

18.2 Plot the probability density f(v) of molecular speeds versus speed for oxygen at 25 °C.

SOLUTION

$$F(v) = 4\pi v^2 \left(\frac{M}{2\pi RT}\right)^{3/2} e^{-Mv^2/2RT}$$

$$= 4\pi v^2\left[\frac{32 \times 10^{-3}\ \text{kg mol}^{-1}}{2\pi(8.314\ \text{J K}^{-1}\ \text{mol}^{-1})(298\ \text{K})}\right]$$

$$\times\ e^{-(32 \times 10^{-3}\ \text{kg mol}^{-1})v^2/2(8.324\ \text{J K}^{-1}\ \text{mol}^{-1})(298\ \text{K})}$$

$$= v\ (3.704 \times 10^{-8}\ \text{s m}^{-1})\ e^{-6.458 \times 10^{-6} v^2}$$

where v is in m s^{-1}

$v/10^2$ m s^{-1}	$f(v)/10^{-4}$ s m^{-1}
1	3.47
3	18.64
5	18.42
7	7.67
10	0.58

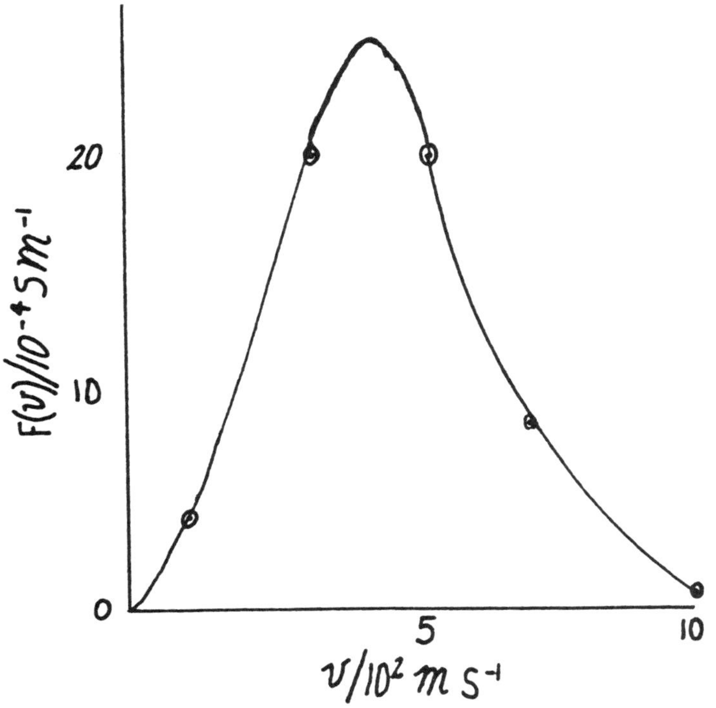

18.3 What is the ratio of the probability that gas molecules have two times the mean speed to the probability that they have the mean speed?

SOLUTION

$F(\langle v \rangle) \quad \propto \langle v \rangle^2 \, e^{-\langle v \rangle^2 m/2kT}$

$F(2\langle v \rangle) \quad \propto 4\langle v \rangle^2 \, e^{-4\langle v \rangle^2 m/2kT}$

$$\frac{F(2\langle v\rangle)}{F(\langle v\rangle)} = 4\, e^{-3\langle v\rangle^2 m/2kT}$$

Since $\langle v\rangle^2 = \dfrac{8kT}{\pi m}$

$$\frac{F(2\langle v\rangle)}{F(\langle v\rangle)} = 4e^{-12/\pi} = 0.0877$$

18.4 Calculate the mean speed and the root-mean-square speed for the following set of molecules: 10 molecules moving 5×10^2 m s^{-1}, 20 molecules moving 20×10^2 m s^{-1}, and 5 molecules moving 15×10^2 m s^{-1}.

SOLUTION

$$\langle v\rangle = \frac{\Sigma N_i v_i}{\Sigma N_i} = \frac{10(500) + 20(1000) + 5(1500)}{35}$$

$$= 928 \text{ m s}^{-1}$$

$$\langle v^2\rangle^{1/2} = \left(\frac{\Sigma N_i v_i^2}{\Sigma N_i}\right)^{1/2} = \left[\frac{10(500)^2 + 20(1000)^2 + 5(1500)^2}{35}\right]^{1/2}$$

$$= 982 \text{ m s}^{-1}$$

18.5 Calculate the most probable, mean, and root-mean-square speeds for oxygen molecules at 25 °C.

SOLUTION

$$v_p = \left(\frac{2RT}{M}\right)^{1/2} = \left[\frac{(2)(8.314 \text{ J K}^{-1} \text{ mol}^{-1})(298 \text{ K})}{32 \times 10^{-3} \text{ kg mol}^{-1}}\right]^{1/2}$$

$$= 393 \text{ m s}^{-1}$$

$$\langle v \rangle = \left(\frac{8RT}{\pi M}\right)^{1/2} = \left[\frac{(8)(8.314 \text{ J K}^{-1} \text{ mol}^{-1})(298 \text{ K})}{(3.1416)(32 \times 10^{-3} \text{ kg mol}^{-1})}\right]^{1/2}$$

$$= 444 \text{ m s}^{-1}$$

$$\langle v^2 \rangle^{1/2} = \left[\frac{(3)(8.314 \text{ J K}^{-1} \text{ mol}^{-1})(298 \text{ K})}{32 \times 10^{-3} \text{ kg mol}^{-1}}\right]^{1/2} = 481 \text{ m s}^{-1}$$

18.6 Derive equations for $\bar{U}$ and $\bar{C}_V$ for any monatomic gas from kinetic theory.

SOLUTION

$$\bar{U} = \frac{N_A}{2} m \int_0^\infty v^2 \, F(v) dv$$

$$= \frac{N_A}{2} m \langle v^2 \rangle = \frac{N_A}{2} m \left(\frac{3kT}{m}\right) = \frac{3}{2} RT$$

$$\bar{C}_V = \left(\frac{\partial \bar{U}}{\partial T}\right)_V = \frac{3}{2} R$$

18.7 Calculate the velocity of sound in nitrogen gas at 25 °C. (See Section 18.4.)

SOLUTION

$$v = \left(\frac{\bar{C}_P RT}{\bar{C}_V M}\right)^{1/2}$$

$$= [\frac{(29.125\ J\ K^{-1}\ mol^{-1})(8.3144\ J\ K^{-1}\ mol^{-1})(298.15\ K)}{(20.811\ J\ K^{-1}\ mol^{-1})(2)(14.0067\ x\ 10^{-3}\ kg\ mol^{-1})}]^{1/2}$$

$$= 352\ m\ s^{-1}$$

$$= \frac{(352\ m\ s^{-1})(10^2\ cm\ m^{-1})\text{-}(60\ s\ min^{-1})(60\ min\ hr^{-1})}{(2.54\ cm\ in^{-1})(12\ in\ ft^{-1})(5280\ ft\ mile^{-1})}$$

$$= 787\ miles\ hr^{-1}$$

18.8 (a) Calculate the collision frequency for a nitrogen molecule in nitrogen at 1 bar pressure and 25 °C. (b) What is the collision density? What is the effect on the collision density (c) of doubling the absolute temperature at constant pressure, and (d) of doubling the pressure at constant temperature?

SOLUTION

(a) $$z_{1(1)} = 2^{1/2}\ \rho\pi d^2\ \langle v\rangle$$

$$\rho = \frac{PN_A}{RT} = \frac{(10^5\ N\ m^{-2})(6.022\ x\ 10^{23}\ mol^{-1})}{(8.324\ J\ K^{-1}\ mol^{-1})(298.15\ K)}$$

$$= 2.429\ x\ 10^{25}\ m^{-3}$$

$$\langle v\rangle = (\frac{8RT}{\pi M})^{1/2} = [\frac{8(8.314)(298.15)}{\pi(28x10^{-3})}]^{1/2} = 475\ m\ s^{-1}$$

$$z_{1(1)} = 2^{1/2}\ (2.429\ x\ 10^{25})\pi(0.375\ x\ 10^{-9})^2(475)$$

$$= 7.21\ x\ 10^9\ s^{-1}$$

(b) $$Z_{11} = \frac{1}{2^{1/2}}\rho^2\pi d^2\langle v\rangle$$

$$= 2^{-1/2}\ (2.429\ x\ 10^{25})^2\pi(0.375\ x\ 10^{-9})^2(475)$$

$$= 8.75\ x\ 10^{34}\ m^{-3}\ s^{-1}$$

$= (8.75 \text{ x } 10^{34} \text{ m}^{-3} \text{ s}^{-1})(10^{-2} \text{ m cm}^{-1})^3$

$= 8.75 \text{ x } 10^{28} \text{ cm}^2 \text{ s}^{-1}$

(c) $\rho \propto T^{-1}$ $\langle v \rangle \propto T^{1/2}$

$Z_{11} \propto \rho^2 \langle v \rangle\ T^{-2}\ T^{1/2} \propto T^{-1.5}$

$$\frac{Z_{11}(2T)}{Z_{11}(T)} = \frac{T^{1.5}}{(2T)^{1.5}} = \frac{1}{2^{1.5}} = 0.354$$

(d) $Z_{11} \propto P^2$

$$\frac{Z_{11}(2P)}{Z_{11}(P)} = 4$$

18.9 (a) Calculate the mean free path for hydrogen gas (d = 0.247 nm) at 1 bar and 0.1 Pa at 25 °C. (b) Repeat the calculation for chlorine gas (d = 0.496 nm).

SOLUTION

At 1 bar

$$\rho = \frac{N_A P}{RT} = \frac{(6.022 \text{ x } 10^{23} \text{ mol}^{-1})(10^5 \text{ N m}^{-2})}{(8.314 \text{J K}^{-1} \text{ mol}^{-1})(298)\text{K}} = 2.44 \text{ x } 10^{25} \text{ m}^{-3}$$

At 0.1 Pa

$$\rho = \frac{N_A P}{RT} = \frac{(6.022 \text{ x } 10^{23} \text{ mol}^{-1})(0.1 \text{ N m}^{-2})}{(8.314 \text{J K}^{-1} \text{ mol}^{-1})(298)\text{K}} = 2.44 \text{ x } 10^{19} \text{ m}^{-3}$$

(a) $\lambda = \dfrac{1}{2^{1/2}\pi d^2 \rho}$

$$= \frac{1}{2^{1/2}\,\pi(0.247 \times 10^{-9}\ \text{m})^2(2.44 \times 10^{25}\ \text{m}^{-3})}$$

$= 1.52 \times 10^{-7}$ m at 1 bar

$$\lambda = \frac{1}{2^{1/2}\,\pi(0.247 \times 10^{-9}\ \text{m})^2(2.43 \times 10^{25}\ \text{m}^{-3})}$$

$= 0.152$ m at 0.1 Pa

$$\text{(b)}\ \lambda = \frac{1}{2^{1/2}\,\pi(0.496 \times 10^{-9}\ \text{m})^2(2.44 \times 10^{25}\ \text{m}^{-3})}$$

$= 3.77 \times 10^{-8}$ m at 1 bar

$$\lambda = \frac{1}{2^{1/2}\,\pi(0.496 \times 10^{-9}\ \text{m})^2(2.43 \times 10^{19}\ \text{m}^{-3})}$$

$= 0.037$ m at 0.1 Pa

18.10 The pressure in interplanetary space is estimated to be of the order of 10^{-14} Pa. Calculate (a) the average number of molecules per cubic centimeter, (b) the collision frequency, and (c) the mean free path in miles. Assume that only hydrogen atoms are present and that the temperature is 1000 K. Assume d = 0.2 nm.

SOLUTION

$$\text{(a)}\ \rho = \frac{N_A P}{RT} = \frac{(6.022 \times 10^{23}\ \text{mol}^{-1})(10^{-14}\ \text{N m}^{-2})}{(8.314\ \text{J K}^{-1}\ \text{mol}^{-1})(10^{3}\ \text{K})}$$

$= 0.724 \times 10^{6}\ \text{m}^{-3} = 0.724\ \text{cm}^{-3}$

$$\text{(b)}\ \langle v \rangle = \left(\frac{8RT}{\pi M}\right)^{1/2} = \left[\frac{8(8.314\ \text{J K}^{-1}\ \text{mol}^{-1})(10^{3}\ \text{K})}{\pi(1 \times 10^{-3}\ \text{kg mol}^{-1})}\right]$$

$= 4600 \text{ m s}^{-1}$

$z_1 = 2^{1/2} \rho\pi d^2 \langle v \rangle$

$= 2^{1/2} (0.724 \times 10^6 \text{ m}^{-3})(0.2 \times 10^{-9} \text{ m})(4600 \text{ m s}^{-1})$

$= 5.92 \times 10^{-10} \text{ s}^{-1}$

(c) $\lambda = \dfrac{1}{2^{1/2}\pi d^2 \rho} = \dfrac{1}{2^{1/2}\pi(0.2 \times 10^{-9} \text{ m})^2(0.724 \times 10^6)}$

$= 7.77 \times 10^{12} \text{ m}$

$= \dfrac{(7.77 \times 10^{12} \text{ m})(10^2 \text{ cm m}^{-1})}{(2.54 \text{ cm in}^{-1})(12 \text{ in ft}^{-1})(5280 \text{ ft mile}^{-1})}$

$= 4.83 \times 10^9$ miles

18.11 Calculate the collision frequency $z_{1(1)}$ and the collision density Z_{11} for molecular chlorine at 25 °C and 1 bar. The collision diameter is 0.544×10^{-9} m.

SOLUTION:

$$\rho = \frac{N_A P}{RT} = \frac{(6.022 \times 10^{23} \text{ mol}^{-1})(10^5 \text{ Pa})}{(8.314 \text{ JK}^{-1}\text{mol}^{-1})(298\text{K})}$$

$= 2.43 \times 10^{25} \text{ m}^{-3}$

$$\langle v \rangle = \left(\frac{8RT}{\pi M}\right)^{1/2}$$

$$= \left(\frac{8\ (8.314 \text{ JK}^{-1}\text{mol}^{-1})(298\text{K})}{\pi\ (70.9 \times 10^{-3} \text{ kg mol}^{-1})}\right)^{1/2}$$

$= 298 \text{ m s}^{-1}$

$$z_{1(1)} = 2^{1/2}\rho\pi d^2 \langle v \rangle$$

$$= 2^{1/2}(2.43 \times 10^{25} \text{ m}^{-3})\pi(0.544 \times 10^{-9} \text{ m})^2(298 \text{ m s}^{-1})$$

$$= 9.52 \times 10^9 \text{ s}^{-1}$$

$$Z_{11} = 2^{-1/2}\rho^2\pi d^2 \langle v \rangle$$

$$= 2\ (2.43 \times 10^{25} \text{ m}^{-3})^2\ \pi\ (0.544 \times 10^{-9}\text{m})^2\ (298 \text{ m s}^{-1})$$

$$= 1.157 \times 10^{25} \text{ s}^{-1} \text{ m}^{-3}$$

$$= (1.157 \times 10^{25} \text{ s}^{-1} \text{ m}^{-3})(10^{-3} \text{ m}^{-3}\text{L}^{-1})/(6.022 \times 10^{23} \text{ mol}^{-1})$$

$$= 1.921 \times 10^8 \text{ mol L}^{-1} \text{ s}^{-1}$$

18.12 A gas mixture contains H_2 at 0.666 bar and O_2 at 0.333 bar at 25 °C. (a) What is the collision frequency $z_{1(2)}$ of a hydrogen molecule with an oxygen molecule? (b) What is the collision frequency $z_{2(1)}$ of an oxygen molecule with a hydrogen molecule? (c) What is the collision density Z_{12} between hydrogen molecules and oxygen molecules in mol L^{-1} s^{-1}? The collision diameters of H_2 and O_2 are 0.272 nm and 0.360 nm, respectively.

<u>SOLUTION</u>:

(a) $\langle v_{12} \rangle = \left(\frac{8kT}{\pi\mu}\right)^{1/2} = 1824 \text{ m s}^{-1}$ from Ex 18.4

$$\rho_1 = \frac{N_A P_1}{RT} = \frac{P_1}{kT} = \frac{0.666 \times 10^5 \text{Pa}}{(1.381 \times 10^{-23}\text{JK}^{-1})(298\text{K})} = 16.18 \times 10^{23} \text{ m}^{-3}$$

$$\rho_2 = \frac{N_A P_2}{RT} = \frac{P_2}{kT} = \frac{0.333 \times 10^5 \text{Pa}}{(1.381 \times 10^{-23}\text{JK}^{-1})(298\text{K})} = 8.092 \times 10^{23} \text{ m}^{-3}$$

d_{12} = (1/2)(0.272 + 0.360) x 10 $^{-9}$m = 0.316 x 10^{-9} m

(b) $z_{1(2)} = \rho_2 \pi d_{12}^2 <v_{12}>$

= (8.092 x 10 ^{24} m^{-3}) π (0.316 x 10^{-9} m)^2 (1824 m s^{-1})

= 4.630 x 10^9 s^{-1}

$z_{2(1)} = \rho_1 \pi d_{12}^2 <v_{12}>$

= (16.18 x 10 ^{24} m^{-3}) π (0.316 x 10^{-9} m)^2 (1824 m s^{-1})

= 9.258 x 10^9 s^{-1}

(c) $Z_{12} = \rho_1 \rho_2 \pi d_{12}^2 <v_{12}> = \rho_1 z_{1(2)}$

= (16.18 x 10 ^{24} m^{-3})(4.630 x 10^9 s^{-1})

= 7.491 x 10^{34} s^{-1} m^{-3}

= (7.491 x 10^{34} s^{-1} m^{-3})(10^{-3} m^3L^{-1})/(6.022 x 10^{23} mol^{-1})

= 1.244 x 10^8 mol L^{-1} s^{-1}

18.13 For O_2(g), d = 0.361 nm. Θ_r = 2.079 K. Θ_v = 2273.64 K, and M = 31.9988 g mol^{-1}. At 1 bar and 25 °C what is the average time between collisions? How many vibrational oscillations will have occurred during this time?

SOLUTION:

From Example 18.4, <v> = 444 m s^{-1}

From Example 18.5, l = 7.11 x 10^{-8} m

$\Delta t = l/<v>$ = (7.11 x 10^{-8} m)/(444 m s^{-1})

= 1.58 x 10^{-10} s

$$\omega = \frac{k}{h}\Theta_v = \frac{(1.381 \times 10^{-23} \text{ J K}^{-1})(2274 \text{ K})}{6.626 \times 10^{-34} \text{ J s}})$$

= 4.74 x 10^{13} s^{-1}

The average number of oscillations between collisions is

$(1.60 \times 10^{-10}\ s)(4.74 \times 10^{13}\ s^{-1}) = 7590$

18.14 (a) How many molecules of H_2 strike the wall per unit area per unit time at one bar at 298 K? 100 K? (b) How many molecules of O_2 strike the wall per unit area per unit time at one bar at 298 K? 1000 K?

SOLUTION

(a) For H_2 at 1 bar and 298 K.

$$m_{H_2} = \frac{2.016 \times 10^{-3}\ kg\ mol^{-1}}{6.022 \times 10^{23}\ mol^{-1}} = 3.348 \times 10^{-27}\ kg$$

$$J_N = \frac{P}{(2\pi mkT)^{1/2}}$$

$$= \frac{10^5\ N\ m^{-2}}{[(2\pi)(3.348 \times 10^{-27}\ kg)(1.381 \times 10^{-23}\ J\ K^{-1})(298\ K)]^{1/2}}$$

$$= 1.075 \times 10^{28}\ m^{-2}\ s^{-1}$$

At 1000 K, $J_N = (1.075 \times 10^{28}\ m^2\ s^{-1})(298/1000)^{1/2}$

$$= 5.866 \times 10^{27}\ m^{-2}\ s^{-1}$$

(b) For O_2 at 1 bar and 298 K

$$m_{O_2} = \frac{32 \times 10^{-3}\ kg\ mol^{-1}}{6.022 \times 10^{23}\ mol^{-1}} = 5.314 \times 10^{-26}\ kg$$

$$J_N = \frac{10^5\ N\ m^{-2}}{[(2\pi)(5.314 \times 10^{-26}\ kg)(1.381 \times 10^{-23}\ J\ K^{-1})(298)]^{1/2}}$$

$$= 2.698 \times 10^{27}\ m^{-2}\ s^{-1}$$

At 1000 K, $J_N = (2.698 \times 10^{27}\ m^{-2}\ s^{-1})(298/1000)^{1/2}$

$= 1.473 \times 10^{27}\ m^{-2}\ s^{-1}$

18.15 A Knudsen cell containing crystalline benzoic acid ($M = 122\ g\ mol^{-1}$) is carefully weighed and placed in an evacuated chamber thermostated at 70 °C for 1 hr. The circular hole through which effusion occurs is 0.60 mm in diameter. Calculate the sublimation pressure of benzoic acid at 70 °C in Pa from the fact that the weight loss is 56.7 mg.

SOLUTION

$$P = \frac{w}{tA}\left(\frac{2\pi RT}{M}\right)^{1/2}$$

$$= \frac{56.7\text{x}210^{-6}\ kg}{(60\text{x}60\ s)\pi(0.3\text{x}10^{-3}\ m)^2}\left[\frac{2\pi(8.314\ J\ K^{-1}\ mol^{-1})(343\ K)}{122\text{x}10^{-3}\ kg\ mol^{-1}}\right]^{1/2}$$

$= 21.3\ N\ m^{-2}$

$= 21.3$ Pa

18.16 R. B. Holden, R. Speiser, and H. L. Johnston [*J. Am. Chem. Soc.* **70**, 3897 (1948)] found the rate of loss of weight of a Knudsen effusion cell containing finely divided beryllium to be $19.8 \times 20^{-7}\ g\ cm^{-2}\ s^{-1}$ at 1320

K and 1210×10^{-7} g cm^{-2} s^{-1} at 1537 K. Calculate ΔH_{sub} for this temperature range.

<u>SOLUTION</u>

According to equations 18.45 and 18.46 the two vapor pressures are proportional to $\Delta w T^{1/2}$.

$$\Delta H_{sub} = \frac{RT_1T_2}{(T_2 - T_1)} \ln \frac{P_2}{P_1}$$

$$= \frac{(8.314)(1320)(1537)}{(217)} \ln \frac{1210(1537)^{1/2}}{19.8(1320)^{1/2}}$$

$$= 326 \text{ kJ mol}^{-1}$$

18.17 A 5 mL container with a hole 10 μ in diameter is filled with hydrogen. This container is placed in an evacuated chamber at 0 °C. How long will it take for 90% of the hydrogen to effuse out?

<u>SOLUTION</u>

$$J_N = \frac{\rho\langle v\rangle}{4} \text{ may be written } -\frac{1}{A}\frac{dN}{dt} = \frac{N}{V}\frac{\langle v\rangle}{4}$$

$$\langle v\rangle = 1.69 \times 10^3 \text{ m s}^{-1} \text{ from Example 18.3}$$

$$-\frac{dN}{dt} = \frac{\pi(5 \times 10^{-6} \text{ m})^2(1.69 \times 10^3 \text{ m s}^{-1})N}{(5 \times 10^{-6} \text{ m}^3)(4)}$$

$$= (6.64 \times 10^{-3}\ s^{-1})N$$

$$\int_{N_o}^{N_t} \frac{dN}{N} = -(6.64 \times 10^{-3}\ s^{-1}) \int_0^t dt$$

$$ln \frac{N_t}{N_o} = -(6.64 \times 10^{-3}\ s^{-1})t$$

$$t = \frac{ln\ 0.1}{-6.64 \times 10^{-3}\ s^{-1}} = 347\ s$$

18.18 The vapor pressure of naphthalene ($M = 128.16\ g\ mol^{-1}$) is 17.7 Pa at 30 °C. Calculate the weight loss in a period of 2 hr of a Knudsen cell filled with naphthalene and having a round hole 0.50 mm in diameter.

SOLUTION

$$J_N = \frac{P}{(2\pi mkT)^{1/2}} = \frac{\Delta w}{mtA}$$

$$A = \pi r^2 = \pi(0.025\ cm)^2(0.01\ m\ cm^{-1})^2 = 1.96 \times 10^{-7}\ m^2$$

$$m = \frac{128.16\ g\ mol^{-1}}{6.022 \times 10^{23}\ mol^{-1}} = 2.128 \times 10^{-22}\ g$$

$$= 2.128 \times 10^{-25}\ kg$$

$$\Delta w = \frac{mtAP}{(2\pi mkT)^{1/2}}$$

$$= \frac{(2.128 \times 10^{-25}\ kg)(2 \times 60 \times 60 s)(1.96 \times 10^{-7}\ m^2)(17.7\ Pa)}{[2\pi(2.128 \times 10^{-25}\ kg)(1.381 \times 10^{-23}\ J\ K^{-1})(303\ K)]^{1/2}}$$

$$= 0.0711\ g$$

18.19 The viscosity of helium is 1.88×10^{-5} Pa s at ^{o}C. **Calculate (a) the collision diameter, and (b) the diffusion coefficient at 1 bar.**

SOLUTION

(a) $$d = \left[\frac{5}{16}\frac{(\pi mkT)^{1/2}}{\pi\eta}\right]^{1/2}$$

$$m = \frac{4.0026 \times 10^{-3}\ \text{kg mol}^{-1}}{6.022 \times 10^{23}\ \text{mol}^{-1}} = 6.647 \times 10^{-27}\ \text{kg}$$

$$d = \left[\frac{5}{16}\frac{(\pi 6.647 \times 10^{-27} \times 1.38 \times 10^{-23} \times 273)^{1/2}}{\pi(1.88 \times 10^{-5})}\right]^{1/2}$$

$$= 0.217\ \text{nm}$$

(b) $$\rho = \frac{PN_A}{RT} = \frac{(10^5\ \text{Pa m}^{-2})(6.022 \times 10^{23}\ \text{mol}^{-1})}{(8.314\ \text{J K}^{-1}\ \text{mol}^{-1})(273\ \text{K})}$$

$$= 2.65 \times 10^{25}\ \text{m}^{-3}$$

$$D = \frac{3}{8}\frac{(\pi mkT)^{1/2}}{\pi d^2 \rho m}$$

$$= \frac{3}{8}\frac{[\pi(6.647 \times 10^{-27}\text{kg})(1.38 \times 10^{-23}\text{J K}^{-1})(273\ \text{K})]^{1/2}}{\pi(0.217 \times 10^{-9}\ \text{m})^2(2.65 \times 10^{25}\ \text{m}^{-3})(6.647 \times 10^{-27}\ \text{kg})}$$

$$= 1.28 \times 10^{-4}\ \text{m}^2\ \text{s}^{-1}$$

18.20 What is the self-diffusion coefficient of radioactive CO_2 in ordinary CO_2 at 1 bar and 25 °C? The collision diameter is 0.40 nm.

SOLUTION

$$m = \frac{44 \times 10^{-3} \text{ kg mol}^{-1}}{6.022 \times 10^{23} \text{ mol}^{-1}} = 7.307 \times 10^{-26} \text{ kg}$$

$$\rho = \frac{N}{V} = \frac{PN_A}{RT} = \frac{(1 \text{ bar})(6.022 \times 10^{23} \text{ mol}^{-1})(10^3 \text{ L mol}^{-1})}{(0.083\,14 \text{ L bar K}^{-1} \text{ mol}^{-1})(298 \text{ K})}$$

$$= 2.43 \times 10^{25} \text{ m}^{-3}$$

$$D = \frac{3}{8}\frac{(\pi mkT)^{1/2}}{\pi d^2 \rho m}$$

$$= \frac{3[\pi(7.307 \times 10^{-26} \text{ kg})(1.38 \times 10^{-23} \text{ J K}^{-1})(298 \text{ K})]^{1/2}}{\pi(0.4 \times 10^{-9} \text{ m})^2(2.43 \times 10^{25} \text{ m}^{-3})(7.307 \times 10^{-26}\text{kg})}$$

$$= 1.26 \times 10^{-5} \text{ m}^2 \text{ s}^{-1}$$

Answers to Second Set

18.21	0.03878 eV
18.23	0.199
18.24	4.18×10^2, 4×10^2, 4.4×10^2 m s^{-1}
18.25	7.9×10^2 m s^{-1}, 802 K
18.26	281 m s^{-1}
18.27	(a) 1016, (b) 351.9 m s^{-1}
18.28	8.314 51 J K^{-1} mol^{-1}
18.29	(a) 1.6×10^{-10}, (b) 1.60×10^{-4}, (c)160 s
18.30	1.4×10^{-10} s
18.31	(a) 4.75×10^{29} m^{-3} s^{-1}, (b) 2.86×10^{-5} m
18.32	(a) 1.908×10^9 s^{-1}, 1.908×10^9 s^{-1}
	(b) 3.87×10^7 mol L^{-1} s^{-1}
18.33	(a) 6.24×10^6 s^{-1} (b) 1.26×10^2 mol L^{-1} s^{-1}
	(c) 160×10^{-7} s
18.34	7.1 m, 62.6 s^{-1}
18.35	2.697×10^{23} cm^{-2} s^{-1}
18.36	(a) 6590 m, (b) 2.88×10^{16} m^{-2} s^{-1}
18.37	(a) 1.136×10^{26} m^{-2} s^{-1}, (b) 20.39 g cm^{-2} min^{-1}
18.38	2.0×10^{-10} kg

19.1 Nitrogen pentoxide (N_2O_5) gas decomposes according to the reaction

$$2N_2O_5 = 4NO_2 + O_2$$

At 328 K, the rate of reaction v under certain conditions if 0.75 x 10^{-4} mol L^{-1} s^{-1}. Assuming that none of the intermediates have appreciable concentrations, what are the values of $d[N_2O_5]/dt$, $d[N_2]/dt$, and $d[O_2]/dt$?

SOLUTION

v = 0.75 x 10^{-4} mol l^{-1} s^{-1}

$$= \frac{1}{2}\frac{d[N_2O_5]}{dt} = \frac{1}{4}\frac{d[NO_2]}{dt} = \frac{d[O_2]}{dt}$$

$$\frac{d[N_2O_5]}{dt} = -2v = -1.5 \times 10^{-4} \text{ mol L}^{-1} \text{ s}^{-1}$$

$$\frac{d[NO_2]}{dt} = 4v = 3.0 \times 10^{-4} \text{ mol L}^{-1} \text{ s}^{-1}$$

$$\frac{d[O_2]}{dt} = v = 0.75 \times 10^{-4} \text{ mol L}^{-1} \text{ s}^{-1}$$

19.2 In studying the decomposiiton of ozone

$$2O_3(g) = 3)_2(g)$$

in a 2 L reaction vessel, it is found that $d[O_3]/dt$ = - -1.5 x 10^{-2} mol L^{-1} s^{-1}. (a) What is the rate of reaction v? (b) What is the rate of conversion $d\xi/dt$? (c) What is the value of $d[O_2]/dt$?

SOLUTION

(a) v $= -\frac{1}{2}\frac{d[O_3]}{dt} = 0.75 \times 10^{-2} \text{ mol L}^{-1} \text{ s}^{-1}$

(b) $\frac{d\xi}{dt} = Vv = (2\ L)(0.75 \times 10^{-3}\ mol\ L^{-1}\ s^{-1})$

(c) $\frac{d[O_2]}{dt} = 3v = 3(0.75 \times 10^{-2}\ mol\ L^{-1}\ s^{-1})$

$$= 2.25 \times 10^{-2}\ mol\ L^{-1}\ s^{-1}$$

19.3 The decomposition of N_2O_5

$$2N_2O_5 = 4NO_2 + O_2$$

is studied by measuring the concentration of oxygen as a function of time, and it is found that

$$d[O_2]/dt = (1.5 \times 10^{-4}\ s^{-1})[N_2O_5]$$

at constant temperature and pressure. Under these conditions the reaction goes to completion to the right. What is the half life of the reaction under these conditions?

<u>SOLUTION</u>

$$-\frac{1}{2}\frac{d[N_2O_5]}{dt} = \frac{d[O_2]}{dt} = (1.5 \times 10^{-4}\ s^{-1})[N_2O_5]$$

Thus k in

$$-\frac{d[A]}{dt} = k[A]$$

is $3.0 \times 10^{-4}\ s^{-1}$ and

$$t_{1/2} = \frac{0.6793}{3.0 \times 10^{-4}\ s^{-1}} = 2310\ s$$

19.4 The following data were obtained on the rate of hydrolysis of 17% sucrose in 0.099 mol L^{-1} HCl aqueous solutions at 35 °C.

t/min	9.82	59.60	93.18	142.9	294.8	5894
Sucrose remaining,%	96.5	80.3	71.0	59.1	32.8	11.1

What is the order of the reaction with respect to sucrose and the value of the rate constant k?

SOLUTION

$ln \frac{[A]}{[A]_o} = - kt$ slope $= -k = -3.76 \times 10^{-3} \text{ m}^{-1}$

$k = 3.76 \times 10^{-3} \text{ m}^{-1}$

$= (3.76 \times 10^{-3} \text{ m}_{-}^{1})((\frac{1}{60} \text{ m s}^{-1})$

$= 6.27 \times 10^{-5} \text{ s}^{-1}$

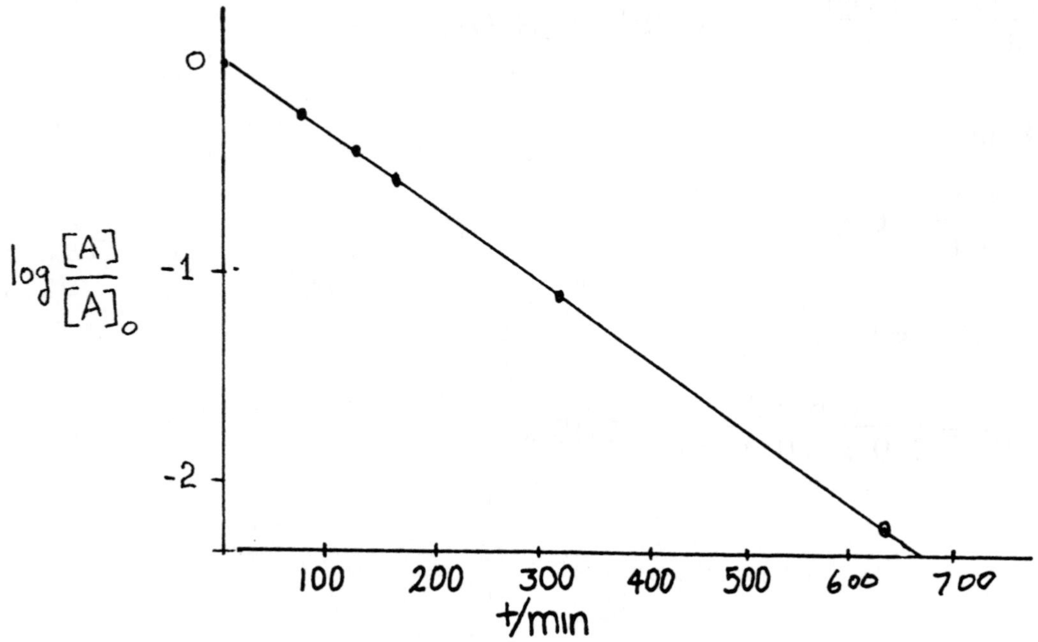

19.5 Methyl acetate is hydrolyzed in approximately 1 mol L^{-1} HCl at 25 °C. Aliquots of equal volume are removed at intervals and titrated with a solution of NaOH. Calculate the first-order rate constant form the following experimental data.

t/s	339	1242	2745	4546	∞
V/cm^3	26.34	27.80	29.70	31.81	39.81

SOLUTION

Any quantity proportional to the concentration of reactant A that remains may be used in equation

$ln[A] = -kt + \log[A]_o$

In this case the concentration of A remaining is proportional to 39.81 cm^3 - V. Therefore $ln(V\text{-}39.81\ cm^3)$ is plotted versus t.

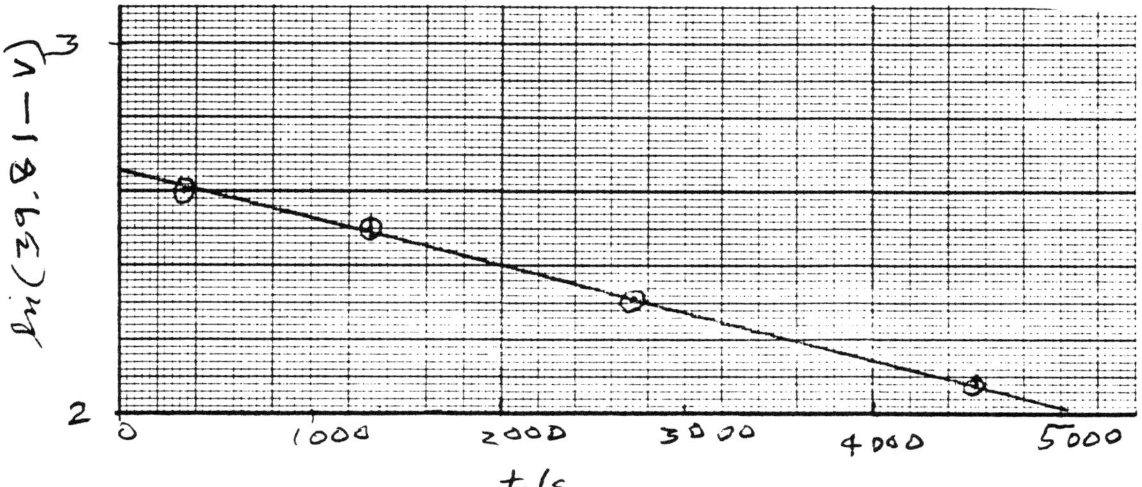

$$\text{slope} = -k = -\frac{0.66}{5000} = -1.32 \times 10^{-4}\ s^{-1}$$

$$k = 1.32 \times 10^{-4}\ s^{-1}$$

19.6 Prove that in a first-order reaction, where dn/dt = -kn, the average life, that is the average life expectancy of the molecules, is equal to 1/k.

SOLUTION

$$\text{Average life} = \frac{\int_0^\infty n(t)dt}{n_0}$$

$$\text{Average life} = \frac{-\frac{1}{k}\int_0^\infty \frac{dn}{dt}dt}{n_0}$$

$$= \frac{-\frac{1}{k}(0-n_o)}{n_o}$$

$$= \frac{1}{k}$$

The average life is also referred to as the relaxation time.

19.7 The hydrolysis of 1-chloro-1methylcycloundecane in 80% ethanol has been studied by H. C. Brown and M. Borkowski (*J. Chem. Soc.* **74**, 1894 (1952)) at 25 ^oC. The extent of hydrolysis was measured by titrating the acid formed after measured intervals of time with a solution of NaOH. The data are as follows:

t/h	0	1.0	3.0	5.0	9.0	12	∞
x/cm^3	0.035	0.295	0.715	1.055	1.505	1.725	2.197

(a) What is the order of the reaction? (b) What is the value of the rate constant? (c) What fraction of the 1-chloro-1-methylcycloundecane will be left unhydrolyzed after 8 hr?

SOLUTION

(a)

t/h	0	1.0	3.0	5.0	9.0	12
$\ln(x_\infty - x)$	0.771	0.643	0.393	0.133	-0.268	-0.751

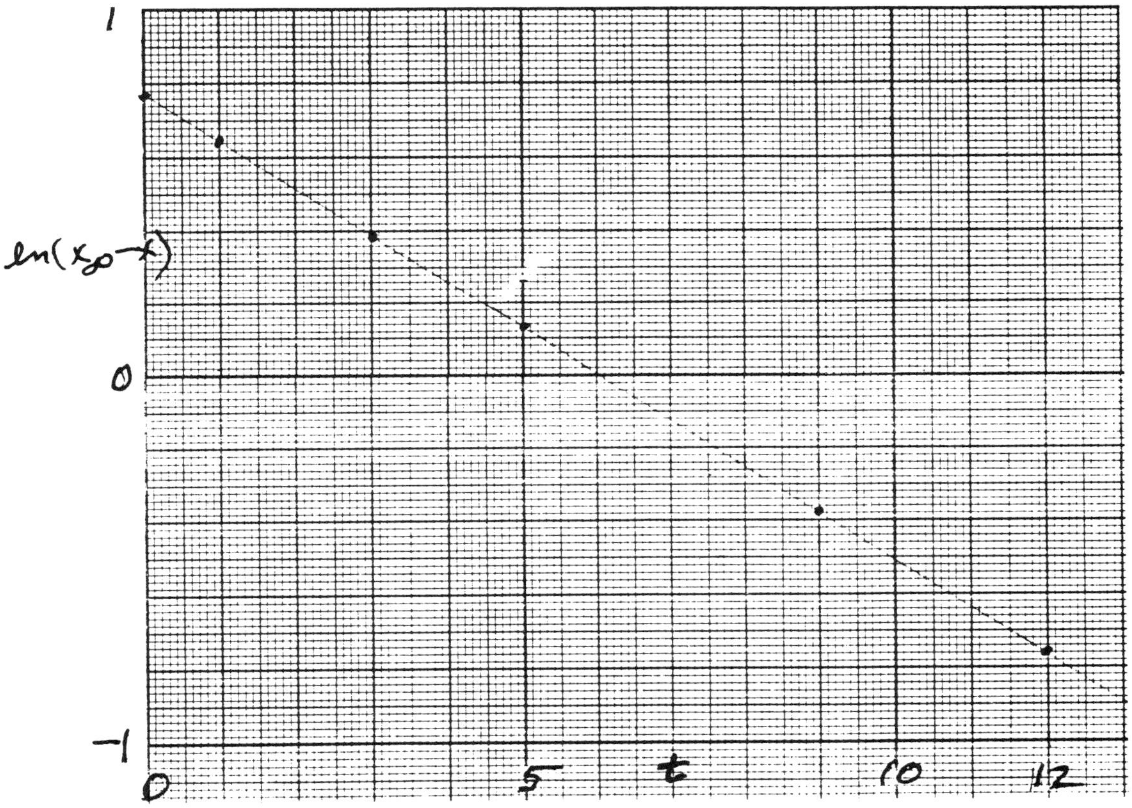

This plot is linear, and so the reaction is first order.

(b) $$\text{slope} = \frac{-0.51 - 0.77}{10} = -0.128 \text{ hr}^{-1} = -k$$

since $\ln c = -kt - \text{const} = -kt + 0.771$

$$k = 0.128 \quad \text{hr}^{-1}$$

(c) At 8 hours, $\ln c = \ln (x_\infty - x) = -(0.128\ \text{hr}^{-1})(8\text{hr}) + 0.771$

$x_\infty - x = 0.776$ $\qquad = -0.253$

$$\text{Fraction unhydrolyzed} = \frac{2.197-0.776}{2.197-0.035} = 0.325$$

19.8 The following values of percent transmission are obtained with a spectrophotometer at a series of times during the decomposition of a substance absorbing light at a particular wavelength. Calculate k, $t_{1/2}$, and t assuming the reaction is first order.

t	*Percent Transmission*
5 min	14.1
10 min	57.1
∞	100

Beer's law: log 100/T = abc, where T = percent transmission, a = absorbancy index, b = cell thickness, and c = concentration.

SOLUTION

The concentration of the reactant is proportional to log(100/T)

t/min	log(100/T)
5	0.851
10	0.243
∞	0

$$\ln \frac{c_1}{c_2} = k(t_2 - t_1)$$

$$\ln \frac{0.851}{0.243} = k5$$

$k = 0.251\ min^{-1}$

$$t_{1/2} = \frac{0.693}{0.251} = 2.76\ min$$

$$t = \frac{1}{0.251} = 3.98\ min$$

19.9 Since radioactive decay is a first-order process, the decay rate for a particular nuclide is commonly given as the half life. Given that potassium contains 0.0118% ^{40}K, which has a half life of 1.27×10^9 years, how many disintegrations per second are there in a gram of KCl?

SOLUTION

$$t_{1/2} = (1.27 \times 10^9\ y)(365\ d\ y^{-1})(24\ h\ d^{-1})(60\ min\ h^{-1})(60\ s\ min^{-1})$$

$$= 4.01 \times 10^{16}\ s$$

$$\frac{dN}{dt} = \frac{0.693}{5_{1/2}} N$$

$$= \frac{(0.6930)[(39.1/76.6)g](6.02 \times 10^{23}\ mol^{-1})(1.18 \times 10^4)}{(4.01 \times 10^{16}\ s)(40.0\ g\ mol^{-1})}$$

$$= 15.7\ s^{-1}$$

19.10 The decomposition of HI to $H_2 + I_2$ at 508 ^oC has a half-life of 135 min when the initial pressure of HI is 0.1 atm and 13.5 min when the pressure is 1 atm. (a) Show that this proves that the reaction is second order. (b) What is the value of the rate constant in L mol^{-1} s^{-1}? (c) What is the value of the rate constant in bar^{-1} s^{-1}? (d) What is the value of the rate constant in cm^3 s^{-1}?

SOLUTION

(a) The reaction studied is

$$2HI = H_2 + I_2$$

If the reaction is second order,

$$v = k[HI]^2 = -\frac{1}{2}\frac{d[HI]}{dt}$$

According to eq. 19.23

$$2kt = \frac{1}{[HI]} - \frac{1}{[HI]_o}$$

$$2kt_{1/2} = \frac{2}{[HI]_o} - \frac{1}{[HI]_o} = \frac{1}{[HI]_o}$$

$$t_{1/2} = \frac{1}{2k[HI]_o}$$

This is in agreement with the fact that the half life is reduced by a factor of 10 when the pressure is increased by a factor of 10.

(b) $$k = \frac{1}{2t_{1/2}[HI]_o}$$

$$P_{HI} = [HI]RT$$

$$[HI]_o = P_{HI}/RT = (1\ bar)/(0.08314\ L\ bar\ K^{-1}\ mol^{-1}(781\ K)$$

$$k = \frac{1}{2(13.5 \times 60\ s)(1.54 \times 10^{-2}\ mol\ L^{-1})}$$

$$= 4.01 \times 10^{-2}\ L\ mo^{-1}\ s^{-1}$$

(c) $$k = \frac{1}{2t_{1/2}P_o} = \frac{1}{2(13.5 \times 60\ s)(1.013\ bar)}$$

$$= 6.17 \times 10^{-4}\ bar^{-1}\ s^{-1}$$

(d) $k = \frac{(4.01 \times 10^{-2}\ L\ mol^{-1}\ s^{-1})(10^3\ cm^3\ L^{-1})}{(6.022 \times 1^{-23}\ mol^{-1})}$

$= 6.56 \times 10^{-22}\ cm^3\ s^{-1}$

19.11 The reaction between propionaldehyde and hydrocyanic acid has been studied at 25 °C by W. J. Svirbely and J. F. Roth [*J. Am. Chem. Soc.* **75**, 3106 (1953)]. In a certain aqueous solution at 25 °C the concentrations at various times were as follows.

t/min	2.78	5.33	8.17	15.23	19.80	∞
$\frac{[HCN]}{mol\ L^{-1}}$	0.0990	0.0906	0.0830	0.0706	0.0653	0.0424
$\frac{[C_3H_7CHO]}{mol\ L^{-1}}$	0.0566	0.0482	0.0406	0.0282	0.0229	0.0000

What is the order of the reaction and the value of the rate constant k?

SOLUTION

The data do not give a linear log versus t plot, and so the data are tested in the integrated equation for a second order reaction. Equation 19.26 is

$$\frac{1}{([A]_0 - [B]_0)} \ln \frac{[A][B]_0}{[A]_0[B]} = kt$$

In order to get $[A]_0$ and $[B]_0$ for t = 0, the clock is started at the first experimental point. This yields the following data to be plotted.

t/m	0	2.55	5.39	12.45	17.02
$\frac{1}{([A]_0-[B]_0)} \ln \frac{[A][B]_0}{[A]_0[B]}$	0	1.695	3.677	8.458	11.523

The slope of this plot, 0.675 L mol^{-1} min^{-1}, is the second order rate constant.

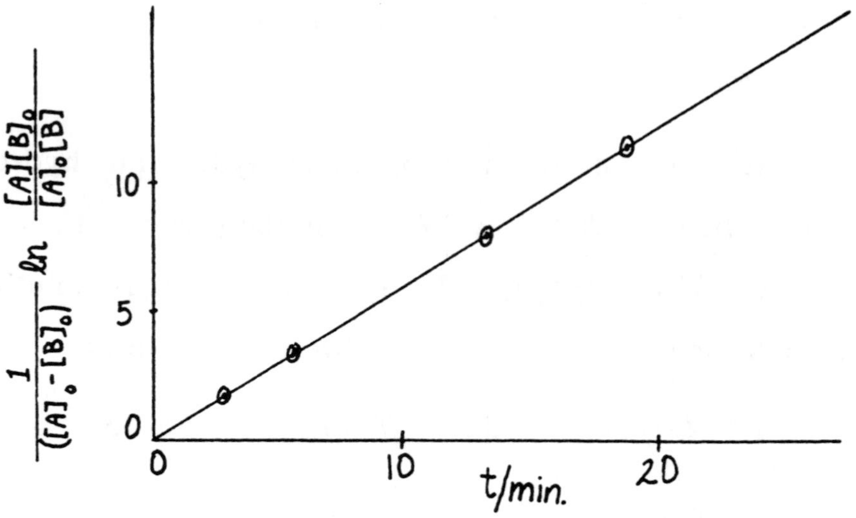

19.12 Hydrogen peroxide reacts with thiosulfate ion in slightly acidic solution as follows:

$$H_2O_2 + 2S_2O_3^{2-} + 2H^+ \rightarrow 2H_2O + S_4O_6^{2-}$$

This reaction rate is independent of the hydrogen-ion concentration in the pH range 4 to 6. The following data were obtained at 25 °C and pH 5.0.

Initial concentrations: $[H_2O_2]$ = 0.036 80 mol L^{-1};
$[S_2O_3^{2-}]$ = 0.020 40 mol L^{-1}

t/min	16	36	43	52
$\frac{[S_2O_3^{2-}]}{10^{-3}\ mol\ L^{-1}}$	10.30	5.18	4.16	3.13

(a) What is the order of reaction? (b) What is the rate constant?

SOLUTION

(a) In the first 16 minutes, $[S_2O_3^{2-}]$ is approximately halved. In the next 10 minutes, $[S_2O_3^{2-}]$ is approximately halved. In the next 16 minutes, $[S_2O_3^{2-}]$ is considerably less than halved. Therefore, the order is higher than first. The next section shows that the reaction is first order in H_2O_2, first order in $S_2O_3^{2-}$, and second order overall.

(b) Let A = H_2O_2 and B = $S_2O_3^{2-}$. Equation 19.26 becomes

$$kt = \frac{1}{(2[A]_o - [B]_o)} \ln \frac{[A][B]_o}{[A]_o[B]}$$

$$\ln \frac{[A]}{[B]} = \ln \frac{[A]_o}{[B]_o} + (2[A]_o - [B]_o)\, kt$$

t/min	0	16	36	43	52
$\frac{[B]}{10^{-3}\ mol\ L^{-1}}$	20.40	10.30	5.18	4.16	3.13
$\frac{[A]}{10^{-3}\ mol\ L^{-1}}$	36.80	31.75	29.19	28.68	28.17
$\ln \frac{[A]}{[B]}$	0.590	1.126	1.729	1.931	2.197

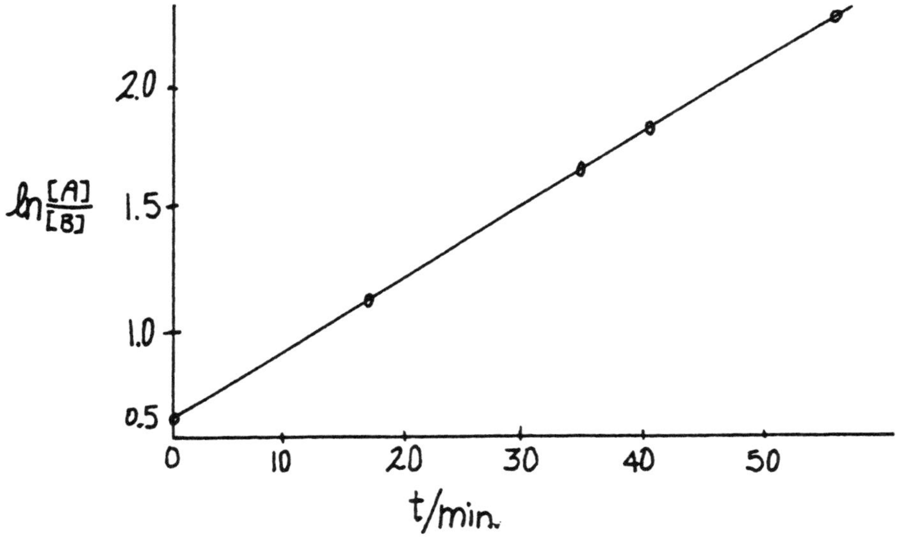

$$\text{slope} = \frac{2.46-0.59}{60 \text{ min}} = 0.0312 \text{ min}^{-1}$$

$$= (2[A]_o - [B]_o)\, k$$

$$k = \frac{0.0312 \text{ min}^{-1}}{2(36.80 \times 10^{-3} \text{ mol L}^{-1}) - 20.40 \times 10^{-3} \text{ mol L}^{-1}}$$

$$= 0.59 \text{ L mol}^{-1} \text{ min}^{-1}$$

19.13 The reaction A=B is nth order (where n=1/2, 3/2, 2, 3, ···) and goes to completion to the right. Derive the expression for the half life in terms of k, n, and $[A]_o$.

SOLUTION

$$\frac{1}{[A]^{n-1}} - \frac{1}{[A]_o{}^{n-1}} = (n-1)kt.$$

If $[A] = \frac{[A]_o}{2}$, $t = t_{1/2}$.

$$\frac{2^{n-1}}{[A]_o{}^{n-1}} - \frac{1}{[A]^{n-1}} = (n-1)kt_{1/2}$$

$$t_{1/2} = \frac{2^{n-1} - 1}{(n-1)k[A]_o{}^{n-1}}$$

19.14 A gas reaction 2A=B is second order in A and goes to completion in a reaction vessel of constant volume and temperature with a half life of 1 hour. If the initial pressure of A is 1 bar, what are the partial pressures of A, of B, and the total pressure at 1 hour, 2 hours, and at equilibrium?

SOLUTION.

$$2kt = \frac{1}{[A]} - \frac{1}{[A]_o}$$

$$2kt_{1/2} = \frac{2}{[A]} - \frac{1}{[A]_o} = \frac{1}{[A]_o}$$

$$t_{1/2} = \frac{1}{2k[A]_o} = 1 \text{ hr or } k = \frac{1}{2t_{1/2}[A]_o}$$

Eliminating k from the first equation,

$$\frac{t}{t_{1/2}[A]_o} = \frac{1}{[A]} - \frac{1}{[A]_o}$$

or

$$[A] = \frac{[A]_o}{t/t_{1/2} + 1}$$

t/h	P_A/bar	P_B/bar	P/bar
1	0.50	0.25	0.75
2	0.33	0.33	0.66
∞	0	0.50	0.50

19.15 The rate constant for the reaction

$$I + I + Ar \rightarrow I_2 + Ar$$

is 0.59×10^{16} cm^6 mol^{-2} s^{-1} at 293 K (G. Porter and J. A. Smith, *Proc. R. Soc. London Set. A* **261**, 28 (1961)). What is the half life of I if $[I]_o = 2 \times 10^{-5}$ mol L^{-1} and $[Ar] = 5 \times 10^{-3}$ mol L^{-1}?

SOLUTION

The rate constant given is defined by

$$-\frac{d[I]}{dt} = (0.59 \times 10^{16} \text{ cm}^6 \text{ mol}^{-1} \text{ s}^{-1})[I]^2[Ar]$$

The authors did not follow the more recent convention that

$$v = k[I]^2[Ar] = -\frac{1}{2}\frac{d[I]}{dt} = \frac{d[I_2]}{dt}$$

The k in this equation is 0.30×10^{16} cm^6 mol^{-2} s^{-1}.

$$[Ar] = (5 \times 10^{-3} \text{ mol L}^{-1})(10^{-3} \text{ L cm}^{-3})$$

$$= 5 \times 10^{-6} \text{ mol cm}^{-3}$$

$$-\frac{d[I]}{dt} = (3.0 \times 10^{10} \text{ cm}^3 \text{ mol}^{-1} \text{ s}^{-1})[I]^2$$

$$t_{1/2} = \frac{1}{k[I]_o} = \frac{1}{(3.0 \times 10^{10} \text{ cm}^3 \text{ mol}^{-1} \text{ s}^{-1})(10^{-3} \text{ L cm}^3)(2 \times 10^{-5} \text{ mol L}^{-1})}$$

$$= 1.70 \times 10^{-3} \text{ s}$$

19.16 A solution of A is mixed with an equal volume of a solution of B containing the same number of moles, and the reaction A + B = C occurs. At the end of 1 hr A is 75% reacted. How much of A will be left unreacted at the end of 2 hr if the reaction is (a) first order in A and zero order in B; (b) first order in both A and B; and (c) zero order in both A and B?

SOLUTION

(a) If the reaction is first order in A and zero order in B, the half time is 1/2 hour, and the following table may be constructed doing calculations in one's head.

t/hours	0	1/2	1	$1\frac{1}{2}$	2
% Unreacted	100	50	25	12.5	6.25
% Reacted	0	50	75	8.75	93.75

(b) If the reaction is first order in both A and B, and the initial concentrations are equal, and the stoichiometry is 1:1, the concentration of A will follow

$$k = \frac{1}{t}\left(\frac{1}{[A]} - \frac{1}{[A]_0}\right) = \left(\frac{[A]_0}{[A]} - 1\right)$$

$$k[A]_0 = \frac{1}{t}\left(\frac{[A]_0}{[A]} - 1\right) = \frac{1}{1\ hr}\left[\frac{100}{25} - 1\right] = 3\ hr^{-1}$$

After 2 hr

$$3\ hr^{-1} = \frac{1}{2hr}\left[\frac{100}{[A]} - 0\right]$$

$$(A) = \frac{100}{7} = 14.3\%$$

(c) If the reaction is zero order in both A and B, both will be completely gone in $1\frac{1}{3}$ hr.

19.17 Show that for a first order reaction R $\rightarrow$ P the concentration of product can be represented as a function of time by [P] = a + bt + ct^2 ··· and express a, b, and c in terms of $[R]_o$ and k.

SOLUTION

$$[P] = [R]_o - [R] = [R]_o(1 - e^{-kt})$$

Expanding e^{-kt} as a power series gives

$$[P] = [R]_o\left(kt - \frac{(kt)^2}{2!} + \frac{(kt)^3}{3!} - \cdots\right)$$

Thus a = 0, b = $[R]_ok$, and c = $\frac{[R]_ok^2}{2}$.

19.18 For a reaction A $\rightarrow$ X, the following concentrations of A were found in a single kinetics experiment

[A]/mol L^{-1}	1.000	0.952	0.909	0.870	0.833	0.800
t/hours	0	0.05	0.10	0.15	0.20	0.25

What is the rate v of this reaction at [A] = 1.000 mol L^{-1}?

SOLUTION

$$v = -\frac{d[A]}{dt}$$

As a first approximation $v = \frac{0.048}{0.05} = 0.960$ mol L^{-1} hr^{-1}

A better value may be obtained by using

$$\frac{[P]}{t} = b + ct$$

for the first 10% of the reaction.

$[X]/mol\ L^{-1}$	0.048	0.091	0.130	0.167	0.200
t/hr	0.05	0.10	0.15	0.10	0.25

$$\frac{0.048}{0.05} = b + c\ (0.05) = 0.960$$

$$\frac{0.091}{0.10} = b + c\ (0.10) = 0.910$$

Taking the difference between these equations

$0.050 = -0.05\ c$

$c = -1.00, \qquad b = 1.010 = v$

More experimental points can be included by least squaring.

Least squaring through 13% reaction yields

$v = 1.005\ mol\ L^{-1}\ hr^{-1}$

Least squaring through 16.7% reaction yields

$v = 0.998\ mol\ L^{-1}\ hr^{-1}$

Least squaring through 20% reaction yields

$v = 0.993\ mol\ L^{-1}\ hr^{-1}$

As the extent of reaction increases, this truncated series expansion becomes less adequate, and so the initial velocity is 1.00 ± 0.005 mol $L^{-1}\ hr^{-1}$, rather than 0.96 mol $L^{-1}\ hr^{-1}$, obtained as a first approximation.

19.19 The following table gives kinetic data [Y. T. Chia and R. E. Connick, *J. Phys. Chem.* **63**, 1518 (1959)] for the following reaction at 25 °C.

$$OCl^- + I^- = OI^- + Cl^-$$

$[OCl^-]$/mol L^{-1}	$[I^-]$/mol L^{-1}	$[OH^-]$/mol L^{-1}	$\frac{d[IO^-]/dt}{10^{-4}\ mol\ L^{-1}\ s^{-1}}$
0.0017	0.0017	1.00	1.75
0.0034	0.0017	1.00	3.50
0.0017	0.0034	1.00	3.50
0.0017	0.0017	0.5	3.50

What is the rate law for the reaction and what is the value of the rate constant?

SOLUTION

When other concentrations are held constant, doubling $[OCl^-]$ doubles the rate, doubling $[I^-]$ doubles the rate, and halving $[OH^-]$ doubles the rate. Therefore the rate law is

$$\frac{d[OI^-]}{dt} = \frac{k[OCl^-][dI^-]}{[dOH^-]}$$

Substituting the values for the first experiment

$$1.75 \times 10^{-4}\ mol\ L^{-1}\ s^{-1} = \frac{k(0.0017\ mol\ L^{-1})(0.017\ mol\ L^{-1})}{(1\ mol\ L^{-1})}$$

$$k = 61\ s^{-1}$$

19.20 For a reversible first-order reaction

$$A \underset{k_2}{\overset{k_1}{\rightleftarrows}} B$$

$k_1 = 10^{-2}\ s^{-1}$ and $[B]_{eq}/[A]_{eq} = 4$. If $[A]_o = 0.01$ mol L^{-1} and $[B]_o = 0$, what will be the concentration of B and 30 s?

SOLUTION

$$[B] = \frac{k_1[A]_o}{k_1 = k_2}\left[1 - e^{-(k_1+k_2)t}\right]$$

$$\frac{[B]_{eq}}{[A]_{eq}} = 4 = \frac{k_1}{k_2}$$

$$k_2 = 2.5 \times 10^{-3}\ s^{-1}$$

$$[B] = \frac{(10^{-2})(10^{-2})}{1.25 \times 10^{-2}}\left[1 - e^{-(1.25 \times 10^{-2})(30)}\right]$$

$$= 2.50 \times 10^{-3}\ mol\ L^{-1}$$

19.21 The first three steps in the decay of ^{238}U are

$$^{238}U \underset{4.5 \times 10^9\ y}{\overset{\alpha}{\rightarrow}} {}^{234}Th \underset{24.1\ d}{\overset{\beta}{\rightarrow}} {}^{234}Pa \underset{1.14\ m}{\overset{\beta}{\rightarrow}} {}^{234}U$$

If we start with pure ^{238}U, what fraction will be ^{234}Th after 10, 20, 3, and 80 days.

SOLUTION

$$k_1 = \frac{0.693}{(4.5 \times 10^9\ y)(365\ d\ y^{-1})} = 4.2 \times 10^{-13}\ d^{-1}$$

$$k_2 = \frac{0.693}{24.1\ d} = 0.0288\ d^{-1}$$

$$F = \frac{[^{234}Th]}{[^{238}U]_o} = \frac{k_1}{k_2 - k_1}(e^{-k_1 t} - e^{-k_2 t})$$

t/d	10	20	40	80
F	3.6×10^{-12}	6.39×10^{-12}	9.98×10^{-12}	13.14×10^{-12}

19.22 The initial rate of the reaction

$$BrO_3^- + 3SO_3^{2-} = Br^- + 3SO_4^{2-}$$

was found by F. S. Williamson and E. L. King (*J. Am. Chem. Soc.* **79**, 5397 (1953)) to be given by $[BrO_3^-][SO_3^{2-}][H^+]$. Give one of the thermodynamically possible rate laws for the backward motion.

<u>SOLUTION</u>

The net rate is given by

$$-\frac{d[BrO_3^-]}{dt} = k_f[BrO_3^-][SO_3^{2-}][H^+]$$

$$- k_b[Br^-]^\alpha[SO_4^{2-}]^\beta[BrO_3^-]^\gamma[SO_3^{2-}]^\delta[H^+]^\varepsilon$$

At equilibrium this rate is equal to zero

$$\frac{k_f}{k_b} = \frac{[Br^-]^{\alpha}[SO_4^{2-}]^{\beta}[BrO_3^-]^{\gamma}[SO_3^{2-}]^{\delta}[H^+]^{\varepsilon}}{[BrO_3^-][BrO_3^{2-}][H^+]}$$

$$= \frac{[Br^-][O_4^{2-}]^3}{[BrO_3^-][SO_3^{2-}]^3} \quad \text{from thermodynamics.}$$

Thus $\alpha = 1$, $\beta = 3$, $\gamma = 0$, $\delta = -2$, and $\varepsilon = 1$.

Thus the rate law for the backward reaction is

$$-\frac{d[Br^-]}{dt} = k_b[Br^-][SO_4^{2-}]^3[H^+][SO_3^{2-}]^{-2}$$

A different reverse rate law would be obtained if the reaction was multiplied by 2.

19.23 Suppose the transformation of A to B occurs by both a reversible first-order reaction and a reversible second-order reaction involving hydrogen ion.

$$A \underset{k_2}{\overset{k_1}{\rightleftarrows}} B \qquad A + H^+ \underset{k_4}{\overset{k_3}{\rightleftarrows}} B + H^+$$

What is the relationship between these four rate constants?

SOLUTION

$$\frac{[B]_{eq}}{[A]_{eq}} = \frac{k_1}{k_2}$$

$$\frac{[B]_{eq}[H^+]_{eq}}{[A]_{eq}[H^+]_{eq}} = \frac{k_3}{k_4} \quad \text{or} \quad \frac{[B]_{eq}}{[A]_{eq}} = \frac{k_3}{k_4}$$

$$\text{Therefore} \quad \frac{k_1}{k_2} = \frac{k_3}{k_4} \quad \text{or} \quad k_1k_4 = k_2k_3$$

19.24 The hydrolysis of $(CH_2)_6C\begin{matrix}Cl\\CH_3\end{matrix}$ in 80% ethanol follows the first-order rate equation. The values of the specific reaction-rate constants, as determined by H.C. Brown and M. Borkowski [*J. Am. Chem. Soc.* **74**, 1896 (1952)], are as follows:

$t/°C$	0	25	35	45
k/s^{-1}	1.06×10^{-5}	3.19×10^{-4}	9.86×10^{-4}	2.92×10^{-3}

(a) Plot log k against 1/T; (b) calculate the activation energy; (c) calculate the pre-exponential factor.

SOLUTION

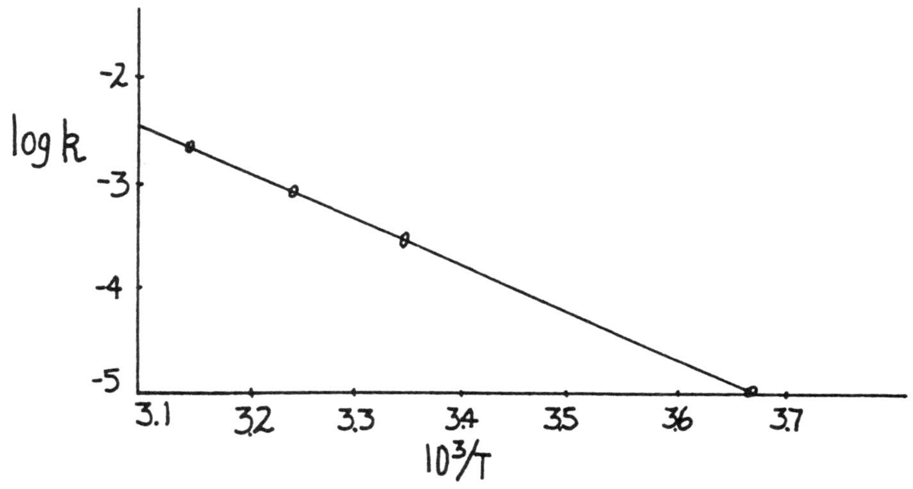

(b) slope $= \dfrac{-E_a}{2.303\ R} = -400\ K$

$E_a = (4600\ K)(2.303)(8.314\ K^{-1}\ mol^{-1})$

$= 88.1\ kJ\ mol^{-1}$

(c) Taking the 35 °C point and $E_a = 88.1\ kJ\ mol^{-1}$

$$\log k = \frac{-E_a}{2.303\ RT} + \log A$$

$$-3.006 = \frac{-88100\ J\ mol^{-1}}{2.303(8.314\ K^{-1}\ mol^{-1})(308.15\ K)} + \log A$$

$A = 8.4 \times 10^{11}\ s^{-1}$

(If the data are fitted by linear regression using a hand-held calculator, the following parameters are obtained: $A = 1.99 \times 10^{12}\ s^{-1}$, $E_a = 90.3\ kJ\ mol^{-1}$.)

19.25 If a first-order reaction has an activation energy of 104,600 J mol^{-1} and a pre-exponential factor A of $5 \times 10^{13}\ s^{-1}$, at what temperature will the reaction have a half-life of (a) 1 min, and (b) 30 days?

SOLUTION:

$$k = \frac{0.693}{t_{1/2}} = Ae^{-E_a/RT}$$

$$T = \frac{-E_a}{R\ln\left(\frac{0.693}{t_{1/2}}\right)}$$

$$\text{(a) } T = \frac{-104600\ J\ mol^{-1}}{(8.314\ J\ K^{-1}\ mol^{-1})\ln\left[\frac{0.693}{(5 \times 10^{13}\ s^{-1})(60\ s)}\right]}$$

$= 349$ K or 76 °C

$$\text{(b) } T = \frac{-104600\ J\ mol^{-1}}{(8.314\ J\ K^{-1}\ mol^{-1})\ln\left[\frac{0.693}{(5 \times 20^{13})(60)^2(24)(30)}\right]}$$

$= 270$ K or -3 °C.

19.26 Isopropenyl allyl ether in the vapor state isomerizes to allyl acetone according to a first-order rate equation. The following equation gives the influence of temperature on the rate constant (in s^{-1}) $k = 5.4 \times 10^{11} e^{-123,000/RT}$ where the activation energy is expressed in J mol^{-1}. At ·150 °C, how long will it take to build up a partial pressure of 0.395 bar of allyl acetone, starting with 1 bar of isopropenyl allyl ether [L. Stein and G. W. Murphy, *J. Am. Chem. Soc.* **74**, 1041 (1952)]?

SOLUTION

$$k = (5.4 \times 10^{11}\ s^{-1})\, e^{-123,000/(8.314)(423.15)}$$

$$= 3.53 \times 10^{-4}\ s^{-1}$$

$$t = \frac{1}{k} ln \frac{P^0}{P} = \frac{1}{3.53 \times 10^{-4}\ s^{-1}} ln \frac{1\ bar}{0.605\ bar}$$

$$= 1420\ s$$

19.27 The pre-exponential factor for the trimolecular reaction $2NO + O_2 \rightarrow 2NO_2$ is 10^9 cm^6 mol^{-2} s^{-1}. What is the value in L^2 mol^{-2} s^{-1} and cm^6 s^{-1}?

SOLUTION

$$(10^9\ cm^6\ mol^{-2}\ s^{-1})(10^{-1}\ dm\ cm^{-1})^6 = 10^3 dm^6\ mol^{-2}\ s^{-1}$$

$$= 10^3\ L^2\ mol^{-2}\ s^{-1}$$

$$= \frac{(10^3\ L^2\ mol^{-2}\ s^{-1})(10^3\ cm^3\ L^{-1})^2}{(6.02 \times 10^{23}\ mol^{-1})^2}$$

$$= 2.76 \times 10^{-39}\ cm^6\ s^{-1}$$

19.28 A reaction A + B + C + → D follows the mechanism

$$A + B \rightleftarrows AB \qquad AB + C \rightarrow D$$

in which the first step remains essentially in equilibrium. **Show that** the dependence of rate on temperature is given by

$$k = Ae^{-(E_a+\Delta H)/RT}$$

where ΔH is the enthalpy change for the first reaction.

SOLUTION:

For A + B = AB

$$K = \frac{[AB]}{[A][B]} = e^{\Delta S^o/R} \; e^{-\Delta H^o/RT}$$

$$[AB] = [A][B] \; e^{\Delta S^o/R} \; e^{-\Delta H^o/RT}$$

$$\frac{d[D]}{dt} = k'[AB][C] = k'[A][B][C] \; e^{\Delta S^o/R} \; e^{-\Delta H^o/RT}$$

$$= k[A][B][C]$$

where

$$k = k'e^{\Delta S^o/R} \; e^{-\Delta H^o/RT}$$

Assuming

$$k' = A'e^{-E_a/RT}$$

$$k = A'e^{\Delta S^o/R} \; e^{-(E_a+\Delta H^o)/RT}$$

$$= Ae^{-(E_a+\Delta H^o)/RT}$$

19.29 For the mechanism

$$A + B \underset{k_2}{\overset{k_1}{\rightleftarrows}} C \qquad\qquad C \xrightarrow{k_3} D$$

(a) Derive the rate law using the steady state approximation to eliminate the concentration of C. (b) Assuming that $k_3 << k_2$, express the pre-exponential factor A and E_a for the apparent second-order rate constant in terms of A_1, A_2, and A_3 and E_a, E_{a2}, and E_{a3} for the three steps.

<u>SOLUTION</u>

(a) $$\frac{d[C]}{dt} = k_1[A][B] - (k_2 + k_3)[C] = 0$$

$$\frac{d[D]}{dt} = k_3[C] = \frac{k_1k_3[A][B]}{k_2 + k_3}$$

(b) For $k_2 >> k_3$,

$$k_{app} = \frac{k_1k_3}{k_2} = \frac{A_1{}^{-E_{a1}/RT}A_3{}^{-E_{a3}/RT}}{A_2e^{-E_{a2}/RT}}$$

$$= \frac{A_1A_3}{A_2} e^{-(E_{a1} + E_{a3} - E_{a2}/RT}$$

$$A_{app} = \frac{A_1A_3}{A_2} \qquad\qquad E_{app} = E_{a1} + E_{a3} - E_{a2}$$

19.27 For the two parallel reactions $A \xrightarrow{k_1} B$ and $A \xrightarrow{k_2} C$, show that the activation energy E' for the disappearance of A is given in terms of the activation energies E_1 and E_2 for the two paths by

$$E' = \frac{k_1E_1 + k_2E_2}{k_1 + k_2}$$

SOLUTION

The rate equation for A is

$$\frac{d[A]}{dt} = k_1[A] + k_2[A] = (k_1 + k_2)[A] = k'[A]$$

where $k' = k_1 + k_2 = Ae^{-E'/RT}$

$$\frac{d \ln k'}{dt} = \frac{E'}{RT^2} = \frac{d \ln(k_1 + k_2)}{dT} = \frac{d(k_1 + k_2)}{(k_1 + k_2)dT} = \frac{1}{(k_1 + k_2)}\left(\frac{dk_1}{dT} + \frac{dk_2}{dT}\right)$$

$$= \frac{1}{(k_1 + k_2)} k_1 \left[\frac{d \ln k_1}{dT} + k_2\frac{d \ln k_2}{dT}\right]$$

$$= \frac{1}{(k_1 + k_2)}\left[\frac{k_1E_1}{RT^2} + \frac{k_2E_2}{RT^2}\right]$$

$$E' = \frac{k_1E_1 + k_2E_2}{k_1 + k_2}$$

19.31 Set up the rate expressions for the following mechanism

$$A \underset{k_2}{\overset{k_1}{\rightleftarrows}} B \qquad B + C \xrightarrow{k_3} D$$

If the concentration of B is small compared with the concentrations of A, C, and D, the steady-state approximation may be used to derive the rate law. Show that this reaction may follow the first-order equation at high pressures and the second-order equation at low pressure.

SOLUTION

Since the concentration of B is small, it is assumed to be in a steady state.

$$\frac{d[B]}{dt} = k_1[A] - (k_2 + k_3[C])[B] = 0$$

$$[B] = \frac{k_1[A]}{k_2 + k_3[C]}$$

$$\frac{d[D]}{dt} = k_3[B][C] = \frac{k_1k_3[A][C]}{k_2 + k_3[C]}$$

At high pressure, $k_3[C] >> k_2$, $\frac{d[D]}{dt} = k_1[A]$

At low pressure, $k_2[C]$, $\frac{d[D]}{dt} = \frac{k_1k_3}{k_2}[A][C]$

19.32 The reaction $NO_2Cl = NO_2 + \frac{1}{2} Cl_2$ is first order and apears to follow the mechanism

$$NO_2Cl \xrightarrow{k_1} NO_2 + Cl \qquad\qquad NO_2Cl + Cl \xrightarrow{k_2} NO_2 + Cl_2$$

(a) Assuming a steady state for the chlorine atom concentration, show that the empirical first-order rate constant can be identified with $2k_1$. (b) The following data were obtained by H. F. Cordes and H. S. Johnston [*J. Am. Chem. Soc.* **76**, 4264 (1954)] at 180 °C. In a single experiment the reaction is first order, and the empirical rate constant is represented by k. Show that the reaction is second order at these low gas pressures and calculate the second-order rate constant.

$c/10^{-8}$ mol cm^{-3}	5	10	15	20
$k/10^{-4}$ s^{-1}	1.7	3.4	5.2	6.9

SOLUTION

(a) $$\frac{d[Cl]}{dt} = k_1[NO_2Cl] - k_2[NO_2Cl][Cl] = 0$$

Therefore $[Cl] = \frac{k_1}{k_2}$

$$\frac{-d[NO_2Cl]}{dt} = k_1[NO_2Cl] + k_2[NO_2Cl][Cl]$$

$$= k_1[NO_2Cl] + k_2[NO_2Cl]k_1/k_2$$
$$= 2k_1[NO_2Cl]$$

(b) At low pressures the rate a determining step is the first step, and the reaction becomes second order.

$$NO_2Cl + NO_2Cl \xrightarrow{k_1} NO_2Cl + N)_2 + Cl$$

$$\frac{-d[NO_2Cl]}{dt} = k_1'[NO_2Cl]^2 \text{ assuming } [NO_2Cl] \text{ is approximately constant}$$

$$k_{obs} = k_1'[NO_2Cl]$$

$$1.7 \times 10^{-4} = k_1'(5 \times 10^{-8})$$

$$k_1' = 3.4 \times 10^3 \text{ cm}^3 \text{ mol}^{-1} \text{ s}^{-1}$$

19.33 The reaction

$$2\ SO_2 + O_2 = 2\ SO_4$$

is catalyzed by the mechanism

$$2\ NO + O_2 \underset{k_{-1}}{\overset{k_1}{\rightleftarrows}} 2NO_2$$

$$NO_2 + SO_2 \underset{k_{-2}}{\overset{k_2}{\rightleftarrows}} NO + SO_3$$

In order to obtain the overall reaction from this mechanism, the second step has to be taken twice, and so the stoichiometric number s_2 of the second step is said to be 1. The equilibrium constant K_c for an overall reaction is related to the rate constants for the individual steps, k_i and k_{-i}, by

$$K_c = \prod_{i-1}^{s} \left(\frac{k_i}{k_{-i}}\right)^{s_i}$$

where s_i is the stoichiometric coefficient of the ith step and S is the number of steps. Verify this relation for the above mechanism.

SOLUTION

The standard Gibbs energy for the overall reaction is given by

$$\Delta G^o = \Delta G_1^o + 2\Delta G_2^o$$

$$-RT\mathit{ln}K_c = -RT\mathit{ln}K_1 - RT\mathit{ln}K_2^2$$

$$= -RT\mathit{ln}\frac{k_1}{k_{-1}} - RT\mathit{ln}\left(\frac{k_2}{k_{-2}}\right)$$

where the equilibrium constants of the elementary reactions have been expressed in terms of the rate constants. The last form of the equation may be written

$$K_c = \left(\frac{k_1}{k_{-1}}\right)\left(\frac{k_2}{k_{-2}}\right)$$

which may be generalized to

$$K_c = \prod_{i=1}^{s} \left(\frac{k_i}{k_{-i}}\right)^{s_i}$$

19.34 What is the rate constant for the following reaction at 500 K?

$$H + HCl \rightarrow Cl + H_2$$

The data required are to be found in Tables 19.3 and C.2.

SOLUTION

The rate constant for the backward reaction is given by

$k_b \quad = (10^{10.9}\ L\ mol^{-1}\ s^{-1})e^{-23,000/8.314 \times 500}$

$\quad\quad = 3.1 \times 10^8\ L\ mol^{-1}\ s^{-1}$

$\Delta G^o \quad = \Delta_f G^o(Cl) - \Delta_f G^o(H) - \Delta_f G^o(HCl)$

$\quad\quad = 94.203 - 192.957 + 97.166$

$\quad\quad = -1.588\ kJ\ mol^{-1}$

$K_P \quad = e^{1588/8.324 \times 500} = 1.47$

$K_p \quad = K_c = \frac{k_f}{k_b}$

$k_f \quad = (1.47)(3.1 \times 10^8\ L\ mol^{-1}\ s^{-1}) = 4.5 \times 10^8\ L\ mol^{-1}\ s^{-1}$

19.35 For the gas reaction $O + O_2 + M \underset{k_{-1}}{\overset{k_1}{\rightleftarrows}} O_3 + M$

When $M = O_2$, Benson and Axworthy [*J. Chem. Phys.* **26**, 1718 (1957)] obtained $k = (6.0 \times 20^7\ L^2\ mol^{-2}\ s^{-1})\ e^{2.5/RT}$ where the activation energy is in $kJ\ mol^{-1}$. Calculate the values of the parameters in the Arrhenius equation for the reverse reaction assuming ΔH^o and ΔS^o are independent of temperature.

SOLUTION

$\Delta H^o = 142.7 - 249.170 = -106.5$ kJ mol^{-1}

$\Delta S^o = 238.82 - 160.946 - 205.029 = -127.16$ J K^{-1} mol^{-1}

$K_P = e^{-\Delta H^o/RT} e^{-\Delta S^o}$

$= e^{106,500/RT} e^{-127.16/R}$

$$K_c = K_p \left(\frac{P^o}{c^oRT}\right)^{\Sigma v_i} = K_P\left(\frac{1}{24.46}\right)^{-1} \text{ L mol}^{-1} = \frac{k}{k'}$$

$$k' = \frac{k}{24.46\ K_P}$$

$$k' = \frac{6 \times 10^7\ e^{2500/RT}}{24.46\ e^{106500/RT}\ e^{-127.16/R}}$$

$$= \left[\frac{6 \times 10^7}{24.46} e^{127.16R}\right] e^{(2500\ -\ 106,500)/RT}$$

$$= (1.1 \times 10^{13} \text{ L mol}^{-1} \text{ s}^{-1})\ e^{-104,000/RT}$$

19.36 For the mechanism

$$H_2 + X_2 \underset{k_{-1}}{\overset{k_1}{\rightleftarrows}} 2HX \qquad X + H_2 \underset{k_{-3}}{\overset{k_3}{\rightleftarrows}} HX + H$$

$$X_2 \underset{k_{-2}}{\overset{k_2}{\rightleftarrows}} 2X \qquad H + X_2 \underset{k_{-4}}{\overset{k_4}{\rightleftarrows}} 2X + 4$$

show that the steady-state rate law is

$$\frac{d[HX]}{dt} = 2k_1[H_2][X_2]\left(1 - \frac{[HX]^2}{K[H_2][X_2]}\right)\left(1 + \frac{\frac{k_3}{k_1}\sqrt{\frac{2k_2}{k_{-2}[X_2]}}}{1 + \frac{k_{-3}[X]}{k_4[X_2]}}\right)$$

SOLUTION

There are five species, but there are two conservation equations.

$[H] + [HX] + 2[H_2] = \text{const.}$

$[X] + [HX] + 2[X_2] = \text{const.}$

Therefore, there are only three independent rate equations.

(1) $$\frac{d[HX]}{dt} = 2k_1[H_2][X_2] + k_3[X][H_2] + k_4[H][X_2]$$
$$- k_{-1}[HX]^2 - k_{-3}[HX][H] - k_{-4}[HX][X]$$

(2) $$\frac{d[X]}{dt} = 2k_2[X_2] + k_{-3}[HX][H] + k_4[H][X_2] - k_{-2}[X]^2$$
$$- k_3[X][H_2] - k_{-4}[HX][X] = 0$$

(3) $$\frac{d[H]}{dt} = k_3[X][H_2] + k_{-4}[HX][X] - k_{-3}[HX][H] - k_4[H][X_2] = 0$$

Adding equations 2 and 3 $[X] = \sqrt{\frac{2k_2}{k_{-2}}}[X_2]$

Substituting in equation 3 yields

$$[H] = \frac{k_3[H_2] + k_{-4}[HX]}{k_{-3}[HX] + k_4[X_2]}\sqrt{\frac{2k_2[X_2]}{k_{-2}}}$$

Substituting in equation 1

$$\frac{d[HX]}{dt} = 2k_1[H_2][X_2] - k_{-1}[HX]^2(k_4[X_2] - k_3[HX])$$

$$\times \frac{(k_3[H_2] + k_{-4}[HX])\sqrt{\frac{2k_2[X_2]}{k_{-2}}}}{k_{-3}[HX] + k_4[X_2]}$$

$$+ (k_3[H_2] - k_{-4}[HX] \sqrt{\frac{2k_2[X_2]}{k_{-2}}} \frac{(k_{-3}[HX] + k_4[X_2])}{(k_3[HX] + k_4[X_2])}$$

$$\frac{d[HX]}{dt} = 2k_1[H_2][X_2] - 2k_{-1}[HX]^2 + k_3k_4[H_2][X_2]$$

$$+ k_4k_{-4}[X_2][HX] - k_3k_{-3}[HX][H_2] - k_{-3}k_{-4}[HX]^2$$

$$+ k_3k_{-3}[H_2][HX] + k_3k_4[H_2][X_2] - k_{-3}k_{-4}[HX]^2$$

$$- k_4k_{-4}[HX][X_2] \times \frac{\sqrt{\frac{2k_2[X_2]}{k_{-2}}}}{k_{-3}[HX] + k_4[X_2]}$$

$$= 2k_1[H_2][X_2] - 2k_{-1}[HX]^2 + (2k_3k_4[H_2][X_2] - 2k_{-3}k_{-4}[HX]^2)$$

$$\times \frac{(\frac{2k_2[X_2]}{k_{-2}})^{1/2}}{k_{-3}[HX] + k_4[X_2]}$$

$$= 2k_1[H_2] (1 - \frac{k_1[HX]^2}{k_1[H_2][X_2]}) + 2k_3k_4[H_2][X_2] (1 - \frac{k_{-3}k_{-4}[HX]^2}{k_3k_4[H_2][X_2]})$$

$$\times \frac{(\frac{2k_2[X_2]}{k_{-2}})^{1/2}}{k_3[HX] + k_4[X_2]}$$

$$\frac{d[HX]}{dt} = 2k_1[H_2][X_2] (1 - \frac{[HX]^2}{K[H_2][X_2]}) (1 + \frac{k_3k_4}{k_1} \frac{\sqrt{\frac{2k_2[X_2]}{k_{-2}}}}{k_{-3}[HX] + k_4[X_2]})$$

$$= 2k_1[H_2][X_2] (1 - \frac{[HX]^2}{K[H_2][X_2]}) \left(1 + \frac{\frac{k_3}{k_1} \sqrt{\frac{2k_2}{k_{-2}[X_2]}}}{\frac{k_{-3}[X_2]}{k_4[X_2]}}\right)$$

19.37 The mechanism of the pyrolysis of acetaldehyde at 520 °C and 0.2 bar is

$$CH_3CHO \xrightarrow{k_1} CH_3 + CHO$$

$$CH_3 + CH_3CHO \xrightarrow{k_2} CH_4 + CH_3CO$$

$$CH_3CO \xrightarrow{k_3} CO + CH_3$$

$$CH_3 + CH_3 \xrightarrow{k_4} C_2H_6$$

What is the rate law for the reaction of acetaldehyde, using the usual assumptions? (As a simplification further reactions of the radical CHO have been omitted and its rate equation may be ignored.)

SOLUTION

(1) $$\frac{d[CH_3CHO]}{dt} = -(k_1 + k_2[CH_3])[CH_3CHO] = 0$$

(2) $$\frac{d[CH_3]}{dt} = k_1[CH_3CHO] - k_2[CH_3][CH_3CHO] + k_3[CH_3CO] - 2k_4[CH_3]^2 = 0$$

(3) $$\frac{d[CH_3CO]}{dt} = k_2[CH_3][CH_3CHO] - k_3[CH_3CO] = 0$$

Equation 3 yields $[CH_3CO] = k_2[CH_3][CH_3CHO]/k_3$

Substituting this in equation 2 yields

$$[CH_3] = \left(\frac{k_1[CH_3CHO]}{2k_4}\right)^{1/2}$$

Substituting this in equation 1 yields

$$\frac{d[CH_3CHO]}{dt} = -\left[k_1 + k_2\left(\frac{k_1}{2k_4}\right)^{1/2}[CH_3CHO]^{1/2}\right][CH_3CHO]$$

If k_1 is small,

$$\frac{d[CH_3CHO]}{dt} = -k_2(k_1/2k_4)^{1/2}[CH_3CHO]^{3/2}$$

19.38 (a) 1.2×10^{-3} mol L^{-1} s^{-1}, (b) 2.4×10^{-3} mol L^{-1} s^{-1},

(c) 3.6×10^{-3} mol s^{-1}, (d) 7.2×10^{-3} mol

19.39 (a) 0.05 mol L^{-1} s^{-1}, (b) -0.05 mol L^{-1} s^{-1}, (c) -0.15 mol L^{-1} s^{-1}

19.40 $v = k[H_2][Br_2]^{1/2}$

19.41 1.56%

19.42 0.0396, 0.0421 min^{-1}

19.43 14.6% reacted

19.44 (b) 20.0 min.

19.45

t/min	P_A/bar	P_B/bar
10	0.50	1.00
20	0.25	1.50
∞	0	2.00

19.46 (a) 1.38×10^{-10}%, (b) 6950 g

19.47 7160 y

19.48 3989 s

19.49 (a) 221 s, (b) 82.4 s

19.50 26.3 min

19.51 3.73×10^9 L mol^{-1} s^{-1}

19.52 (a) 0.107 L mol^{-1} s^{-1}, (b) 2.85×10^3 s

19.53 $[A]_0^{1/2} - [A]^{1/2} = (k/2)t$

$t_{1/2} = (\sqrt{2}/k)(\sqrt{2} - 1)[A]_0^{1/2}$

19.54 $[A]^{-1/2} - [A]_0^{-1/2} = kt/2$ $\quad t_{1/2} = 2(\sqrt{2} - 1)/k[A]_0^{1/2}$

19.55 (a) 80%, (b) 67.1%, (c) 61.5%

19.56 $-d[H_2Se_3]/dt = k[H_2SeO_3][H^+]^2[I^-]^3$

19.57 5.78×10^{-4} s^{-1}

19.58 $[B] = k_1[A]_0 t\, e^{-k_1 t}$

$[C] = [A]_0[1 - e^{-k_1 t}(1 + k_1 t)]$

19.59 $k[B][C]/[A]$, $k[B]^{1/2}[C]^{1/2}$

19.60 (a) 97.0 kJ mol^{-1}, (b) 8.99 x 10^{13} s^{-1}, (c) 1.71 s

19.61 (a) 721 K, (b) 663 K

19.62 (a) 4.6 x 10^6 s, (b) 2430 s

19.63 (a) 1105 K, (b) 736 K, (c) The rate of reaction with the higher activation energy increases more rapidly with increasing temperature than the rate of the reaction with the lower activation energy.

19.64 $k = (5.98 \times 10^{10}\ L^2\ mol^{-2}\ s^{-1}) \exp[5730\ J\ mol^{-1}/(8.314\ JK^{-1}\ mol^{-1}) \times (T/K)]$

19.65 (a) 15.1 kJ mol^{-1}, (b) 2.27

19.66 $d[NO_2]/dt = k'[NO]^2[O_2]$

19.67 $d[I_2]/dt = k_1K[I]^2[M] - k_{-1}[I_2][M]$

$[I_2]/[I]^2 = k_1K/k_{-1}$

19.68 $d[A_2]/dt = k_1k_2[M][A]^2/(k_{-1} + k_2[M])$

19.70 E_a = 18,931 J mol^{-1}, A = 4.74 x 10^{10} L mol^{-1} s^{-1}

19.71 3.0×10^{16} L mol^{-1} s^{-1}

19.72 $d[O_2]/t = 2k_1[NO][O_3]$

19.73 $$\frac{d[COCl_2]}{dt} = \frac{k_2k_3(2k_1/k_{-1})^{1/2}[Cl_2]^{3/2}[CO]}{k_2 + k_3[Cl_2]}$$

20.1 According to the Boltzmann distribution, the number of molecules $dN(\varepsilon)$ in the energy range dE is given by

$$dN(\varepsilon) = Ke^{-\varepsilon/kT}\, d\varepsilon$$

If the density of states is independent of E, (a) Derive the expression for the probability that these molecules have energies greater than ε_a.

(b) What is the probability that the molecules have energies greater than 20 kJ mol^{-1} at 298 ? (c) What is the probability that the molecules have energies greater than 20 kJ mol^{-1} at 100 K?

SOLUTION

(a) $N = \int_0^\infty dN(\varepsilon) \qquad = K\int_0^\infty e^{-\varepsilon/kT}\, d\varepsilon = kTK$

so that $K = N/kT$.

$P(\varepsilon > \varepsilon_a) = (1/N)\int_{\varepsilon_a}^\infty Ke^{-\varepsilon/kT}\, d\varepsilon = e^{-\varepsilon_a/kT} = e^{-\varepsilon_a/kT}$

where ε_a is expressed energy per mole.

(b) $P(E > 20 \text{ kJ mol}^{-1})$

$= \exp[\ (-20{,}000 \text{ J mol}^{-1}) / \frac{(8.314 \text{ J k mol}^{-1})}{298 \text{ K}}]$

$= 3.12 \times 10^{-4}$

(c) $P(E > 20 \text{ kJ mol}^{-1}) = \exp[\ (-20{,}000 \text{ J mol}^{-1}) / \frac{(8.314 \text{ J k mol}^{-1})}{1000 \text{ K}}]$

$= 0.0902$

20.2 Use equation 19.68 to derive the expression for the activation energy for a reaction following simple collision theory.

SOLUTION

$E_a \quad = RT^2\ (d\mathit{ln}k/dT)$

$k \quad = N_A\ \pi d_{AB}^2(8k_BT\pi\mu)^{1/2}\ e^{-\varepsilon_o/k_BT}$

$E_a \quad = RT^2 \frac{d}{dT}\left\{ \mathit{ln}\left[\ N_A\ \pi d_{AB}^2\ (8k_B/\pi\mu)^{1/2}\right] + \frac{1}{2}\mathit{ln}\ T - \varepsilon_0/k_BT\right\}$

$= RT^2\ (\frac{1}{2T} + \frac{\varepsilon_0}{k_BT^2})$

$= N_A\varepsilon_0 + \frac{1}{2}RT$

20.3 Use the pre-exponential factor A = 3 x 10^{13} cm^3 mol^{-1} s^{-1} for the reaction Br + $H_2 \rightarrow$ HBr + H to calculate the cross section and collision diameter for this reaction at 400 K.

SOLUTION

$$\mu N_A = \frac{1}{\frac{1}{79.904 \times 10^{-3}\ kg} + \frac{1}{2.0158 \times 10^{-3}\ kg}}$$

$$= 1.966 \times 10^{-3}\ kg\ mol^{-1}$$

$$(\frac{8RT}{\pi\mu N_A})^{1/2} = \left[\frac{8(8.314\ J\ K^{-1}\ mol^{-1})(400\ K)}{\pi(1.966 \times 10^{-3}\ kg\ mol^{-1})}\right]^{1/2}$$

$$= 2075\ m\ s^{-1}$$

$$A = N_A \pi d_{12}^2 \left(\frac{8RT}{\pi \mu N_A}\right)^{1/2}$$

$$d_{12} = \left[\frac{A}{N_A \pi \left(\frac{8RT}{\pi \mu N_A}\right)^{1/2}}\right]^{1/2}$$

$$= \left[\frac{(3 \times 10^{13}\ cm^3\ mol^{-1}\ s^{-1})(10^{-2}\ m\ cm^{-1})^3}{\pi(6.02 \times 10^{23}\ mol^{-1})(2075\ m\ s^{-1})}\right]^{1/2} = 87.4\ pm$$

The cross section is $\pi d_{12}^2 = 2.40 \times 10^{-2}\ m^2$.

20.4 (a) Calculate the second-order rate constant for collisions of dimethyl ether molecules with each other at 777 K. It is assumed that the molecules are spherical and have a radius of 0.25 nm. If every collision was effective in producing decompositon what would be the half-life of the reaction (b) at 1 bar pressure, and (c) at a pressure of 0.13 Pa?

SOLUTION

(a) $k = 2N_A \pi\ d_{12}^2 \left(\frac{8RT}{\pi \mu N_A}\right)^{1/2}$ since A_1 and A_2 are identical

$$\mu N_A = \frac{M}{2} = \frac{46 \times 10^{-3}\ kg\ mol^{-1}}{2} = 23 \times 10^{-3}\ kg\ mol^{-1}$$

$$d_{12} = 0.5 \times 10^{-9}\ m$$

$$k = 2(6.02 \times 10^{23}\ mol^{-1})\pi(0.5 \times 10^{-9}\ m)^2$$

$$\left[\frac{8(8.314\ J\ K^{-1}\ mol^{-1})(777\ K)}{(3.1416)(23 \times 10^{-3}\ kg\ mol^{-1})}\right]^{1/2}$$

$$= 8.00 \times 10^8\ m^3\ mol^{-1}\ s^{-1}$$

$$= (8.00 \times 10^8\ m^3\ mol^{-1}\ s^{-1})(10^3\ L\ m^{-3})$$

$$= 8.00 \times 10^{11}\ L\ mol^{-1}\ s^{-1}$$

(b) $$t_{1/2} = \frac{1}{k[A]_0}$$

$$[A]_0 = \frac{P}{RT} = \frac{1\ bar}{(0.08314\ L\ bar\ k^{-1}\ mol^{-1})(777\ K)}$$

$$= 0.0155\ mol\ L^{-1}$$

$$t_{1/2} = \frac{1}{(8.00 \times 10^{11}\ L\ mol^{-1}\ s^{-1})(0.0155\ mol\ L^{-1})}$$

$$= 8.1 \times 10^{-11}\ s$$

(c) $$P = 0.13 \times 10^{-5}\ bar$$

$$[A]_0 = \frac{P}{RT} = \frac{(0.13 \times 10^{-5})}{0.08314 \times 777}$$

$$= 2.0 \times 10^{-8}\ mol\ L^{-1}$$

$$t_{1/2} = \frac{1}{(8.00 \times 10^{11}\ L\ mol^{-1}\ s^{-1})(2.0 \times 10^{-8}\ mol\ L^{-1})}$$

$$= 6.2 \times 10^{-5}\ s$$

20.5 Show that the transition-state theory yields the simple collision theory result when it is applied to the reaction of two rigid spherical molecules.

SOLUTION

Assuming that the molecules react on their first collision (that is, the activation energy is zero), transition state theory yields

$$k = \frac{RT}{h} \frac{q''_{AB^\ddagger}}{q'_A q'_B} \quad (1)$$

where q'_A and q'_B are molecular partition fucntions without the volume factor, and $q''_{AB}{}^\ddagger$ is the molecular partition function for the transition state without the volume factor and without the vibrational factor.

$$q'_A = \left(\frac{2\pi m_A kT}{h^2}\right)^{3/2} \quad (2) \qquad q'_B = \left(\frac{2\pi m_B kT}{h^2}\right)^{3/2} \quad (3)$$

$$q''_{AB} = \left[\frac{2\pi(m_A + m_B)kT}{h^2}\right]^{3/2} \frac{8\pi^2\mu(R_A + R_B)^2 kT}{h^2} \quad (4)$$

$$\mu = \frac{m_A m_B}{m_a + m_B} \quad (5)$$

Substituting equations, 2, 3, 4 and 5 in equation 1 yields

$$k = N_A\left(\frac{8\pi kT}{\mu}\right)^{1/2} (R_A + R_B)^2$$

which may be compared with equation 20.2

$$k = \pi d_{12}^2 \left(\frac{8RT}{\pi\mu N_A}\right)^{1/2} \qquad = \left(\frac{8\pi kT}{\mu}\right)^{1/2} (R_A + R_B)^2$$

The difference between these expressions of a factor of N_A is simply a matter of units.

20.6 What is the expression for the pre-exponential factor of a second order gas reaction according to the thermodynamic formulation of transition-state theory? What is the value of the pre-exponential factor at 500 K if the entropy of activation is zero? When this value is compared with values in Table 19.3, what do you conclude about the sign of $\Delta S^{\ddagger}$?

SOLUTION

$$A = \frac{e^2RT}{c^oN_Ah}\frac{e^{\Delta S^{\ddagger}}}{R} = \frac{(2.718)^2(8.314\ \text{J K}^{-1}\ \text{mol}^{-1})(500\ \text{K})}{(1\ \text{mol L}^{-1})(6.022 \times 10^{23}\ \text{mol}^{-1})(6.63 \times 10^{-34}\ \text{J s})}$$

$$= 7.70 \times 10^{13}\ \text{L mol}^{-1}\ \text{s}^{-1}$$

$\log\ (A/\text{L mol}^{-1}\ \text{s}^{-1}) = 13.9$

$\Delta S^{\ddagger}$ is negative for all of these reactions, and is more negative for metathesis reactions not involving atoms. This means that the transition state has a lower entropy than the reactants.

20.7 F. W. Schuler and G. W. Murphy [*J. Am. Chem. Soc.* **72**, 3155 (1950)] studied the thermal rearrangement of vnyl allyl ether to allyl acetaldehyde in the range 150 to 200 °C andi found that

$$k = 5 \times 10^{11}\ e^{-128,000/RT}$$

where k is in s^{-1} and the activation energy is in J mol^{-1}. Calculate (a) the enthalpy of activation, and (b) the entropy of activation, and (c) give an interpretation of the latter.

SOLUTION

The values of $\Delta H^{\ddagger}$ and $\Delta S^{\ddagger}$ are calculated for the mean temperature of the range 273 + 175 = 448 K.

(a) $\Delta H^{\ddagger} = E_a - RT$

$= 128.0 \text{ kJ mol}^{-1} - (8.314 \times 10^{-3} \text{ kJ K}^{-1} \text{ mol}^{-1})(448 \text{ K})$

$= 124.3 \text{ kJ mol}^{-1}$

(b) $\Delta S^{\ddagger} = R \, ln \, (AN_Ah/eRT)$

$$\frac{AN_Ah}{eRT} = \frac{(5 \times 10^{11} \text{ s}^{-1})(6.02 \times 10^{23} \text{ mol}^{-1})(6.63 \times 10^{-34} \text{ J s})}{(2.718)(8.314 \text{ J K}^{-1} \text{ mol}^{-1})(448 \text{ K})}$$

$= 1.97 \times 10^{-2}$

$\Delta S^{\ddagger} = (8.314 \text{ J K}^{-1} \text{ mol}^{-1}) \, ln \, 1.97 \times 10^{-2}$

$= -32.6 \text{ J K}^{-1} \text{ mol}^{-1}$

(c) The transition state has a more organized structure than the reactant, perhaps a ring structure.

20.8 Table 20.1 shows that the pre-exponential factor for the reaction

$$CD_3 + CH_4 \rightarrow CH_3H + CH_3$$

is 1×10^{11} cm^3 mol^{-1} s^{-1}. Calculate $\Delta S^{\ddagger}$ for this elementary bimolecular reaction at 300 K. How do you account for the sign of $\Delta S^{\ddagger}$?

SOLUTION

For a bimolecular reaction

$$A = \frac{e^2RT}{N_Ahc^o} e^{\Delta S^{\ddagger}/R}$$

$$\Delta S^{\ddagger} = R \ln \frac{AN_Ahc^o}{e^2RT}$$

The safest way to make the calculation is to convert A and c^o to SI base units.

$$A = (1 \times 10^{11} \text{ cm}^3 \text{ mol}^{-1} \text{ s}^{-1})(10^{-2} \text{ m cm}^{-1})^3$$

$$= 1 \times 10^5 \text{ m}^3 \text{ mol}^{-1}\text{s}^{-1}$$

$$c^o = (1 \text{ mol L}^{-1})(10^3 \text{ L m}^{-3}) = 10^3 \text{ mol m}^{-3}$$

$$\frac{AN_Ahc^o}{RTe^2} = \frac{(10^5 \text{ m}^3 \text{ mol}^{-1} \text{ s}^{-1})(6.62 \times 10^{-34} \text{ J s})(10^3 \text{ mol m}^{-3})}{(8.314 \text{ J K}^{-1} \text{ mol}^{-1})(300 \text{ K})(2.718)^2}$$

$$= 2.16 \times 10^{-6}$$

$$\Delta S^{\ddagger} = 1(8.314 \text{ J K}^{-1} \text{ mol}^{-1}) \ln 2.16 \times 10^{-6}$$

$$= -108 \text{ J K}^{-1} \text{ mol}^{-1}$$

There is a decrease in entropy when two molecules of gas come together to make a transition complex.

20.9 For the unimolecular dissociation of ethane to methyl radicals

$$C_2H_6(g) \underset{k_b}{\overset{k_f}{\rightleftarrows}} 2CH_3(g)$$

In the temperature range 800-1000 K and about 1 bar, it has been found that

$$-\frac{d[C_2H_6]}{dt} = k_f[C_2H_6] - k_b[CH_3]^2$$

k_f = $(10^{16}\ s^{-1})\ \exp\ (-360\ kJ\ mol^{-1}/RT)$

k_b = $1.19 \times 10^{10}\ L\ mol^{-1}\ s^{-1}$

At 1000 K, calculate $\Delta S^{\ddagger}$ for (a) the forward reaction and (b) the backward reaction. (c) How do you interpret these values? (d) Calculate ΔS° for the overall reaction. This is different from the value of ΔS° you would calculate from data on $K_P = (P_{CH_3}/P^{\circ})^2/(P_{C_2H_6}/P^{\circ})$ at a series of temperatures. Why?

<u>SOLUTION</u>

(a) For a first order reaction

$$\Delta S^{\ddagger} = R \ln \left(\frac{AN_Ah}{eRT}\right)$$

$$= (8.314\ J\ K^{-1}\ mol^{-1})\ ln \frac{(10^{16})(6.02 \times 10^{23})(6.63 \times 10^{-34})}{(2.718)(8.315)(1000\ )}$$

$$= 43\ J\ K^{-1}\ mol^{-1}$$

(b) For a second order reaction

$$\Delta S^{\ddagger} = R \ln \frac{AN_Ahc^{o}}{e^2RT}$$

$$= (8.314\ J\ K^{-1}\ mol^{-1})\ ln \frac{(1.19 \times 10^{10})(6.02 \times 10^{23})(6.63 \times 10^{-34})}{(2.718)^2(8.314)(1000)}$$

$$= -79\ J\ K^{-1}\ mol^{-1}$$

Note that c^o has been taken as 1 mol L^{-1}, which cancels the liter unit in the rate constant.

(c) There is an increase in S in forming the transition state from C_2H_6 in otherwords, a loosening of the structure. There is a decrease in entropy in forming the transition state from two gas molecules; in otherwords, a loss in degrees of freedom.

(d) $C_2H_6 \rightarrow C_2H_6^{\ddagger}$ $\Delta S^{\ddagger} = 43$ J K^{-1} mol^{-1}

$2CH_3 \rightarrow C_2H_6^{\ddagger}$ $\Delta S^{\ddagger} = -79$ J K^{-1} mol^{-1}

$C_2H_6 = 2CH_3$ $\Delta S^{\circ} = 122$ J K^{-1} mol^{-1}

This is the correct entropy change when the standard state is taken to be 1 mol L^{-1}.

20.10 A certain photochemical reaction requires an excitation energy of 12 kJ mol^{-1}. To what value does this correspond in the following unit (a) frequency of light, (b) wave number, (c) wavelength in nanometer and (d) electron volts?

SOLUTION

$$\text{(a) } \nu = \frac{E}{h} = \frac{1.26 \times 10^{5} \text{ J mol}^{-1}}{(6.63 \times 10^{-34} \text{ J s})(6.02 \times 10^{23} \text{ mol}^{-1})}$$

$$= 3.16 \times 10^{14} \text{ s}^{-1}$$

$$\text{(b) } \tilde{\nu} = \frac{\nu}{c} = \frac{(3.16 \times 10^{14} \text{ s}^{-1})(10^{-2} \text{ m cm}^{-1})}{2.998 \times 10^{8} \text{ m s}^{-1}}$$

$$= 10{,}500 \text{ cm}^{-1}$$

$$\text{(c) } \lambda = \frac{1}{\tilde{\nu}} = \frac{1}{10500 \text{ cm}^{-1}} = 9.52 \times 10^{-5} \text{ cm} = 925 \text{ nm}$$

(d) $\dfrac{126 \text{ kJ mol}^{-1}}{96485 \text{ C mol}^{-1}} = 1.31 \text{ eV}$

20.11 How may moles of photons does a laser with an intensity of 0.1 watt at 560 nm produce in one hour?

SOLUTION

$$\frac{E\lambda}{N_A hc} = \frac{(0.1 \text{ J s}^{-1})(60 \times 60 \text{ s})(560 \times 10^{-9} \text{ m})}{(6.022 \times 10^{23} \text{ mol}^{-1})(6.626 \times 10^{-34} \text{ J s})(2.998 \times 10^{8} \text{ m s}^{-1})}$$

$= 1.7 \times 10^{-3}$ mol

20.12 A sample of gaseous acetone is irradiated with monochromatic light having a wavelength of 313 nm. Light of this wavelength decomposed the acetone according to the equation

$$(CH_3)_2CO \rightarrow C_2H_6 + CO$$

The reaction cell used has a volume of 59 cm^3. The acetone vapor absorbs 91.5% of the incident energy. During the experiment the following data are obtained.

Temperature of reaction	= 56.7 °C
Initial pressure	= 102.16 kPa
Final pressure	= 104.42 kPa
Time of radiation	= 7 hr
Incident energy	= 48.1×10^{-4} J s^{-1}

What is the quantum yield?

SOLUTION

Since PV = nRT, $V\Delta P = RT\Delta n$

$$\Delta n = \frac{V\Delta P}{RT} = \frac{(59 \times 10^{-6} \text{ m}^{-3})[(104.42 - 102.16) \times 10^{3} \text{ Pa}]}{(8.314 \text{ J K}^{-1} \text{ mol}^{-1})(329.85 \text{ K})}$$

$= 4.86 \times 10^{-5}$ mol

$= (4.86 \times 10^{-5}$ mol$)(6.02 \times 10^{23}$ mol$^{-1})$

$= 2.93 \times 10^{19}$ molecules reacting

$$E = \frac{hc}{\lambda} = \frac{(6.626 \times 10^{-34} \text{ J s})(2.988 \times 10^{8} \text{ m s}^{-1})}{313 \times 10^{-9} \text{ m}})$$

$= 6.36 \times 10^{-19}$ J

Number of quanta absorbed

$$= \frac{48.6 \times 10^{-5} \text{ mol}}{(7 \text{ hr})(60 \text{ min hr}^{-1})(60 \text{ s min}^{-1})(0.059 \text{ L})} = 3.27 \times 10^{-8} \text{ mol L}^{-1} \text{ s}^{-1}$$

$$I_a = \frac{(4.81 \times 10^{-3} \text{ J s}^{-1})(0.915)}{(6.36 \times 10^{-19} \text{ J})(6.022 \times 10^{23} \text{ mol}^{-1})(0.059 \text{ L})}$$

$= 1.95 \times 10^{-7}$ mol L^{-1} s^{-1}

$\phi = v/I_a = (3.7 \times 10^{-8}$ mol L^{-1} s$^{-1})/(1.95 \times 10^{-7}$ mol L^{-1} s^{-1}

$= 0.167$

20.13 A 100-cm^3 vessel containing hydrogen and chlorine was irradiat with light of 400 nm. Measurements with a thermopile showed that x 10^{-7} J of light energy was absorbed by the chlorine per secon During an irradiation of 1 min the partial pressure of chlorine, determined by the absorption of light and the application of Beer's la decreased from 27.3 to 20.8 kPa (corrected to 0 °C). What is t quantum yield?.

SOLUTION

$$E = \frac{hc}{\lambda} = \frac{(6.626 \times 10^{-34}\ J\ s)(2.998 \times 10^{8}\ m\ s^{-1})}{400 \times 10^{-9}\ m}$$

$$= 4.97 \times 10^{-19}\ J$$

$$I_a = \frac{(11 \times 10^{-7}\ J\ s^{-1})}{(4.97 \times 10^{-19}\ J)\ (6.02 \times 10^{23}\ mol^{-1})(0.100\ L)}$$

$$= 3.68 \times 10^{-11}\ mol\ L^{-1}\ s^{-1}$$

Since $PV = nRT$, $V\Delta P = RT\Delta n$

$$N_A \Delta N = \frac{V\Delta P}{RT} = \frac{(6.02 \times 10^{23}\ mol^{-1})(0.1 \times 10^{-3}\ m^3)[(27.3 - 20.8) \times 10^3\ Pa]}{(8.314\ J\ K^{-1} mol^{-1})(273\ K)}$$

$= 1.72 \times 10^{20}$ molecules of Cl_2 react.

Since $H_2 + Cl_2 = 2\ HCl$

$$\Delta n = V\Delta P/RT$$

$$= \frac{(0.1 \times 10^{-3}\ m^3)[(27.3 - 20.8) \times 10^3\ Pa]}{(8.314\ J\ K^{-1}\ mol^{-1})(273\ K)}$$

$$= 2.86 \times 10^{-4}\ mol$$

$$v = \frac{(2.86 \times 10^{-4}\ mol)}{(60\ s)(0.100\ L)} = 4.77 \times 10^{-5}\ mol\ L^{-1}\ s^{-1}$$

$$\phi = \frac{v}{I_a} = \frac{4.77 \times 10^{-5}\ mol\ L^{-1}\ s^{-1}}{3.68 \times 10^{-11}\ mol\ L^{-1}\ s^{-1}} = 1.3 \times 10^{6}$$

$$= -\frac{1}{I_a}\frac{d[Cl_2]}{dt} = \frac{1}{2I_a}\frac{d[HCl]}{dt}$$

20.14 Show that if a solute follows the Beer-Lambert law, the intensity of absorbed radiation I_a in moles of photons per unit volume per second is given by

$$I_a = \frac{I_a}{\ell N_A h\nu}(1-e^{-\kappa cl})$$

where ℓ is the length of the cell in the direction of the incident monochromatic radiation and I_o is in energy per unit area per unit time.

SOLUTION

Consider that monochromataic radiation of intensity I_o is perpendicular to the face of a reaction cell of area A and optical path length ℓ. The rate that energy enters the cell is I_oA, and the rate that energy leaves the cell is IA, where $I = I_o \exp(-\kappa cl)$. The rate I_a of absorption of electromagnetic radiation in moles of photons per unit volume per second is given by

$$I_a = \frac{I_oA - IA}{VN_Ah\nu} = \frac{I_o}{\ell N_Ah\nu}(1 - e^{-\kappa cl})$$

If the cell is thick enough that the radiation is essentially all absorbed, $I_a = I_o/\ell N_a h\nu$.

20.15 When CH_3I molecules in the vapor state absorb 253.7 nm light, they dissociate into methyl radicals and iodine atoms. The energy required to rupture the C-I bond is 209 kJ mol^{-1}. What are the velocities of the iodine atom and the methyl radical, assuming all of the excess energy goes into translational motion?

SOLUTION

$$E = \frac{hc}{\lambda} = \frac{(6.626 \times 10^{-34}\ \text{J s})(2.998 \times 10^{8}\ \text{m s}^{-1})}{(253.7 \times 10^{-9}\ \text{m})}$$

$$= 7.83 \times 10^{-19}\ \text{J}$$

$$\text{Excess energy} = 7.83 \times 10^{-19} - \frac{209000}{6.022 \times 10^{23}}$$

$$= 4.36 \times 10^{-19}\ \text{J}$$

$$= \frac{1}{2} m_I v_I^2 + \frac{1}{2} m_{CH_3} v_{CH_3}^2$$

But the momenta of the fragments have to be equal and opposite.

$$m_I v_I = m_{CH_3} v_{CH_3}$$

$$4.36 \times 10^{-19} = \frac{1}{2} m_I c_I^2 + \frac{1}{2} m_{CH_3} \left(\frac{m_I}{m_{CH_3}}\right)^2 v_I^2$$

$$= \frac{1}{2} m_I c_I^2 \left(1 + \frac{m_I}{m_{CH_3}}\right)$$

$$m_I = \frac{126.9 \times 10^{-3}\ \text{kg mol}^{-1}}{6.022 \times 10^{23}\ \text{mol}^{-1}} = 2.107 \times 10^{-25}\ \text{kg}$$

$$m_{CH_3} = \frac{15.0 \times 10^{-3}\ \text{kg mol}^{-1}}{6.022 \times 10^{23}\ \text{mol}^{-1}} = 2.491 \times 10^{-26}\ \text{kg}$$

$$1 + \frac{m_I}{m_{CH_3}} = 9.459$$

$$v_I = \left[\frac{2(4.36 \times 10^{-19}\ \text{J})}{(2.107 \times 10^{-25}\ \text{kg})(9.459)}\right]^{1/2} = 662\ \text{m s}^{-1}$$

$$v_{CH_3} = 8.459\ v_I = 5595\ \text{m s}^{-1}$$

20.16 The phosphorescence of butyrophenone in acetonitrile is quenched by 1,3-pentadiene (P). The following quantum yields were measured at 25 °C (see N. J. Turro, *Modern Molecular Photochemistry*, The Benjamin/Cummings Publ. Co., Menlo Park, CA (1978), p 248).

$[P]/10^{-3}$ mol L^{-1}	0	1.0	2.0
ϕ/ϕ_o	1	0.61	0.43

Assuming that the quenching reaction is diffusion controlled and the rate constant has a value of 10^{10} L mol^{-1} s^{-1}, what is the lifetime of the triplet state?

SOLUTION

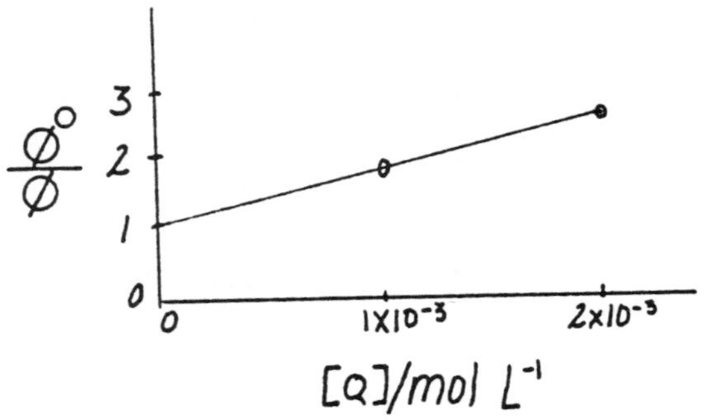

$$\text{slope} = k_Q\tau_{T_1} = \frac{1.33}{2 \times 10^{-3}\ \text{mol L}^{-1}} = 640\ \text{L mol}^{-1}$$

$$\tau_{T_1} = \frac{640\ \text{L mol}^{-1}}{10^{10}\ \text{L mol}^{-1}\ \text{s}^{-1}} = 6.4 \times 10^{-8}\ \text{s}$$

20.17 Biacetyl triplets have a quantum yield of 0.25 for phosphorescence and a measured lifetime of the triplet state of 10^{-3} s. If its phosphorescence is quenched by a compound Q with a diffusion-controlled rate (10^{10} L mol^{-1} s^{-1}), what concentration of Q is required to cut the phosphorescence yield by half?

SOLUTION

$$\frac{I_P^o}{I_P} = 1 + k_Q \tau_{T_1}[Q]$$

$$2 = 1 + (10^{10}\ \text{L mol}^{-1}\ \text{s}^{-1})(10^{-3}\ \text{s})\ [Q]$$

$$[Q] = 10^{-7}\ \text{mol L}^{-1}$$

20.18 For 900 s, light of 436 nm was passed into a carbon tetrachloride solution containing bromine and cinnamic acid. The average power absorbed was 19.2 x 10^{-4} J s^{-1}. Some of the bromine reacted to give cinnamic acid dibromide, and in this experiment the total bromine content decrease by 3.83 x 10^{19} molecules. (a) What was the quantum yield? (b) State whether or not a chain reaction was involved.

SOLUTION

(a) $$E = \frac{h}{\lambda} = \frac{(6.63 \times 10^{-34}\ \text{J s})(3 \times 10^{8}\ \text{m s}^{-1})}{(436 \times 10^{-9}\ \text{m})}$$

$$= 4.56 \times 10^{-19}\text{J}$$

$$I_a = \frac{19.2 \times 10^{4}\ \text{J s}^{-1}}{(4.56 \times 10^{-19}\ \text{J})(6.02 \times 10^{13}\ \text{mol}^{-1})(0.500\ \text{L})}$$

$$= 1.40 \times 10^{-8}\ \text{mol L}^{-1}\ \text{s}^{-1}$$

$$v = \frac{6.36 \times 10^{-5}\ \text{mol}}{(900\ \text{s})(0.500\ \text{L})}$$

$$\phi = \frac{v}{I_a} = \frac{1.41 \times 10^{-7} \text{ mol L}^{-1} \text{ s}^{-1}}{1.40 \times 10^{-8} \text{ mol L}^{-1} \text{ s}^{-1}} = 10.1$$

(b) Since $\phi > 1$, a chain reaction is involved.

20.19 The following calculations are made on a uranyl oxalate actinometer, on the assumption that the energy of all wavelengths between 254 and 435 nm is completely absorbed. The actinometer contains 20 cm^3 of 0.05 mol L^{-1} oxalic acid, which also is 0.01 mol L^{-1} with respect to uranyl sulfate. After 2 hr of exposure to ultraviolet light, the solution required 34 cm^3 of potassium permanganate, $KMnO_4$, solution to titrate the undecomposed oxalic acid. The same volume, 20 cm^3, of unilluminated solution required 40 cm^3 of the $KMnO_4$ solution. If the average energy of the quanta in this range may be taken as corresponding to a wavelength of 350 nm, how many joules were absorbed per second in this experiment? (ϕ = 0.57)

SOLUTION

Moles of oxalic acid decomposed $= (0.02 \text{ L})(0.05 \text{ mol L}^{-1}) \left(\frac{40-34}{40}\right)$

$= 1.5 \times 10^{-4}$ mol

Moles of photons required per second $= \frac{1.5 \times 10^{-4} \text{ mol}}{(2 \times 60 \times 60 \text{ s})(0.57)}$

$= 3.65 \times 10^{-8}$ mol s^{-1}

$$E = \frac{hc}{\lambda} = \frac{(6.266 \times 10^{-34}\ \text{J s})(2.998 \times 10^{8}\ \text{m s}^{-1})}{3.5 \times 10^{-7}\ \text{m}}$$

$= 5.68 \times 10^{-19}$ J

Energy flux $= (3.65 \times 10^{-8}\ \text{mol}^{-1}\ \text{s}^{-1})(6.02 \times 10^{23}\ \text{mol}^{-1})(5.68 \times 10^{-19}\ \text{J})$

$= 1.25 \times 10^{-2}$ J s^{-1}

20.20 A solution of a dye is irradiated with 400 nm light to produce a steady concentration of triplet state molecules. If the triplet state quantum yield is 0.9, and the triplet state lifetime is 20 x 10^{-6} s, what light intensity, expressed in watts per liter, is required to maintain a steady triplet concentration of 5 x 10^{-6} mol L^{-1} in a liter of solution. Assume that all of the light is absorbed.

<u>SOLUTION</u>

$$\frac{d[T_1]}{dt} = 0.9\ I - \frac{1}{20 \times 10^{-6}}[T_1] = 0$$

$$I = \frac{(5 \times 10^{-6}\ \text{mol L}^{-1})}{(20 \times 10^{-6}\ \text{s})(0.9)} = 0.278\ \text{mol L}^{-1}\ \text{s}^{-1}$$

$$E = \frac{N_A hc}{\lambda} = \frac{(6.022 \times 10^{23}\ \text{mol}^{-1})(6.62 \times 10^{-34}\ \text{J s})(3 \times 208\ \text{m s}^{-1})}{400 \times 10^{-9}\ \text{m}}$$

$= 2.99 \times 10^{5}$ J mol^{-1}

$I = (0.278\ \text{mol L}^{-1}\ \text{s}^{-1})(2.99 \times 10^{5}\ \text{J mol}^{-1}) = 83\ \text{kW L}^{-1}$

20.21 The photochemical chlorination of chloroform

$$CHCl_3 + Cl_2 = CCl_4 + Hl$$

is believed to proceed by the following mechanism.

$$Cl_2 + h\nu \xrightarrow{I_a} 2Cl$$

$$Cl + CHCl_3 \xrightarrow{k_1} CCl_3 + HCl$$

$$CCl_3 + Cl^2 \xrightarrow{k_2} CCl_4 + Cl$$

$$2CCL_3 + Cl_2 \xrightarrow{k_3} 2CCl_4$$

Derive the steady state rate law for the production of carbon tetrachloride.

SOLUTION

$$\frac{d[CCl_4]}{dt} = k_2[CCl_3][Cl_2 + 2k_3[CCl_3]^2[Cl_2]$$

$$\frac{d[CCl_3]}{dt} = k_1[Cl][CHCl_3] - k_2[CCl_3][Cl_2] - 2k_3[CCl_3]^2[Cl_2] = 0$$

$$\frac{d[Cl]}{dt} = 2I_a - k_1[Cl][CHCl_3] + k_2[CCl_3][Cl_2] = 0$$

Adding the two steady state expressions yields

$$I_a = k_3[CCl_3]^2[Cl_2]$$

This makes it possible to eliminate $[CCl_3]$ from the first equation to obtain

$$\frac{d[CCl_4]}{dt} = k_2 I_a^{1/2} [Cl_2]^{1/2}/k_3^{1/2} + 2I_a)$$

20.22 Sunlight between 290 and 313 nm can produce sunburn (erythema) in 30 min. The intensity of radiation between these wavelengths in summer and at 45° latitude is about 50 μW cm^{-2}. Assuming that 1 photon produces chemical change in 1 molecule, how many molecules in a square centimeter of human skin must be photochemically affected to produce evidence of sunburn?

SOLUTION

$$E \frac{hc}{\lambda} = \frac{(6.63 \times 10)(3 \times 10^{8})}{3 \times 10^{-7}} = 6.63 \times 10^{-19} \text{ J}$$

$$(50 \times 10^{-6} \text{ J s}^{-1} \text{ cm}^{-2})(30 \times 60 \text{ s}) = 0.09 \text{ J cm}^{-2}$$

$$\frac{0.09}{6.63 \times 10^{-19}} = 1.36 \times 10^{17} \text{ molecules cm}^{-2}$$

20.23	4.87×10^{11} L mol^{-1} s^{-1}, 0.21
20.24	5.69×10^{11} L mol^{-1} s^{-1}
20.26	(a) 159.9 kJ mol^{-1}, (b) 59.4 J K^{-1} mol^{-1}
20.27	(a) 224 kJ mol^{-1}, (b) 18.0 J K^{-1} mol^{-1}
20.28	(a) 6.4×10^{14} s^{-1}, (b) 1.2×10^{-8} s^{-1}
20.29	2.51×10^{-6} mol s^{-1}
20.30	(a) 0.1709 J s^{-1}, (b) 0.3988 J s^{-1}, (c) 0.1709 watt, (d) 0.3988 watt
20.31	(a) $1.55, (b) $0.02
20.32	1.27×10^{-3} mol
20.33	26.9 m
20.34	5×10^{-9} mol s^{-1}
20.35	370 ns
20.36	9×10^{-9} mol L^{-1}
20.37	21.8 hr
20.38	47.1 s
20.39	0.050 J s^{-1}
20.40	(a) 10^{14} quanta, (b) 4.5×10^{-9} g day^{-1}
20.41	617 g m^{-2}
20.42	700 W m^{-2}
20.43	10^{-3}

21.1 A steel ball (ρ= 7.86 g cm^{-3}) 0.2 cm in diameter falls 10 cm through a viscous liquid (ρ_o = 1.50 cm^{-3}) in 25 s. What is the viscosity at this temperature?

SOLUTION

$$\eta = \frac{2r^2(\rho - \rho_o)g}{9\frac{dx}{dt}}$$

$$= \frac{2(1 \times 10^{-3}\ m)^2[(7.86 - 1.50) \times 10^3\ kg\ m^{-3}](9.8\ m\ s^{-2})}{9(\frac{0.10\ m}{25\ s})}$$

$$= 3.46\ Pa\ s$$

21.2 Estimate the rate of sedimentation of water droplets of 1 μm diameter in air at 20 °C. The viscosity of air at this temperature is 1.808 x 10^{-5} Pa s.

SOLUTION

$$\frac{dx}{dt} = \frac{2r^2(\rho - \rho_o)g}{9\eta}$$

$$= \frac{2(0.5 \times 10^{-6}\ m)^2(0.998 \times 10^3\ kg\ m^{-3})(9.8\ m\ s^{-2})}{9(1.808 \times 10^{-5}\ Pa\ s)}$$

$$= 3.01 \times 10^{-5}\ m\ s^{-1}$$

21.3 The viscosity of mercury is 1.661 x 10^{-3} Pa s at 0 °C and 1.476 x 10^{-3} Pa s at 35 °C.. What is the activation energy, and what viscosity is expected at 50 °C?

<u>SOLUTION</u>

$$ln \frac{\eta_1}{\eta_2} = \frac{E_a(T_2 - T_1)}{RT_1T_2}$$

$$E_a = \frac{RT_1T_2}{(T_2-T_1)} \ln \frac{\eta_1}{\eta_2}$$

$$= \frac{(8.314 \text{ J K}^{-1} \text{ mol}^{-1})(273.15 \text{ K})(308.15\text{K})}{35 \text{ K}} ln \frac{1.661}{1.476}$$

$$= 2.36 \text{ kJ mol}^{-1}$$

$$A = \frac{1}{\eta} e^{E_a/RT}$$

$$= \frac{e^{2360/(8.314)(273.15)}}{1.661 \times 10^{-3} \text{ Pa s}}$$

$$= 1.70 \times 10^{3} \text{ Pa}^{-1} \text{ s}^{-1}$$

At 50 °C

$$\eta = \frac{e^{2360/(8.314)(323.15)}}{1.70 \times 10^{3} \text{ Pa}^{-1} \text{ s}^{-1}}$$

$$= 1.41 \times 10^{-3} \text{ Pa s}$$

21.4 A sharp boundary is formed between a dilute aqueous solution of sucrose and water at 25 °C. After 5 hr the standard deviation of the concentration gradient is 0.434 cm. (a) What is the diffusion coefficient for sucrose under these conditions? (b) What will be the standard deviation after 10 hr?

SOLUTION

(a) $$D = \frac{\sigma^2}{2t} = \frac{(4.34 \times 10^{-3}\ \text{m})^2}{2(5 \times 60 \times 60\ \text{s})} = 5.23 \times 10^{-10}\ \text{m}^2\ \text{s}^{-1}$$

(b) $$\sigma = \sqrt{2Dt} = \sqrt{2(5.23 \times 10^{-10}\ \text{m}^2\ \text{s}^{-1})(10 \times 60 \times 60\ \text{s})}$$

$$= 6.14 \times 10^{-3}\ \text{m}$$

21.5 (a) Calculate the time required for the half-width of a freely diffusing boundary of dilute potassium chloride in water to become 0.5 cm at 25 °C ($D = 1.99 \times 10^{-9}$ m^2 s^{-1}). (b) Calculate the corresponding time for serum albumin ($D = 6.15 \times 10^{-11}$ m^2 s^{-1}).

SOLUTION

(a) $$D = \sigma^2/2t$$

$$t = \frac{\sigma^2}{2D} = \frac{(0.005\ \text{m})^2}{2(1.99 \times 10^{-9}\ \text{m}^2\ \text{s}^{-1})}$$

$$= 6.28 \times 10^3\ \text{s} = 1.75\ \text{h}$$

(b) $$t = \frac{\sigma^2}{2D} = \frac{(0.005)^2}{2(6.15 \times 10^{-11}\ \text{m}^2\ \text{s}^{-1})}$$

$= 2.03 \times 10^5 = 56.5$ h

21.6 The standard deviation σ of a freely diffusing boundary between dilute salt solution and water at 25 °C is 3.8 nm after 1 hour. What is the diffusion coefficient of the salt in water? What will the standard deviation be after 2 hours?

SOLUTION

$$D = \frac{\sigma^2}{2t} = \frac{(0.38\ \text{cm})^2}{2(3600\ \text{s})} = 2.00 \times 10^{-5}\ \text{cm}^2\ \text{s}^{-1}$$

$$\sigma = \sqrt{2DT} = \sqrt{2(2.00 \times 10^{-5}\ \text{cm}^2\ \text{s}^{-1})(7200\ \text{s})}$$

$$= 0.54\ \text{cm}$$

21.7 Using a table of the probability integral, calculate enough points on a plot of c versus x (like Fig. 21.2c) to draw in the smooth curve for diffusion of 0.1 mol L^{-1} sucrose into water at 25 °C after 4 hr and 29.83 min ($D = 54.23 \times 10^{-10}\ m^2\ s^{-1}$).

SOLUTION

Equation 21.18

$$c = \frac{c_o}{2}\left[1 + \frac{2}{\sqrt{\pi}} \int_o^{x/2\sqrt{Dt}} e^{-\beta^2}\, d\beta\right]$$

in order to get this equation in the form of the normal probability function let $\beta^2 = t^2/2$. Thus $\beta = t/2^{1/2}$ and $d\beta = dt/2^{1/2}$.

$$c = \frac{c_o}{2}\left[1 + \frac{2}{\sqrt{2\pi}}\int_o^{x/\sqrt{2Dt}} e^{-t^2/2}\, dt\right]$$

$$= \frac{c_o}{2}\left[1 + \frac{1}{\sqrt{2\pi}}\int_{-x/\sqrt{2Dt}}^{x/\,x/\sqrt{2Dt}} e^{-t^2/2}\, dt\right]$$

This area is obtainable from tables of the normal probability integral (N.P.I.)

$$\frac{c_o}{2} = [1 \pm \text{Area of N.P.I.}]$$

The + sign is used when $x > 0$, and the - sign is used when $x < 0$.

At x = 1 cm the Gaussian parameter is

$$\frac{x}{\sqrt{2Dt}} = \frac{10^{-2}\ \text{m}}{\sqrt{2(5.23 \times 10^{-10}\ \text{m}^2\ \text{s}^{-1})(269.83 \times 60\ \text{s})}} = 2.43$$

$\lvert x \rvert$	$\frac{x}{\sqrt{2Dt}}$	Area N.P.I.	$c(x < 0)$	$c(x > 0)$
0	0	0	0.05 mol L^{-1}	0.05 mol L^{-1}
0.1	0.243	0.1922	0.040	0.060
0.2	0.486	0.3732	0.031	0.069
0.3	0.729	0.5340	0.023	0.077
0.4	0.972	0.6690	0.016	0.084
0.6	1.458	0.8556	0.007	0.093
0.8	1.944	0.9480	0.003	0.097

1.0	2.43	0.9850	0.002	0.098
1.2	2.916	0.9964	0.001	0.099

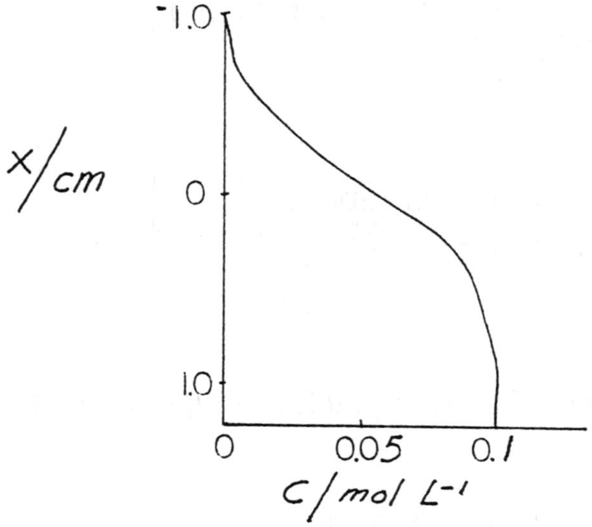

21.8 Calculate the conductivity of 0.001 mol L^{-1} HCl at 25 °C. The limiting ion mobilities may be used for this problem.

SOLUTION

$$\kappa = Fc(u_{H+} + u_{Cl-})$$

$$= (96{,}485\ C\ mol^{-1})(1\ mol\ m^{-3})[(36.25 + 7.91) \times 10^{-8}\ m^2\ V^{-1}\ s^{-1}]$$

$$= 0.0426\ \Omega^{-1}\ m^{-1}$$

21.9 One hundred grams of sodium chloride is dissolved in 10,000 L of water at 25 °C, giving a solution that may be regarded in these calculations as infinitely dilute. (a) What is the conductivity of the solution? (b) This dilute solution is placed in a glass tube of 4-cm diameter provided with electrodes filling the tube and placed 20 cm apart. How much current will flow if the potential drop between the electrodes is 80 V?

SOLUTION

(a) $$c = \frac{(100\ g)/(58.5\ g\ mol^{-1})}{(10^7\ cm^3)(10^{-2}\ m\ cm^{-1})^3}$$

$$= 1.71 \times 10^{-1}\ mol\ m^{-3}$$

$$\kappa = (96{,}485\ C\ mol^{-1})(1.71 \times 10^{-1}\ mol\ m^{-3})(13.105 \times 10^{-8}\ m^2\ V^{-1}\ s^{-1})$$

$$= 2.16 \times 10^{-3}\ \Omega^{-1}\ m^{-1}$$

(b) $$R = \frac{1}{\kappa A} = \frac{0.2\ m}{(2.16 \times 10^{-3}\ \Omega^{-1}\ m^{-1})(2 \times 10^{-2}\ m)^2\pi}$$

$$= 7.36 \times 10^4\ \Omega$$

$$I = \frac{E}{R} = \frac{80\ V}{7.36 \times 10^4\ \Omega} = 1.09 \times 10^{-3}\ A$$

21.10 It is desired to use a conductance apparatus to measure the concentration of dilute solutions of sodium chloride. If the electrodes in the cell are each 1 cm^2 in area and are 0.2 cm apart, calculate the resistance that will be obtained for 1, 10, and 100 ppm NaCl at 25 °C.

SOLUTION

1 ppm $= 1$ g NaCl in 10^6 g H_2O $= \frac{(1\ g)/(58.45\ g\ mol^{-1})}{1\ m^3}$

Using electric mobilities at infinite dilution,

$$\kappa = Fc(u_{Na} + u_{Cl^-})$$

$$= (96{,}485\ C\ mol^{-1})(1.71 \times 10^{-2}\ mol\ m^{-3})[(5.192 + 7.913) \times 10^{-8}\ m\ V^{-1}\ s^{-1}]$$

$= 2.16 \times 10^{-4}\ \Omega^{-1}\ m^{-1}$

$$R = \frac{1}{\kappa A} = \frac{0.2 \times 10^{-2}\ m}{(2.16 \times 10^{-4}\ \Omega^{-1}\ m^{-1})(0.01\ m)^2} = 9.26 \times 10^4\ \Omega$$

For 10 ppm $R = 9.26 \times 10^3\ \Omega$

For 100 ppm $R = 926\ \Omega$

21.11 Show that if A and B can be represented by spheres of the same radius that react when they touch, the second-order rate constant is given by

$$k_a = \frac{8 \times 10^3\ RT}{3\eta}\ L\ mol^{-1}\ s^{-1}$$

where R is in J K^{-1} mol^{-1}. To obtain this result the diffusion coefficient is expressed in terms of the radius of a spherical particle by use of equation 21.12. For water at 25 °C, $\eta = 8.95 \times 20^{-4}$ kg m^{-1} s^{-1}. Calculate k at 25 °C.

SOLUTION

The diffusion coefficient for a spherical particle of radius r is given by

$$D = \frac{RT}{N_A 6\pi\eta r}$$

Substituting in equation 21.30

$$k_a = 4\pi(10^3\ L\ m^{-3})N_A(D_1 + D_2)R_{12}$$

$$= 4\pi(10^3\ L\ m^{-3})(\frac{2RT}{6\pi\eta r})(2r)$$

$$= \frac{8 \times 10^3\ RT}{3\eta}\ L\ mol^{-1}\ s^{-1}$$

where RT is expressed in J mol^{-1} and η is expressed in kg m^{-1} s^{-1}.

For water at 25°C

$$k_a = \frac{8(10^3 \text{ L m}^{-3})(8.314 \text{ J K}^{-1} \text{ mol}^{-1})(298 \text{ K})}{3(8.95 \times 10^{-4} \text{ kg m}^{-1} \text{ s}^{-1})}$$

$= 7.4 \times 10^9 \text{ L mol}^{-1} \text{ s}^{-1}$

21.12 What is the reaction radius for the reaction

$$H^+ + OH^- \xrightarrow{1.4 \times 10^{11} \text{ L mol}^{-1} \text{ s}^{-1}} H_2O$$

at 25 °C given that the diffusion coefficients of H^+ and OH^- at this temperature are 9.1×10^{-9} m^2 s^{-1} and 5.2×10^{-9} m^2 s^{-1}.

SOLUTION

$k_a = 4\pi N_A(D_1 + D_2)R_{12}f$

$$R_{12}f = \frac{k_a}{4\pi N_A(D_1 + D_2)}$$

$$= \frac{(1.4 \times 10^{11} \text{ L mol}^{-1} \text{ s}^{-1})(10^{-3} \text{ m}^3 \text{ L}^{-1})}{4\pi(6.02 \times 10^{23} \text{ mol}^{-1})(14.3 \times 10^{-9} \text{ m}^2 \text{ s}^{-1})}$$

$= 1.29 \times 10^{-9}$ m

To calculate the electrostatic factor f, we have to know R_{12}. Successive approximations have to be made. Let us begin by estimating $R_{12} = 0.9$ nm.

$$x = \frac{z_1 z_2 e^2}{4\pi\varepsilon_0 \kappa k T R_{12}}$$

$$= \frac{-(1.602 \times 10^{-19}\ C)^2 (8.988 \times 10^9\ N\ C^{-2}\ m^2)}{(78.3)(1.38 \times 10^{-23}\ J\ K^{-1})(298\ K)(0.9 \times 10^{-9}\ m)}$$

$$= -0.796$$

$$f = x(e^x - 1)^{-1} = -0.796(e^{-0.796} - 1)^{-1}$$

$$= 1.45$$

Thus $R_{12} = \frac{1.29 \times 10^{-9}\ m}{1.45} = 0.89\ nm$

21.13 For acetic acid in dilute aqueous solution at 25°C, $K = 1.73 \times 10^{-5}$, and the relaxation time is 8.5×10^{-9} seconds for a 0.1 M solution. Calculate k_a and k_d in

$$CH_3CO_2H \underset{k_a}{\overset{k_d}{\rightleftarrows}} CH_3CO_2^- + H^+$$

<u>SOLUTION</u>

$$K = \frac{[H^+][CH_3CO_2^-]}{[CH_3CO_2H]} = \frac{x^2}{0.10} = 1.73 \times 10^{-5} = \frac{k_d}{k_a}$$

$$x = (1.73 \times 10^{-6})^{1/2} = 1.32 \times 10^{-3} = [H^+] = [CH_3CO_2^-]$$

$$\tau = \frac{1}{k_d + k_a([H^+] + [CH_3CO_2^-])}$$

$$8.5 \times 10^{-9}\ s = \frac{1}{1.73 \times 10^{-5}\ k_a + k_a(2.64 \times 10^{-3})}$$

$$k_a = 4.42 \times 10^{10}\ L\ mol^{-1}\ s^{-1}$$

$k_d \quad = (1.73 \times 10^{-5})k_a = 7.65 \times 10^5\ s^{-1}$

21.14 Derive the relation between the relaxation time τ and the rate constants for the reaction $A + B \underset{k_2}{\overset{k_1}{\rightleftarrows}} C + D$, which is subjected to a small displacement from equilibrium.

<u>SOLUTION</u>

$d[C]/dt \quad = k_1[A][B] - k_2[C][D] = 0$

At equilibrium, $0 = k_1[A]_{eq}[B]_{eq} - k_2[C]_{eq}[D]_{eq}$

$[A] \quad = [A]_{eq} - \Delta[C]$

$[B] \quad = [B]_{eq} - \Delta[C]$

$[C] \quad = [C]_{eq} - \Delta[C]$

$[D] \quad = [D]_{eq} - \Delta[C]$

Thus the equation in the first line can be written

$$\frac{d\Delta[C]}{dt} = k_1([A]_{eq} - \Delta[C])([B]_{eq} - \Delta[C]) - k_2([C]_{eq} + \Delta[C])([D]_{eq} + \Delta[C])$$

Multiplying this out, using the equation in the second line, and dropping the $(\Delta[C])^2$ term yields

$$\frac{d\Delta[C]}{dt} = -\frac{\Delta[C]}{\tau}$$

where

$$\frac{1}{\tau} \quad = k_1([A]_{eq} + [B]_{eq}) + k_2([C]_{eq} + [D]_{eq})$$

21.15 Derive the relation between the relaxation time τ and the rate constants for the mechanism.

$$\begin{array}{ccccc} & A & \underset{k_{-1}}{\overset{k_1}{\rightleftarrows}} & B & \\ K_A & \uparrow\downarrow & & \uparrow\downarrow & K_B \\ & A' & \underset{k_{-1}'}{\overset{k_1'}{\rightleftarrows}} & B' & \end{array}$$

which is subjected to a small displacement from equilibrium. It is assumed that the equilibria, $A \rightleftarrows A'$, $K_A = (A')/(A)$, and $B \rightleftarrows B'$, $K_B = (B')/(B)$, are adjusted very rapidly so that these steps remain in equilibrium.

SOLUTION

$[A'] = K_A[A]$, $[B'] = K_B[B]$

$$d[A]/dt = -k_1[A] + k_{-1}[B] - k_1'K_A[A] + k'_{-1}K_B[B]$$

$$= -(k_1 + k_1'K_A)[A] + (k_{-1} + k'_{-1}K_B)[B]$$

At equilibrium, $d[A]/dt = 0 = (-k_1 + k_1'K_A)[A]_{eq} + (k_{-1} + k'_{-1}K_B)[B]_{eq}$

Set $[A] = [A]_{eq} + \Delta[A]$

$[B] = [B]_{eq} - \Delta[A]$

$$\frac{d\Delta[A]}{dt} = -(k_1 + k_1'K_A)\Delta[A] - (k_{-1} + k'_{-1})K_B)\Delta[A] = -\frac{\Delta[A]}{\tau}$$

$$\tau = \frac{1}{k_1 + k_1'K_A + k_{-1} + k'_{-1}K_B}$$

21.16 Calculate the first-order rate constants for the dissociation of the following weak acids: acetic acid, acid form of imidazole $C_3N_2H_5^+$, NH_4^+. The corresponding acid dissociation constants are 1.75 x 10^{-5}, 1.2 x 10^{-7}, and 5.71 x 10^{-10}, respectively. The second-order rate constants for the formation of the acid forms from a proton plus the base are 4.5 x 10^{10}, 1.5 x 10^{10}, and 4.3 x 10^{10} L mol^{-1} s^{-1}, respectively.

<u>SOLUTION</u>

For $HA \underset{k_2}{\overset{k_1}{\rightleftarrows}} H^+ + A^-$

$$K = \frac{[H^+][A^-]}{[HA]} = \frac{k_1}{k_2}$$

For acetic acid $\quad k_1 = k_2K$

$$= (4.5 \times 10^{10}\ L\ mol^{-1}\ s^{-1})(1.75 \times 10^{-5}\ mol\ L^{-1})$$

$$= 7.9 \times 10^5\ s^{-1}$$

For imidazole $\quad k_1 = k_2K$

$$= (1.5 \times 10^{10}\ L\ mol^{-1}\ s^{-1})(1.2 \times 10^{-7}\ mol\ L^{-1})$$

$$= 1.8 \times 10^3\ s^{-1}$$

For NH_4^+ $\quad k_1 = k_2K$

$$= (4.3 \times 10^{10}\ L\ mol^{-1}\ s^{-1})(5.71 \times 10^{-10}\ mol\ L^{-1})$$

$$= 25\ s^{-1}$$

21.17 The solution reaction

$$I^- + OCl^- = OI^- + Cl^-$$

is believed to go by the mechanism

$$OCl^- + H_2O \overset{K_1}{\rightleftarrows} HOCl + OH^- \qquad \text{(fast)}$$

$$I^- + HOCl \overset{k}{\rightarrow} HOI + Cl^- \qquad \text{(slow)}$$

$$HOI + OH^- \overset{K_2}{\rightleftarrows} H_2O + OI^- \qquad \text{(fast)}$$

Derive the rate equation for the forward rate of this reaction that shows the effect of the concentration of OH^-.

SOLUTION

For the slow step,

$$\frac{d[Cl^-]}{dt} = k[I^-][HOCl]$$

Since $K_1 = [HOCl][OH^-]/[OCl^-]$,

$[HOCl] = K_1[OCl^-]/[OH]$

Substituting this in the rate equation for the slow step yields

$$\frac{d[Cl^-]}{dt} = \frac{kK_1[I^-][OCl^-]}{[OH^-]}$$

21.18 The mutarotation of glucose is first order in glucose concentration and is catalyzed by acids (A) and bases (B). The first-order rate constant may be expressed by an equation of the type that is encountered in reactions with parallel paths.

$$k = k_o + k_{H^+}[H^+] + k_A[A] + k_B[B]$$

where k_o is the first-order rate constant in the absence of acids and bases other than water. The following data were obtained by J. H. Brönsted and E. A. Guggenheim [*J. Am. Chem. Soc.* **49**, 2554 (1927)] at 18 °C in a medium containing 0.02 mol L-1 sodium acetate and various concentrations of acetic acid.

$[CH_3CO_2H]$/mol L^{-1}	0.020	0.105	0.199
$k/10^{-4}$ min^{-1}	1.36	1.40	1.46

Calculate k_o and k_A. The term involving k_{H^+} is negligible under these conditions.

SOLUTION

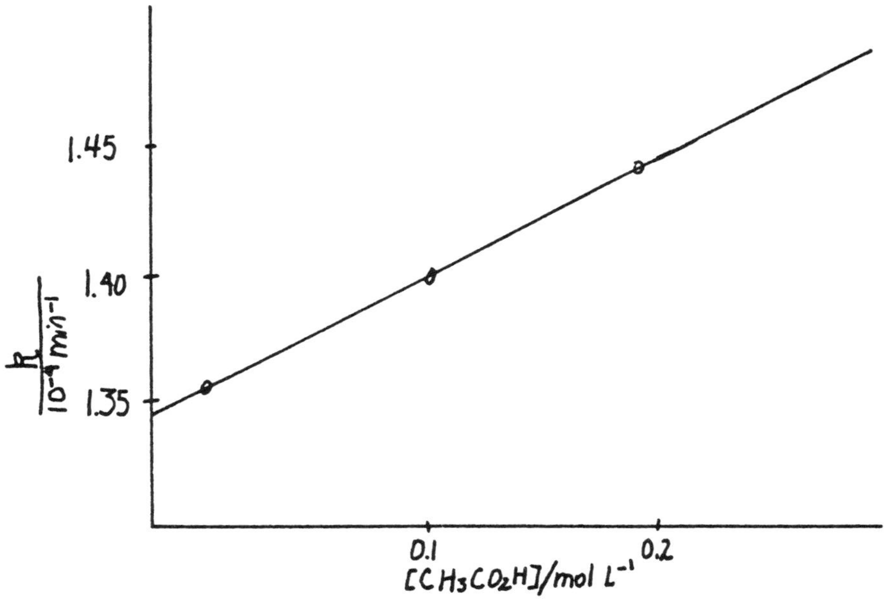

Intercept = k_o = 1.35 x 10^{-4} min^{-1}

Slope $= \dfrac{(1.46 - 1.35) \times 10^{-4}}{0.2 \text{ mol L}^{-1}} = 5.5 \times 10^{-5}$ L mol^{-1} min^{-1} = k_A

21.19 The rate of a reaction between oppositely charged ions is measured at an ionic strength of 0.01 mol L^{-1}. How will the rate be affected if the ionic strength is raised to 0.05 mol L^{-1} if the reaction is (a) $A^+ + B^- \rightarrow$ or (b) $A^{2+} + B^{2+}$?

SOLUTION

(a) $\log k_1 = \log k^o + 2z_Az_B A I_1^{1/2}$

$\log k_2 = \log k^o + 2z_Az_B A I_2^{1/2}$

$\log \frac{k_2}{k_1} = 2z_Az_B A(I_2^{1/2} - I_1^{1/2})$

$\log \frac{k_2}{k_1} = -2[0.509 (\text{mol L}^{-1})^{-1/2}][(0.05 \text{ mol L}^{-1})^{1/2}$

$- [0.01 \text{ mol L}^{-1})^{1/2}]$

$\frac{k_2}{k_1} = 0.748$

Raising the ionic strength reduces k.

(b) $\log \frac{k_2}{k_1} = (2)(2)(2)(0.509)(0.05^{1/2} - 0.01^{1/2})$

$\frac{k_2}{k_1} = 3.20$

Raising the ionic strength increases k.

21.20 Suppose an enzyme has a turnover number of 10^4 min^{-1} and a molar mass of 60,000 g mol^{-1}. How many moles of substrate can be turned over per hour per gram of enzyme if the substrate concentration is twice the Michaelis constant? It is assumed that the substrate concentration is maintained constant by a preceding enzymatic reaction and that products do not accumulate and inhibit the reaction.

SOLUTION

$$v = \frac{V}{1 + K/[S]} = \frac{(10^4\ \text{min}^{-1})[E]_o}{1 + 0.5}$$

$$= \frac{(10^4\ \text{min}^{-1})(1\ \text{g}/60{,}000\ \text{g mol}^{-1})(60\ \text{min hr}^{-1})}{1.5}$$

$$= 6.7\ \text{mol hr}^{-1}$$

21.21 The kinetics of the fumarase reaction fumarate + H_2O = L-malate is studied at 25 °C using an 0.01 ionic strength buffer of pH 7. The rate of reaction is obtained using a recording ultraviolet spectrometer to measure the fumarate concentration. The following rates of the forward reaction are obtained using a fumarase concentration of 5 x 10^{-10} mol L^{-1}.

$[F]/10^{-6}$ mol L^{-1}	$v_F/10^{-7}$ mol L^{-1} s^{-1}
2	2.2
40	5.9

The following rates of the reverse reaction are obtained using a fumarase concentration of 5 x 10^{-10} mol L^{-1}.

$[M]/10^{-6}$ mol L^{-1}	$v_M/10^{-7}$ mol L^{-1} s^{-1}
5	1.3
100	3.6

(a) Calculate the Michaelis constants and turnover numbers for the two substrates. In practice many more concentrations would be studied. (b) Calculate the four rate constants in the mechanism

$$E + F \underset{k_{-1}}{\overset{k_1}{\rightleftarrows}} X \underset{k_{-2}}{\overset{k_2}{\rightleftarrows}} E + M$$

where E represents the catalytic site. There are four catalytic sites per fumarase molecule. (c) Calculate K_{eq} for the reaction catalyzed. The concentration of H_2O is omitted in the expression for the equilibrium constant because its concentration cannot be varied in dilute aqueous solutions.

SOLUTION

(a) $v_F = \dfrac{V_F}{1 + K_F/[F]}.$

$$v_F + \frac{v_F}{[F]} K_F = V_F$$

$$2.2 \times 10^{-7} + 0.110\ K_F = V_F$$

$$5.9 \times 10^{-7} + 0.0148\ K_F = V_F$$

$$-3.7 \times 10^{-7} + 0.0952\ K_F = 0$$

$$K_F = \frac{3.7 \times 10^{-7}}{0.0952} = 3.9 \times 10^{-6} \text{ mol L}^{-1}$$

$$v_F = 2.2 \times 10^{-7} + (0.110)(3.9 \times 10^{-6}) = 6.5 \times 10^{-7} \text{ mol L}^{-1} \text{ s}^{-1}$$

$$v_M = \frac{V_M}{1 + K_M/[M]}$$

$$v_M + \frac{V_M}{[M\}} K_M = V_M$$

$$1.3 \times 10^{-7} + 0.260\ K_M = V_M$$

$$3.6 \times 10^{-7} + 0.0036\ K_M = V_M$$

$$-2.3 \times 10^{-7} + 0.0224\ K_M = 0$$

$$K_M = \frac{2.3 \times 10^{-7}}{0.0224} = 1.03 \times 10^{-5} \text{ mol L}^{-1}$$

$$v_M = 1.3 \times 10^{-7} + 0.026(1.03 \times 10^{-5}) = 4.0 \times 10^{-7} \text{ mol L}^{-1} \text{ s}^{-1}$$

The rate constants should be expressed in terms of the concentration of enzymatic sites and so

$$k_2 = \frac{V_F}{[E]_o} = \frac{6.5 \times 10^{-7} \text{ mol L}^{-1} \text{ s}^{-1}}{4(5 \times 10^{-10} \text{ mol L}^{-1})} = 2.0 \times 10^2 \text{ s}^{-1}$$

$$k_{-1} = \frac{V_M}{[E]_o} = \frac{4.0 \times 10^{-7} \text{ mol L}^{-1} \text{ s}^{-1}}{4(5 \times 10^{-10} \text{ mol L}^{-1})} = 2.0 \times 10^2 \text{ s}^{-1}$$

(b) $$K_F = \frac{k_2 + k_1}{k_1}$$

$$k_1 = \frac{k_2 + k_{-1}}{K_F} = \frac{(3.3 + 2.0) \times 10^2 \text{ s}^{-1}}{3.9 \times 10^{-6} \text{ mol L}^{-1}}$$

$$= 1.4 \times 10^8 \text{ L mol}^{-1} \text{ s}^{-1}$$

$$K_M = \frac{k_2 + k_1}{k_{-2}}$$

$$k_{-2} = \frac{k_2 + k_{-1}}{K_M} = \frac{(3.3 + 2.0) \times 10^2 \text{ s}^{-1}}{1.03 \times 10^{-5} \text{ mol L}^{-1}}$$

$$= 5.1 \times 10^7 \text{ L mol}^{-1} \text{ s}^{-1}$$

(c) $$K = \frac{[M]_{eq}}{[F]_{eq}} = \frac{V_F K_M}{V_M K_F}$$

$$= \frac{(6.5 \times 10^{-7})(1.03 \times 10^{-5})}{(4.0 \times 10^{-7})(3.9 \times 10^{-6})} = 4.3$$

$$K = \frac{k_1 k_2}{k_{-1} k_{-2}} = \frac{(1.4 \times 10^8)(3.3 \times 10^2)}{(2.0 \times 10^2)(5.1 \times 10^7)} = 4.5$$

21.22 Derive the steady-state rate equation for the mechanism

$$E + S \underset{k_2}{\overset{k_1}{\rightleftarrows}} X \xrightarrow{k_3} E + P \qquad\qquad E + I \underset{k_5}{\overset{k_4}{\rightleftarrows}} EI$$

for the case that $[S] >> [E]_o$ and $[I] >> [E]_o$.

SOLUTION

$$[E]_o = [E] + [EI] + [X]$$

$$= [E](1 + [I]/K_I) + [X] \qquad (1)$$

Since $K_I = \frac{[E][I]}{[EI]} = \frac{k_5}{k_4}$

Assuming that X is in a steady state

$$\frac{d[X]}{dt} = k_1[E][S] - (k_2 + k_3)[X] = 0 \qquad (2)$$

Solving equation 1 for [E] and substituting this expression in equation 2 yields

$$\frac{k_1[S][E]_o}{1 + [I]/K_I} = [X]\left[\frac{k_1[S]}{1 + [I]/K_I} + k_2 + k_3\right]$$

$$\frac{d[P]}{dt} = k_3[X] = \frac{k_3[E]_o}{1 + \frac{k_2 + k_3}{k_1[S]}\left(1 + \frac{[I]}{K_I}\right)}$$

$$= \frac{V_s}{1 + \frac{K_s}{[S]}\left(1 + \frac{[I]}{K_I}\right)}$$

21.23 The following initial velocities were determined spectrophotometrically for solutions of sodium succinate to which a constant amount of succinoxidase was added. The velocities are given as the change in absorbancy at 250 nm in 10 s. Calculate V, K_M and K_I for malonate.

[Succinate] 10^{-3} mol L^{-1}	$\frac{A \times 10^3}{10s}$ No Inhibitor	15×10^{-6} mol L^{-1} malonate
10	16.7	14.9
2	14.2	10.0
1	11.3	7.7
0.5	8.8	4.9
0.33	7.1	--

SOLUTION

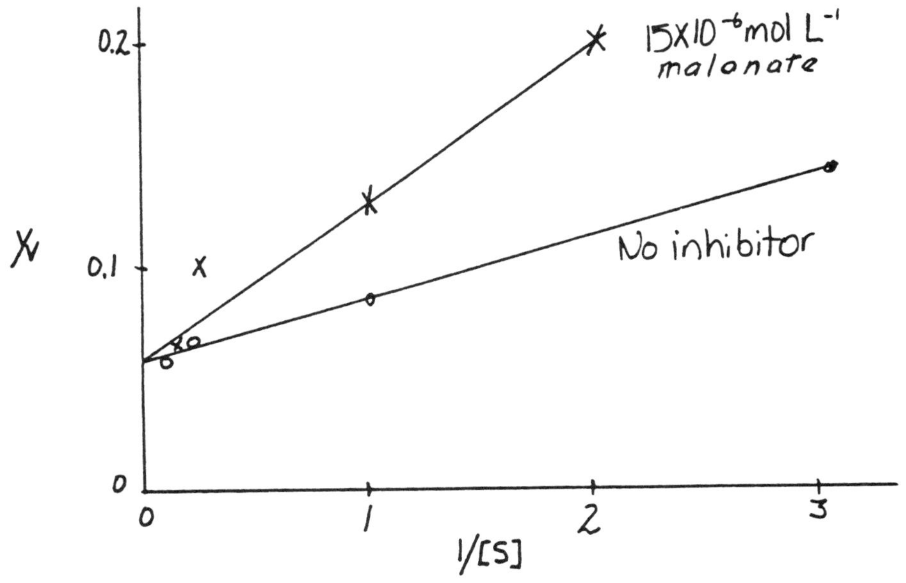

$$v = \frac{d[P]}{dt} = \frac{V}{1 + \frac{K_M}{[S]}(1 + \frac{[I]}{K_I})}$$

$$\frac{1}{v} = \frac{1}{V} + \frac{K_M}{V[S]}(1 + \frac{[I]}{K_I})$$

$$\frac{1}{V} = 0.0662 \qquad V = 15.1$$

Slope without inhibitor

$$= \frac{0.141 - 0.066}{3 \times 10^3 \text{ L mol}^{-1}} = 25 \times 10^{-6} = \frac{K_s}{V_s}$$

$$K_M = (25 \times 10^{-6})(15.1) = 0.38 \times 10^{-3} \text{ mol L}^{-1}$$

Slope with inhibitor

$$= \frac{0.205 - 0.066}{2 \times 10^3 \text{ L mol}^{-1}} = 0.069 = \frac{K_M}{V}[1 + \frac{[I]}{K_I}]$$

$$0.069 = \frac{0.38}{15.1}[1 + \frac{15 \times 10^{-6} \text{ mol L}^{-1}}{K_I}]$$

$$K_I = \frac{15 \times 10^{-6} \text{ mol L}^{-1}}{\frac{(15.1)(0.069)}{0.38} - 1}) = 8.6 \times 10^{-6} \text{ mol L}^{-1}$$

21.24 In the Eadie-Hostee method for determining k_{cat} and K_M for an enzymatic reaction, $v/[E]_o[S]$ is plotted versus $v/[E]_o$. How are the kinetic parameters obtained from this plot?

SOLUTION

The Michaelis-Menten equation can be arranged in the form

$$\frac{v}{[E]_o[S]} = \frac{k_{cat}}{K_M} - \frac{v}{K_M[E]_o}$$

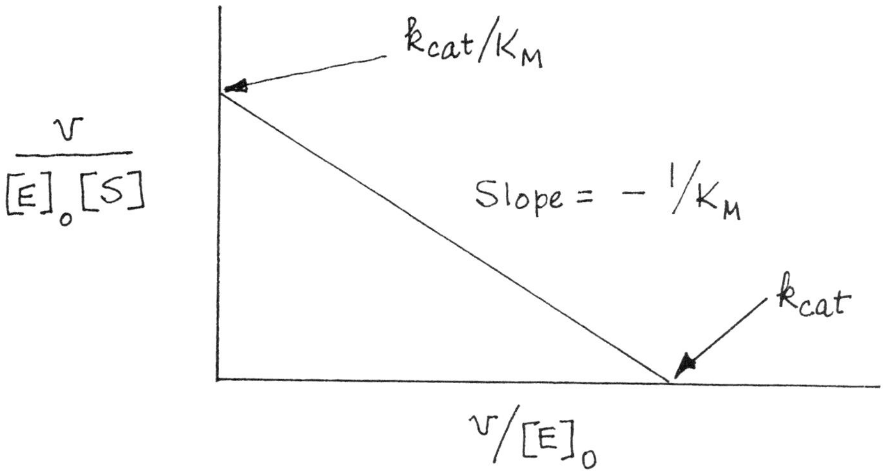

21.25 The maximum initial velocities for an enzymatic reaction are determined at a series of pH values.

pH	6.0	7.0	7.5	8.0	8.5	9.0
V	11	74	129	147	108	53

Calculate the values of the parameters V′, K_a, K_b in

$$V = \frac{V'}{1 + [H^+]/K_a + K_b/[H^+]}$$

Hint: A plot of V versus pH may be constructed and the hydrogen ion concentration at the midpoint on the acid side referred to as $[H^+]_a$

and the hydrogen ion concentration at the midpoint on the basic side is referred to as $[H^+]_b$. Then

$$K_a = [H^+]_a + [H^+]_b - 4\sqrt{[H^+]_a[H^+]_b}$$

$$K_b = \frac{[H^+]_a[H^+]_b}{K_a}$$

SOLUTION

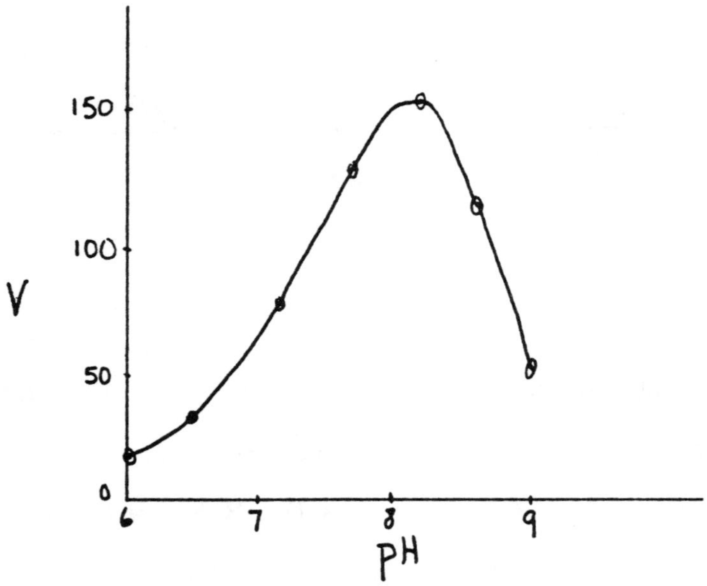

$[H^+]_a = 10^{-7.03} = 9.3 \times 10^{-8}$

$[H^+]_b = 10^{-8.76} = 1.74 \times 10^{-9}$

$$K_a = [H^+]_a + [H^+]_b - 4\sqrt{[H^+]_a[H^+]_b}$$

$$= 9.3 \times 10^{-9} + 1.74 \times 10^{-9} - \sqrt{(9.3 \times 10^{-8})(1.74 \times 10^{-9})}$$

$$= 4.41 \times 10^{-8} \text{ mol L}^{-1}$$

$$pK_a = 7.36$$

$$K_b = \frac{[H^+]_a[H^+]_b}{K_a} = \frac{(9.3 \times 10^{-8})(1.74 \times 10^{-9})}{4.41 \times 10^{-8}}$$

$$= 3.67 \times 10^{-9} \text{ mol L}^{-1}$$

$$pK_b = 8.45$$

$$V = \frac{V'}{1 + [H^+]/K_a + K_b/[H^+]}$$

$$147 = \frac{V'}{1 + (10^{-8})/(4.41 \times 10^{-8}) + (0.367 \times 10^{-8})/(10^{-8})}$$

$$V' = 234$$

21.26 (a) If the anodic transfer coefficient α_a is 0.5 for an oxidation reaction, by what factor will k_{ox} be changed if E is changed by 0.01 V at 25 °C if n = 1? (b) What will the factor be if $\alpha_a = 0.3$?

SOLUTION

(a) $$\frac{k_{ox2}}{k_{ox1}} = \exp[(\alpha_a nF/RT)(E_2 - E_1)]$$

$$= \exp\left[\frac{(0.5)(96{,}485 \text{ C mol}^{-1})(0.01 \text{ V})}{(8.3145 \text{ J K}^{-1} \text{ mol}^{-1})(298 \text{ K})}\right]$$

$$= 1.22$$

(b) $$\frac{k_{ox2}}{k_{ox1}} = 1.12$$

21.27 What is the root-mean-square displacement in the x direction of a molecule of tobacco mosaic virus due to Brownian motion during one minute in water at 20 °C? (See Table 22.4)

SOLUTION

$$\langle(\Delta x)^2\rangle^{1/2} = \sqrt{2Dt} = [(2)(0.53 \times 10^{-11}\ m^2\ s^{-1})(60\ s)]^{1/2}$$

$$= 0.0252\ nm$$

21.28 8.57 minutes

21.29 0.66 s

21.30 17.4 kJ mol^{-1}

21.31 7.09 x 10^{-11} m^2 s^{-1}

21.32 (a) 0.0705 cm (b) 0.141 cm

21.33 1.83, 3.66, 5.49 mm

21.34 3.34 x 10^{-9} m^2 s^{-1}

21.36 0.1058 Ω^{-1} m^{-1}

21.37 1.55 x 10^{-2} Ω^{-1} m^{-1}

21.38 1.99 x 10^{-9} m^2 s^{-1}

21.39 520 nm

21.40 1.4 x 10^{11} L mol^{-1} s^{-1}

21.41 2.7 x 10^{-4} s

21.42 $\tau^{-1} = k_1+k_2$

21.43 k_1 = 1.8 x 10^{10} L mol^{-1} s^{-1}, k_{-1} = 1.11 x 10^3 s^{-1}

21.45 H^+ production: k = 1.75 x 10^5, 1.2 x 10^2, 5.71 s^{-1}; OH^- production: k = 5.71, 0.83 x 10^3, 1.8 x 10^5 s^{-1}. Imidazole can play both roles about equally well.

21.46 k_o = 1.21 x 10^{-4} min^{-1}
k_{H^+} = 3.4 x 10^{-3} L mol^{-1} s^{-1}

21.47 V = 1.2 x 10^{-6} M s^{-1}, K_M = 0.48 x 10^{-3} mol L^{-1}

20.48 (a) k_1 = 1.3 x 10^8 M^{-1} s^{-1}

k_2 = 0.33 x 10^3 s^{-1}

k_{-1} = 0.20 x 10^3 s^{-1}

k_{-2} = 5.3 x 10^7 M^{-1} s^{-1}

(b) -3.4 kJ mol^{-1}

21.49 k_1 = 4.9 x 10^8 M^{-1} s^{-1}, k_2 = 0.8 x 10^3 s^{-1}

$k_{-1} = 2.6 \times 10^3\ s^{-1}$, $k_{-2} = 3.4 \times 10^7\ M^{-1}\ s^{-1}$

21.50 33% inhibition

21.51 $$\frac{v}{[S]} = \frac{V}{K_M(1 + [I]/K_I)} = \frac{v}{K_M(1 + [I]/K_I)}$$

21.52 $V = k_2[E]_o/(1 + [H^+]/K_{EHS})$

$$K_M = \frac{k_{-1} + k_2}{k_1} \frac{[1 + [H^+]/K_{EH}]}{[1 + [H^+]/K_{EHS}\}}$$

21.53 $38.9\ V^{-1}$

21.54 5.41×10^{-4} cm

CHAPTER 22 Macromolecules

22.1 A polymer solution contains 250 molecules of molar mass 75,000 g mol^{-1}, 500 molecules of molar mass 100,000 g mol^{-1} and 250 molecules of molar mass 125,000 g mol^{-1}. Calculate $\bar{M}_n$, $\bar{M}_m$, and the ratio $\bar{M}_m/\bar{M}_n$ (the polydispersity).

SOLUTION

$$\bar{M}_n = \frac{250(75{,}000)}{1000} + \frac{500(100{,}000)}{1000} + \frac{250(125{,}000)}{1000}$$

$$= 100{,}000 \text{ g mol}^{-1}$$

$$\bar{M}_m = \frac{250(75{,}000)^2 + 500(100{,}000)^2 + 250(125{,}000)^2}{250(75{,}000) + 500(100{,}000) + 250(125{,}000)}$$

$$= 103{,}000 \text{ g mol}^{-1}$$

$$\frac{\bar{M}_m}{\bar{M}_n} = 1.03$$

22.2 A beam of sodium D light (589 nm) is passed through 100 cm of an aqueous solution of sucrose containing 10 g sucrose per 100 cm^3. Calculate I/I_o, where I_o is the intensity that would have been obtained with pure water, given that M = 342.30 g mol^{-1} and dn/dc = 0.15 g^{-1} cm^3 for sucrose. The refractive index of water at 20 °C is 1.333 for the sodium D line.

SOLUTION

$$\tau = \frac{32\pi^3 n_o^2 (dn/dc)^2 M c}{3 N_A \lambda^4}$$

$$= \frac{32\pi^3 (1.333)^2 (0.15\ g^{-1}\ cm^3)^2 (342.30\ g\ mol^{-1})(0.1\ g\ cm^{-3})}{3(6.02\ x\ 10^{23}\ mol^{-1})(5.89\ 10^{-5}\ cm^4)}$$

$$= 6.25\ x\ 10^{-5}\ cm^{-1}$$

$$\frac{I}{I_o} = e^{-\tau x} = e^{-(6.25\ x\ 10^{-5}\ cm^{-1})(100\ cm)} = 0.9938$$

22.3 Plot the probability density W(r) for random walk in three dimensions after 1000 steps with a step length of unity. Indicate the root-mean-square end-to-end distance on this plot.

<u>SOLUTION</u>

$$W(r) = (\frac{3}{2\pi n l^2})^{3/2} \quad exp(-\frac{3r^2}{2nl^2})$$

$$= (\frac{3}{2\pi 1000})^{3/2} \quad exp(-\frac{3r^2}{2000})$$

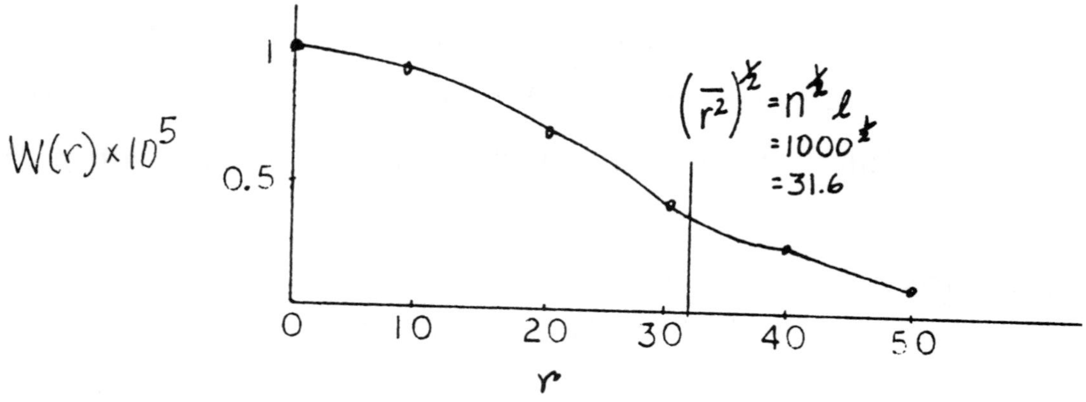

22.4 In polyethylene $H(CH_2—CH_2)_nH$ the bond length 1 is 0.15 nm. What is the root-mean-square end-to-end distance for a molecule with universal joints with a molar mass of 10^5 g mol^{-1}? Taking into account the fact that carbon forms tetrahedral bonds, what is $(\overline{r^2})^{1/2}$?

SOLUTION

$$N = \frac{10^5 \text{ g mol}^{-1}}{14 \text{ g mol}^{-1}}$$

$$(\overline{r^2})^{1/2} = n^{1/2}\ell = \left(\frac{10^4}{14}\right)^{1/2} 0.15 \text{ m}$$

$$= 12.7 \text{ nm}$$

$$(\overline{r^2})^{1/2} = n^{1/2}\ell \times \left(\frac{1 + \cos\theta}{1 - \cos\theta}\right)^{1/2}$$

$$= 12.7 \text{ nm} \left(\frac{1 + \cos 71°}{1 - \cos 71°}\right)^{1/2} = 17.8 \text{ nm}$$

22.5 Adipic acid and 1,10-decamethylene glycol were mixed in equimolar amounts and polymerized at 161 °C and 191 °C [S. D. Hamann, D. H. Solomon, and J. D. Swift. *J. Macromol. Sci. Chem.* **A2**, 153 (1968)]. The extents of reaction, as determined by acid titration, are as follows:

t/min	0	200	400	600	800
p(161 °C)	0.820	0.900	0.927	0.940	0.947

p(191 °C)	0.820	0.937	0.955	0.963	0.968

Time was measured from the point at which there was 82% esterification. Taking $[CO_2H]_0 = 1.25$ mol kg^{-1} at both temperatures, what are the two values of k and the activation energy?

SOLUTION

$$\frac{1}{(1-p)^2} - \frac{1}{(1-0.82)^2} = 2[CO_2H]_0^2 k t$$

At 161 °C

$$k = \frac{(1-p)^{-2} - (1-0.82)^{-2}}{2[CO_2H]_0^2 t}$$

= 0.111, 0.125, 0.132, 0.130 for the 4 successive times

= 0.13 kg^2 mol^{-2} min^{-1}

At 191 °C

k = 0.36 kg^2 mol^{-2} min^{-1}

$$E_a = \frac{RT_1T_2}{T_2 - T_1} \ln \frac{k_2}{k_1} = \frac{(8.314)(459)(489)}{30} \ln \frac{0.36}{0.13}$$

= 63 kJ mol^{-1}

22.6 Adipic acid and diethylene glycol were polymerized at 109 °C using p-toluene sulfuric acid as a catalyst [P. J. Flory, *J. Am. Chem. Soc.* **61**, 3334 (1939)]. The extents of reaction at various times are as follows:

t/min	0	40	80	120

p	0.800	0.909	0.944	0.960

(Note that the time is taken as zero at p = 0.80.) At what time will the number average molar mass reach 10,000 g mol^{-1} for this concentration of reactants and catalyst?

SOLUTION

$$\frac{\frac{1}{1-p} - \frac{1}{1-0.800}}{t} = k_2[CO_2H]_o$$

$$k_2[CO_2H]_o = 0.16 \text{ min}^{-1}$$

$$p = 1 - \frac{M_o}{M_n} = 1 - \frac{216 \text{ g mol}^{-1}}{10^4 \text{ g mol}^{-1}} = 0.9784$$

$$\frac{1}{1-0.9784} - \frac{1}{1-0.800} = (0.16 \text{ min}^{-1})t$$

$$t = 258 \text{ min}$$

22.7 For a condensation polymerization of a hydroxyacid in which 99% of the acid groups are used up, calculate (a) the average number of monomer units in the polymer molecules, (b) the probability that a given molecule will have the number of residues given by this value, and (c) the weight fraction having this particular number of monomer units.

SOLUTION

(a) The number-average degree of polymerization is given by

$$\bar{X}_n = \frac{1}{1 - p} = \frac{1}{1 - 0.99} = 100$$

(b) The probability that a given molecule will have 100 residues is given by

$$\pi_i = p^{i - 1}(1 - p)$$

$$\pi_{100} = 0.99^{100 - 1}(1 - 0.99) = 3.70 \times 10^{-3}$$

(c) The weight fraction having this number of residues is given by

$$W_i = ip^{i-1}(1 - p)^2$$

$$W_{100} = (100)(0.99)^{100 - 1}(1 - 0.99)^2$$
$$= 3.70 \times 10^{-3}$$

22.8 A hydroxyacid $HO\text{-}(CH_2)_5\text{-}CO_2H$ is polymerized and it is found that the product has a number average molar mass of 20,000 g mol^{-1}. (a) What is the extent of reaction p? (b) What is the degree of polymerization $\bar{X}_n$? (c) What is the mass-average molar mass?

SOLUTION

The molar mass of a monomer unit is 114 g $mol^{-1} = M_o$

(a) $\bar{M}_n = M_o/(1 - p)$

$p = 1 - M_o/\bar{M}_n = 1 - 114/20{,}000 = 0.9943$

(b) $\bar{X}_n = \dfrac{1}{1 - p} = \dfrac{1}{1 - 0.9943} = 175$

(c) $\bar{M}_m = M_o\dfrac{1 + p}{1 - p} = 114\dfrac{1.9943}{1 - 0.9943}$

$= 39{,}900 \text{ g mol}^{-1}$

22.9 The relative viscosities of a series of solutions of a sample of polystyrene in toluene were determined with an Ostwald viscometer at 25 °C.

$c/10^{-2}$ g cm^{-3}	0.249	0.499	0.999	1.998
η/η_o	1.355	1.782	2.879	6.090

The ratio η_{sp}/c is plotted against c and extrapolated to zero concentration to obtain the intrinsic viscosity. If the constants in equation 22.78 are $K = 3.7 \times 10^{-2}$ and $a = 0.62$ for this polymer, when concentrations are expressed in g/cm^3, calculate the molar mass.

SOLUTION

$c/10^{-2}$ g cm^{-3}	0.249	0.499	0.999	1.998
$\dfrac{\eta/\eta_o - 1}{c}$	142.6	156.7	188.1	254.8

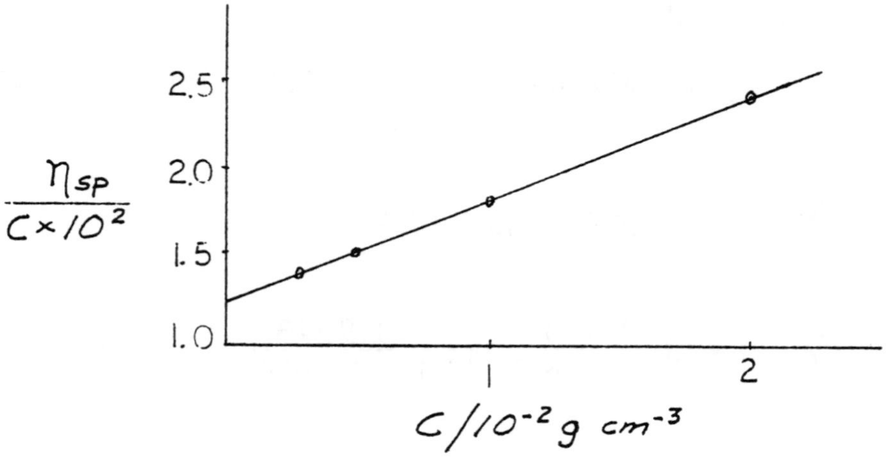

$$ln \frac{[\eta]}{3.7 \times 10^{-2}} = 0.62 \; ln \; M$$

$$M = \exp \left[\frac{1}{0.62} \; ln \; \frac{126}{3.7 \times 10^{-2}}\right] = 500{,}000 \text{ g mol}^{-1}$$

22.10 At 34 °C the intrinsic viscosity of a sample of polystyrene in toluene is 84 $cm^3 \, g^{-1}$. The empirical relation between the intrinsic viscosity of polystyrene in toluene and molar mass is $[\eta] = 1.15 \times 10^{-2} M^{0.72}$. What is the molar mass of this sample?

SOLUTION

$$ln \frac{[\eta]}{1.15 \times 10^{-2}} = 0.72 \; ln \; M$$

$$M = \exp\left[\frac{1}{0.72} \; ln \; \frac{84}{1.15 \times 10^{-2}}\right] = 230{,}000 \text{ g mol}^{-1}$$

22.11 Given that the intrinsic viscosity of myosin is 217 $cm^3\ g^{-1}$, approximately what concentration of myosin in water would have a relative viscosity of 1.5?

SOLUTION

$$\frac{\frac{\eta}{\eta_o} - 1}{c} = 217\ cm^3\ g^{-1} \qquad \frac{0.5}{c} = 217\ cm^3\ g^{-1}$$

$$c = \frac{0.5}{217\ cm^3\ g^{-1}} = 2.30 \times 10^{-3}\ g\ cm^{-3}$$

22.12 The sedimentation coefficient of myoglobin at 20 °C is 2.06×10^{-13} s. What molar mass would it have if the molecules were spherical? Given: $v = 0.749 \times 10^{-3}\ m^3\ kg^{-1}$, $\rho = 0.9982 \times 10^3\ kg\ m^{-3}$, and $\eta = 0.001\ 005$ Pa s.

SOLUTION

$$S = \frac{M(1 - v\rho)}{N_A f} = \frac{M(1 - v\rho)}{N_A 6\pi\eta}\left(\frac{4\pi N_A}{3Mv}\right)^{1/3}$$

$$M = \left(\frac{N_A 6\pi\eta S}{1 - v\rho}\right)^{3/2}\left(\frac{3v}{4\pi N_A}\right)^{1/2}$$

$$M = \left[\frac{(6.022 \times 10^{23}\ mol^{-1})(6\pi)(0.001\ 005)(2.06 \times 10^{-13})}{1 - (0.749 \times 10^{-3})(0.9982 \times 10^3)}\right]^{3/2}$$

$$\times \left(\frac{3 \times 0.749 \times 10^{-3}}{4\pi 6.022 \times 10^{23}}\right)^{1/2}$$

$$= 15.5 \text{ kg mol}^{-1} = 15{,}500 \text{ g mol}^{-1}$$

22.13 The sedimentation and diffusion coefficients for hemoglobin corrected to 20 °C in water are 4.41×10^{-13} s and 6.3×10^{-11} m^2 s^{-1} , respectively. If v = 0.749 cm^3 g^{-1} and ρ_{H_2O} = 0.998 g cm^{-3} at this temperature, calculate the molar mass of the protein. If there is 1 mol of iron per 17,000 g of protein, how many atoms of iron are there per hemoglobin molecule?

SOLUTION

$$M = \frac{RTS}{D(1 - v\rho)}$$

$$= \frac{(8.31 \text{ J K}^{-1} \text{ mol}^{-1})(203 \text{ K})(4.41 \times 10^{-13} \text{ s})}{(6.3 \times 10^{-11} \text{ m}^2 \text{ s}^{-1})[1 - (0.749 \times 10^{-3} \text{ m}^3 \text{ kg}^{-1})(0.998 \times 10^3 \text{ kg m}^{-3})]}$$

$$= 67.6 \text{ kg mol}^{-1} \qquad = 67{,}600 \text{ g mol}^{-1}$$

$$\frac{67000 \text{ g mol}^{-1}}{17000 \text{ g g-atom Fe}} = 4 \text{ g-atom Fe mol}^{-1}$$

$$= 4 \text{ Fe per molecule}$$

22.14 Given the diffusion coefficient for sucrose at 20 °C in water (D = 45.5×10^{-11} m^2 s^{-1}), calculate its sedimentation coefficient. The partial specific volume v is 0.630 cm^3 g^{-1}.

SOLUTION

$$S \quad = \frac{MD(1 - v\rho)}{RT}$$

$$= \frac{(342 \times 10^{-3}\ \text{kg mol}^{-1})(45.5 \times 10^{-11}\ \text{m}^2\ \text{s}^{-1})[1 - (0.630\ \text{cm}^3\ \text{g}^{-1})(A)]}{(8.31\ \text{J K}^{-1}\ \text{mol}^{-1})(293\ \text{K})}$$

$= 0.236 \times 10^{-13}$ s $\qquad\qquad A = 1$ g cm^{-3}

22.15	19,500 g mol^{-1}, 0.9980
22.16	0.57 kg min^{-1} mol^{-1}
22.17	179 hr
22.18	(a) 20, (b) 0.0189, (c) 0.0189
22.19	20 M_o , 39.0 M_o
22.20	(a) 20,000 g mol^{-1}, (b) 39,800 g mol^{-1}
22.21	8.36×10^6 g mol^{-1}
22.22	638,000 g mol^{-1}
22.23	177×10^{-13} s
22.24	0.33 cm
22.25	64,000 g mol^{-1}
22.26	511,000 g mol^{-1}
22.27	100,000 g mol^{-1}
22.28	27.7×10^6 g mol^{-1}

CHAPTER 23 Solid-State Chemistry

23.1 What is the equation for the distances between planes 110 for a crystal with mutually perpendicular axes?

SOLUTION

$$\frac{1}{d} = \left(\frac{1}{a^2} + \frac{1}{b^2}\right)^{+1/2}$$

$$d = \left(\frac{1}{a^2} + \frac{1}{b^2}\right)^{-1/2} = \left(\frac{b^2 + a^2}{a^2b^2}\right)^{-1/2} = \frac{ab}{(a^2 + b^2)^{1/2}}$$

23.2 Calculate the angles at which the first-, second-, and third-order reflections are obtained from planes 500 pm apart, using X rays with a wavelength of 100 pm.

SOLUTION

$$\sin\theta = \frac{\lambda}{2d_{hkl}} = \frac{\lambda}{2(d/n)}$$

$$\sin\theta_1 = \frac{100\ \text{pm}}{2(500\ \text{pm}/1)} \qquad \theta_1 = 5.74°$$

$$\sin\theta_2 = \frac{100\ \text{pm}}{2(500\ \text{pm}/2)} \qquad \theta_2 = 11.54°$$

$$\sin\theta_3 = \frac{100\ \text{pm}}{2(500\ \text{pm}/3)} \qquad \theta_3 = 17.46°$$

23.3 Calculate the structure factor for a cubic unit cell of AB in which the B atoms occupy the body centered position. Which reflections will be strong and which weak?

SOLUTION

$$F_{hkl} = f_A + f_B\, e^{2\pi i(h/2 + k/2 + 1/2)}$$
$$= f_A + f_B\, e^{\pi i(h + k + 1)}$$
$$= f_A + f_B(-1)^{h + k + 1}$$

If h + k + 1 is even, $F_{hkl} = f_A + f_B$, strong reflections.

If h + k + 1 is odd, $F_{hkl} = f_A - f_B$, weak reflections.

23.4 The crystal unit cell of magnesium oxide is a cube 420 pm on an edge. The structure is interpenetrating face centered. What is the density of crystalline MgO?

SOLUTION

$$\frac{4(40.32 \times 10^{-3}\ \text{kg mol}^{-1})}{(6.022 \times 10^{23}\ \text{mol}^{-1})(4.20 \times 10^{-10}\ \text{m})^3} = 3.615 \times 10^3\ \text{kg m}^{-3}$$
$$= 3.615\ \text{g cm}^{-3}$$

23.5 Platinum forms face-centered cubic crystals. If the radius of a platinum atom is 139 pm, what is the length of the side of the unit cell? What is the density of the crystal?

SOLUTION

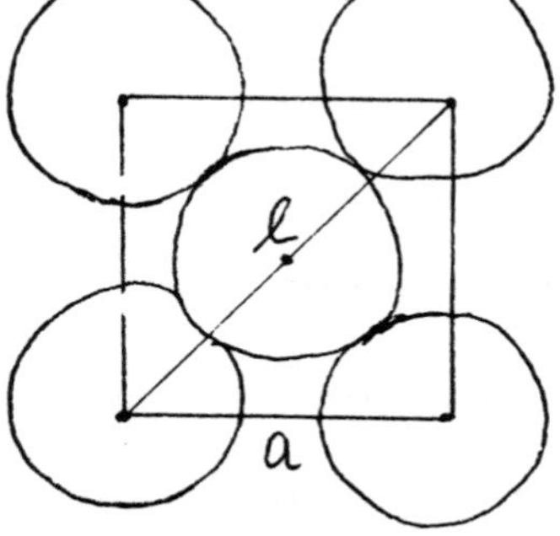

$$l^2 = 2a^2 = [4(139 \text{ pm})]^2$$

$$a = \frac{1}{2^{1/2}} [4(139 \text{ pm})] = 393 \text{ pm}$$

$$d = \frac{4(195.08 \times 10^{-3} \text{ kg mol}^{-1})}{6.022 \times 10^{23} \text{ mol}^{-1} (393 \times 10^{-12} \text{ m})^3}$$

$$= 21.32 \times 10^3 \text{ kg m}^{-3}$$

23.6 Tungsten forms body-centered cubic crystals. From the fact that the density of tungsten is 19.3 g cm^{-3}, calculate (a) the length of the side of this unit cell, and (b) d_{200}, d_{110}, and d_{222}.

SOLUTION

(a) $$19.3 \times 10^3 \text{ kg m}^{-3} = \frac{2(183.85 \times 10^{-3} \text{ kg mol}^{-1})}{(6.022 \times 10^{23} \text{ mol}^{-1})a^3}$$

$$a = 316 \times 10^{-12} \text{ m} = 316 \text{ pm}$$

(b) $$d_{200} = \frac{a}{\sqrt{2^2 + 0 + 0}} = \frac{316 \text{ pm}}{2} = 158 \text{ pm}$$

$$d_{110} = \frac{a}{\sqrt{1 + 1 + 0}} = \frac{316 \text{ pm}}{\sqrt{2}} = 223 \text{ pm}$$

$$d_{222} = \frac{a}{\sqrt{4 + 4 + 4}} = \frac{316 \text{ pm}}{\sqrt{12}} = 91.2 \text{ pm}$$

23.7 Copper forms cubic crystals. When an X-ray powder pattern of crystalline copper is taken using X-rays from a copper target (the wavelength of the Ka line is 154.05 pm), reflections are found at θ = 21.65°, 25.21°, 37.06°, 44.96°, 47.58°, and other larger angles. (a) What type of lattice is formed by copper? (b) What is the length of a side of the unit cell at this temperature? (c) What is the density of copper?

SOLUTION

$$d_{hkl} = \frac{\lambda}{2\sin\theta} = \frac{154.05\ \text{pm}}{2\ \sin\theta} = \frac{a}{\sqrt{h^2 + k^2 + l^2}}$$

(a)

θ	d_{hkl}/pm	Ratio d_{hkl} to largest spacing
21.65°	208.77	1.0000
25.21°	180.84	0.8662
37.06°	127.81	0.6122
44.96°	109.00	0.5221
47.58°	104.34	0.4998

Ratios expected for cubic crystals

$$d_{hkl} = \frac{a}{\sqrt{h^2 + k^2 + l^2}}$$

	Primitive	Body-centered	Face-Centered
100	1.000	--	--
110	0.7071	1.0000	--
111	0.5774	--	1.0000
200	0.5000	0.7071	0.8660
210	0.4472	--	--
211	x	0.5774	--

220	x	0.5000	0.6124
310	x	0.4472	- -
311	x	- -	0.52222
222	x	x	0.5000

-- reflection absent

x not needed for this problem

Thus copper forms face-centered cubic crystals.

(b) $d_{111} = \dfrac{a}{\sqrt{1 + 1 + 1}} = 208.77$ pm

$a = 361.6$ pm

(c) $d = \dfrac{4(63.546 \times 10^{-3}\ \text{kg mol}^{-1})}{(6.022\ 05 \times 10^{23}\ \text{mol}^{-1})(361.6 \times 10^{-12}\ \text{m}^3)^3}$

$= 0.8927 \times 10^3$ kg m^{-3}

23.8 (a) Metallic iron at 20 °C is studied by the Bragg method, in which the crystal is oriented so that a reflection is obtained from the planes parallel to the sides of the cubic crystal, then from planes cutting diagonally through opposite edges, and finally from planes cutting diagonally through opposite corners. Reflections are first obtained at θ = 11° 36', 8° 3', and 20° 26', respectively. What type of cubic lattice does iron have at 20 °C? (b) Metallic iron also forms cubic crystals at 1100 °C, but the reflections determined as described in (a) occur at θ = 9° 8', 12° 57', and 7° 55', respectively. What type of cubic lattice does iron have at 1100 °C? (c) The density of iron at 20 °C is 7.86 g cm^{-3}. What is the length of a side of the unit celll at 30

°C? (d) What is the wavelength of the X-rays used? (e) What is the density of iron at 1100 °C?

SOLUTION

(a)(b) The three orientations will give reflections from planes a00, bb0, and ccc, respectively, where a, b and c are small integers. The smallest suitable integers are: primitive lattice: a=b=c=1; body-centered lattice: b=1, a=c=2 (100 and 111 reflections are missing); face-centered lattice: c=1, a=b=2.

$$\sin\theta = \frac{\lambda}{2d_{hkl}} = \frac{\lambda\sqrt{h^2 + k^2 + l^2}}{2a}$$

Since $\lambda/2a$ is constant for a particular experiment, the three lattice types can be distinguished simply on the basis of the relative magnitudes of the three θ's. Thus the crystal at 20 °C is body-centered since the second angle is smaller than the first and the third is the largest of all (ratio is sin θ is 2: $\sqrt{2}$: $\sqrt{12}$). At 1100 °C the face-centered form is found (ratio of sin θ is 2:$\sqrt{8}$:$\sqrt{3}$). Note that a primitive lattice would give ratios of sin θ of 1:$\sqrt{2}$:$\sqrt{3}$.

(c) $$d = 7.86 \times 10^3 \text{ kg m}^{-3} = \frac{2(55.847 \times 10^{-3} \text{ kg mol}^{-1})}{(6.022 \times 10^{23} \text{ mol}^{-1})a^3}$$

$$a = \left[\frac{(2)(55.847 \times 10^{-3} \text{ kg mol}^{-1})}{(6.022 \times 10^{23} \text{ mol}^{-1})(7.86 \times 10^3 \text{ kg m}^{-3})}\right]^{1/3}$$

$$= 286.8 \times 10^{-12} \text{ m} = 286.8 \text{ pm}$$

(d) $$\lambda = 2d_{hkl} \sin\theta = \frac{2a \sin\theta}{\sqrt{h^2 + k^2 + l^2}}$$

$$\frac{573.6\ \text{pm}}{2} \sin 11° 36' = 57.7\ \text{pm}$$

$$\frac{573.6\ \text{pm}}{\sqrt{2}} \sin 8° 3' = 56.8\ \text{pm}$$

$$\frac{573.6\ \text{pm}}{\sqrt{12}} \sin 20° 26' = 57.8\ \text{pm} \quad \text{Average} = 57.4\ \text{pm}$$

(e) $$d_{hkl} = \frac{\lambda}{2 \sin\theta}$$

$$d_{200} = \frac{57.4\ \text{pm}}{2 \sin 9° 8'} = 180.8\ \text{pm};\ a = d_{200}\sqrt{4} = 361.6\ \text{pm}$$

$$d_{220} = \frac{57.4\ \text{pm}}{2 \sin 12° 57'} = 128.1\ \text{pm};\ a = d_{220}\sqrt{8} = 362.3\ \text{pm}$$

$$d_{111} = \frac{57.4\ \text{pm}}{2 \sin 7° 55'} = 208.4\ \text{pm};\ a = d_{111}\sqrt{3} = 360.9\ \text{pm}$$

Average = 361.6 pm

$$d = \frac{4(55.847 \times 10^{-3}\ \text{kg mol}^{-1})}{(6.022 \times 10^{23}\ \text{mol}^{-1})(361.6 \times 10^{-12}\ \text{m})^3}$$

$$= 7.846 \times 10^3\ \text{kg m}^{-3}$$

23.9 Cesium chloride, bromide, and iodide form interpenetrating simple cubic crystals instead of interpenetrating face-centered cubic crystals like the other alkali halides. The length of the side of the unit cell of CsCl is 412.1 pm. (a) What is the density? (b) Calculate the ionic radius of Cs^+, assuming that the ions touch along a diagonal through the unit cell and that the ion radius of Cl^- is 181 pm.

SOLUTION

(a) $$d = \frac{(168.36 \times 10^{-3}\ \text{kg mol}^{-1})}{(6.022 \times 10^{23}\ \text{mol}^{-1})(412.1 \times 10^{-12}\ \text{m})^3}$$

$= 3.995 \times 10^3$ kg m^{-3}

(b) The diagonal plane through the cubit unit cell is as follows:

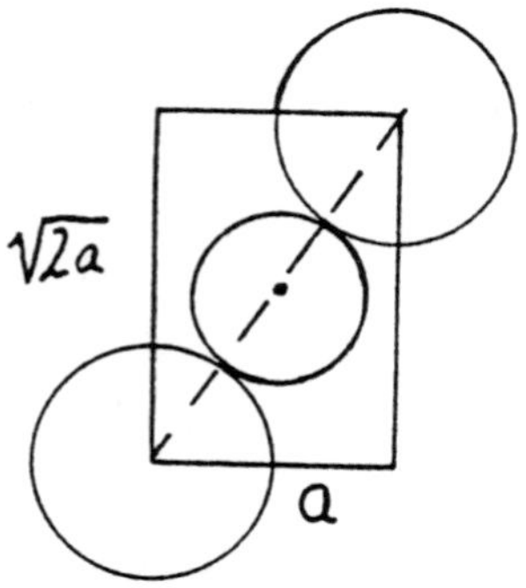

$$a^2 + (\sqrt{2}\ a)^2 = (2r_{Cs} + 2r_{Cl})^2$$

$$r_{Cs} = \frac{1}{2}[a^2 - (\sqrt{2}\ a)^2]^{1/2} - 2(181 \text{ pm})$$

$$= 176 \text{ pm}$$

23.10 Deslattes, et al., *Phys. Rev. Lett.* **33**, 463 (1974) found the following values for a single crystal of very pure silicon at 25 °C: ρ = 2.328 992 g cm^{-3}, a = 543.1066 pm. Silicon has a face-centered cubic lattice like diamond. The atomic mass is 28.085 41 g mol^{-1}. What value of Avogadro's constant is obtained from these values?

SOLUTION

$$\rho = \frac{nM}{N_A a^3}$$

$$N_A = \frac{nM}{\rho a^3}$$

$$= \frac{(8)(28.085\ 41 \times 10^{-3} \text{ kg mol}^{-1})}{(2.328\ 992 \times 10^3 \text{ kg m}^{-3})(543.106\ 6 \times 10^{-12} \text{ m})^3}$$

$$= 6.022\ 093 \times 10^{23} \text{ mol}^{-1}$$

23.11 Insulin forms crystals of the orthohombic type with unit-cell dimensions of 13.0 x 7.48 x 3.09 nm. If the density of the crystal is 1.315 g cm^{-3} and there are six insulin molecules per unit cell, what is the molar mass of the protein insulin?

<u>SOLUTION</u>

$d = 1.315 \times 10^3 \text{ kg m}^{-3}$

$$= \frac{6M}{(6.022 \times 10^{23} \text{ mol}^{-1})(13.0 \times 7.48 \times 3.09 \times 10^{-27} \text{ m}^3)}$$

$$M = \frac{1}{6}(1.315 \times 10^3 \text{ kg m}^{-3})(6.022 \times 10^{23} \text{ mol}^{-1})(13.0 \times 7.48 \times 3.09 \times 10^{-27} \text{ m}^3))$$

$$= 39.7 \text{ kg mol}^{-1} = 39{,}700 \text{ g mol}^{-1}$$

23.12 Molybdenum forms body-centered cubic crystals and, at 20 °C, the density is 10.3 g cm^{-3}. Calculate the distance between the centers of the nearest molybdenum atoms.

<u>SOLUTION</u>

$$10.3 \times 10^3 \text{ kg m}^3 = \frac{2(95.94 \times 10^{-3} \text{ kg mol}^{-1})}{(6.022\,045 \times 10^{23} \text{ mol}^{-1})a^3}$$

$$a = 314 \times 10^{-12} \text{ m} = 314 \text{ pm}$$

Consider a plane bisecting the cube and cutting diagonally through opposite faces.

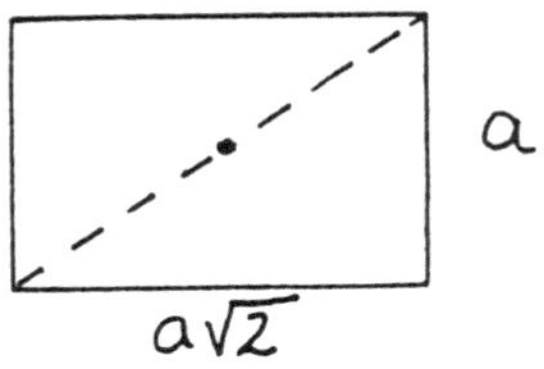

length of diagonal $= \sqrt{a + 2b} = a\sqrt{3}$

distance between atoms $= \frac{a\sqrt{3}}{2} = 273$ pm

23.13 Si has a face-centered cubic structure with two atoms per lattice point, just like diamond. At 25 °C, a = 543.1 pm. What is the density of silicon?

SOLUTION

$$n = 8 = 8 \times \frac{1}{8} + 6 \times \frac{1}{2} + 4$$

= corners + faces + interstitial

$$d = \frac{8(28.0855 \times 10^{-3}\ \text{kg mol}^{-1})}{(6.022 \times 10^{23}\ \text{mol}^{-1})(5.431 \times 10^{-10}\ \text{m})^3}$$

$$= 2.33 \times 10^3\ \text{kg m}^{-3}$$

23.14 The common form of ice has a tetrahedral structure with protons located on the lines between oxygen atoms. A given proton is closer to one oxygen atom than the other and is said to belong to the closer oxygen atom. How many different orientations of a water molecule in space are possible in this lattice?

SOLUTION:

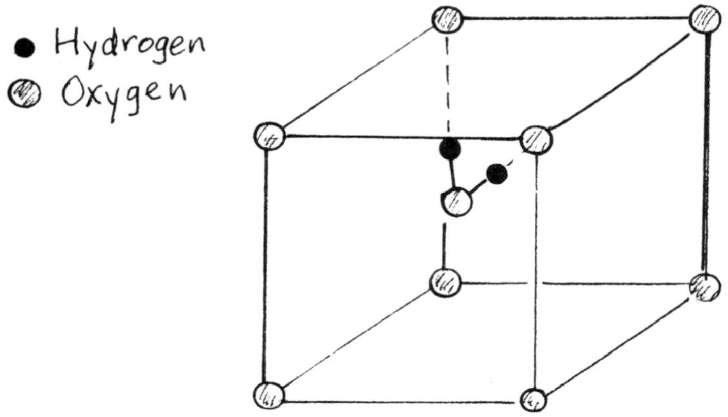

On each of the 6 faces there are 2 pairs of oxygens so that the H's can point to 12 possible orientations

23.15 The diamond has a face-centered cubic crystal lattice, and there are eight atoms in a unit cell. Its density is 3.51 g cm^{-3}. Calculate the first six angles at which reflections would be obtained using an X-ray beam of wavelength 71.2 pm.

SOLUTION

$$d = 3.51 \times 10^3 \text{ kg m}^{-3} = \frac{8(12.011 \times 10^{-3} \text{ kg mol}^{-1})}{(6.022 \times 10^{23} \text{ mol}^{-1})a^3}$$

$$a = 357 \times 10^{-12} \text{ m} = 357 \text{ pm}$$

$$\theta = \sin^{-1} \frac{\lambda}{2a}\sqrt{h^2 + k^2 + l^2} = \sin^{-1} \frac{71.2}{2(357)}\sqrt{h^2 + k^2 + l^2}$$

For face-centered cubic the first six reflections are:

hkl	θ
111	9.95
200	11.50°
220	16.38°
311	19.31°
222	20.21°
400	23.51°

23.16 Calculate the ratio of the radii of small and large spheres for which the small spheres will just fit into octahedral sites in a close-packed structure of the large spheres.

SOLUTION

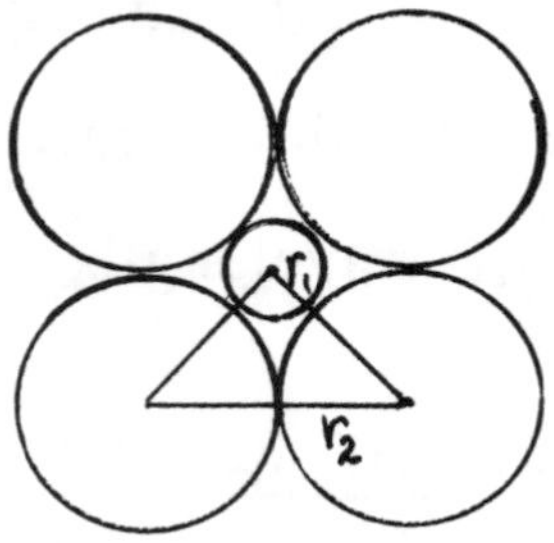

$$2(r_1 + r_2)^2 = (2r_2)^2$$

$$r_1 + r_2 = \frac{2r_2}{\sqrt{2}} = \sqrt{2}\ r_2$$

$$\frac{r_1}{r_2} + 1 = \sqrt{2}$$

$$\frac{r_1}{r_2} = 0.414$$

23.17 A close-packed structure of uniform spheres has a cubic unit cell with a side of 800 pm. What is the radius of the spherical molecule?

SOLUTION

Since the cubic close-packed structure is face-centered cubic, a face diagonal is equal to four radii.

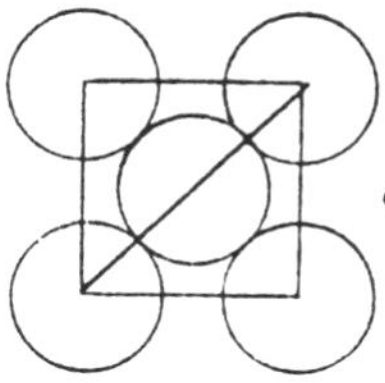

Length of face diagonal = $\sqrt{128 \times 10^4}$

Radius of sphere = $\dfrac{\sqrt{128 \times 10^4}}{4}$ = 282.8 pm

23.18 Titanium forms hexagonal close-packed crystals. Given the atomic radius of 146 pm, what are the unit cell dimensions and what is the density of crystals?

SOLUTION

a = b = 2(156 pm) = 292 pm $c = \sqrt{2}\ 2\ a/\sqrt{3}$ = 477 pm

V = $a^2\ c(1 - \cos^2 \gamma)^{1/2} = a^2\ c \sin\gamma$

= $(292\ \text{pm})^2\ (477\ \text{pm}) \sin 120° = 35.22 \times 10^{-30}\ \text{m}^3$

d $= \dfrac{2(47.90 \times 10^{-3}\ \text{kg mol}^{-1})}{(6.022 \times 10^{23}\ \text{mol}^{-1})(35.22 \times 10^{-30}\ \text{m}^3)}$

$= 4.517 \times 10^3\ \text{kg m}^{-3}$

23.19 What neutron energy in electronvolts is required for a wavelength of 100 pm?

SOLUTION

$$\text{Kinetic energy} = \frac{p^2}{2m_n} = \frac{h^2}{\lambda^2 2m_n} \quad \text{since } \lambda = h/p$$

$$Ee = \frac{h^2}{2\lambda^2 m_n}$$

$$E = \frac{h^2}{2\lambda^2 m_n e}$$

$$= \frac{(6.626 \times 10^{-34}\ \text{J s})^2}{2(10^{-10}\ \text{m})^2(1.675 \times 10^{-27}\ \text{kg})(1.602 \times 10^{-19}\ \text{C})}$$

$= 0.0818$ V or 0.0818 eV

23.20 A solution of carbon in face-centered cubic iron has a density of 8.105 g cm^{-3} and a unit cell edge of 358.3 pm. Are the carbon atoms interstitial, or do they substitute for iron atoms in the lattice? What is the weight percent carbon?

SOLUTION

$$d = \frac{4(55.847 \times 10^{-3}\ \text{kg mol}^{-1})}{(6.022 \times 10^{23}\ \text{mol}^{-1})(358.3 \times 10^{-12}\ \text{m})^3}$$

$= 8.064 \times 10^3$ kg m^{-3}

Since the experimental density is greater, the carbon atoms must be interstitial.

$$\begin{array}{r} 8.105 \times 10^3 \text{kg m}^{-3} \\ -8.064 \times 10^3 \\ \hline 0.041 \times 10^3 \end{array}$$

$$\frac{0.041 \times 10^3}{8.105 \times 10^3} 100 = 0.51\% \text{ by weight}$$

23.21 At 550 °C the conductivity of solid NaCl is $2 \times 10^{-4}\ \Omega^{-1}\ m^{-1}$. Since the sodium ions are smaller than the chloride ions (see Table 23.2), they are responsible for most of the electric conductivity. What is the ionic mobility of Na^+ under these conditions?

SOLUTION

Since there are 4 Na^+ per unit cell of a = 564 pm

$$H = \frac{4}{(564 \times 10^{-12}\ m)^3} = 2.23 \times 10^{28}\ m^{-3}$$

According to equation 23.36

$$u = K/Nq = \frac{2 \times 10^{-4}\ \Omega^{-1}\ m^{-1}}{(2.23 \times 10^{28}\ m^{-3})(1.60 \times 10^{-19}\ C)}$$

$$= 5.61 \times 10^{-14}\ m\ V^{-1}\ s^{-1}$$

23.22 What fraction of the lattice sites of a crystal are vacant at 300 K if the energy required to move an atom from a lattice site in the crystal to a lattice site on the surface is 1 eV? at 1000 K?

SOLUTION

At 300 K $\frac{n}{N} = e^{-E_V/kT}$

$$= e^{-\frac{(1 V)(1.602 \times 10^{-19} C)}{(1.38 \times 10^{-23}\ J\ K^{-1})(300\ K)}} = 1.56 \times 10^{-17}$$

$$\text{At 1000 K} \quad \frac{n}{N} = e^{-\frac{(1V)(1.602 \times 10^{-19} C)}{(1.38 \times 10^{-23}\ J\ K^{-1})(1000\ K)}}$$

$$= 9.08 \times 10^{-6}$$

23.23	236.5, 193.1, 149.6, 136.6 pm
23.24	7
23.25	When h + k is odd, the structure factor F(hkl) is zero and the reflection is extinguished.
23.26	392.4 pm
23.27	(a) body-centered; (b) 314.8; (c) 10.21 g cm^{-3}.
23.28	74.69 g mol^{-1}
23.29	2826 kg m^{-3}
23.30	3.516×10^3 kg m^{-3}
23.31	(a) 2 (b) 328 (c) 164 (d) 232 (e) 94.7 pm
23.32	287, 124 pm
23.33	8.84 g cm^{-3}
23.34	(a) 2.698×10^3 kg m^{-3} (b) 202.5, 143.2, 233.8 pm
23.35	152 pm
23.36	1.295×10^{19}, 9.157×10^{18}, 1.495×10^{19} m^{-2}
23.37	1414.2, 1154.7 pm
23.38	200.3 pm
23.39	184 pm
23.40	3.6×10^{-6}
23.41	490 cm^2 V^{-1} s^{-1}

CHAPTER 24 Surface Dynamics

24.1 In an ultrahigh vacuum chamber ($P = 5 \times 10^{-10}$ torr), how many molecules strike 1 cm^2 of surface in one second at 298 K, if (a) the gas is helium, and (b) if the gas is mercury vapor?

<u>SOLUTION</u>

$J_n \quad = PN_A/(2\pi MRT)^{1/2}$

$P \quad = (5 \times 10^{-10}\text{ torr})(133.3\text{ Pa torr}^{-1}) = 6.66 \times 10^{-8}\text{ Pa}$

$$(a) J_n = \frac{(6.66 \times 10^{-8}\text{ Pa})(6.02 \times 10^{23}\text{ mol}^{-1})}{[2\pi(4.00 \times 10^{-3}\text{ kg mol}^{-1})(8.314\text{ J K}^{-1}\text{ mol}^{-1})(298\text{ K})]^{1/2}}$$

$= (5.08 \times 10^{15}\text{ m}^{-2}\text{ s}^{-1})(0.01\text{ m cm}^{-1})^2$

$= 5.08 \times 10^{11}\text{ cm}^{-2}\text{ s}^{-1}$

(b) $J_n \quad = (5.08 \times 10^{11}\text{ m}^{-2}\text{ s}^{-1})(4/200.6)^{1/2}$

$= 7.17 \times 10^{10}\text{ cm}^{-2}\text{ s}^{-1}$

24.2 A readily oxidized metal surface with 10^{15} metal atoms per square centimeter is exposed to molecular oxygen at 10^{-5} Pa at 298 K. How long will it take to completely oxidize the surface if the oxide formed is MO?

<u>SOLUTION</u>

$$J_n = \frac{(10^{-5}\text{ Pa})(6.02 \times 10^{23}\text{ mol}^{-1})}{[2\pi(32 \times 10^{-3}\text{ kg mol}^{-1})(8.314\text{ J K}^{-1})(298\text{ K})]^{1/2}}$$

$= (2.7 \times 10^{17}\text{ m}^{-2}\text{ s}^{-1})(0.01\text{ m cm}^{-1})^2$

$= 2.7 \times 10^{13}\text{ cm}^{-2}\text{ s}^{-1}$

$$t = \frac{10^{15}\text{ cm}^{-2}}{2(2.7 \times 10^{13}\text{ cm}^{-2}\text{ s}^{-1})} = 18.5\text{ s}$$

24.3 For the adsorption of nitrogen molecules on a certain sample of carbon, the pressure required to half saturate the surface at 298 K is 2×10^{-5} Pa. If the enthalpy of adsorption is -10 kJ mol^{-1} and the sticking coefficient s^* is unity, what is the rate constant of desorption k_d?

SOLUTION

$$k_d = \frac{PN_A}{(2\pi MRT)^{1/2} \exp(\Delta_{ads}H/RT)}$$

$$= \frac{(2 \times 10^{-5}\ \text{Pa})(6.02 \times 10^{23}\ \text{mol}^{-1})\exp(10{,}000/8.314 \times 298\ \text{K})}{[2\pi(28 \times 10^{-3}\ \text{kg mol}^{-1})(8.314\ \text{J K}^{-1}\ \text{mol}^{-1})(298\ \text{K})]^{1/2}}$$

$$= 3.26 \times 10^{19}\ \text{m}^{-2}\ \text{s}^{-1}$$

24.4 The pressure of nitrogen required for adsorption of 1.0 $cm^3\ g^{-1}$ (25 °C, 1.013 bar) of gas on graphitized carbon black are 24 Pa at 77.5 K and 290 Pa at 90.1 K. Calculate the enthalpy of adsorption at this fraction of surface coverage.

SOLUTION

$$\Delta_{ads}H = \frac{RT_1T_2 \ln (P_1/P_2)}{T_2 - T_1}$$

$$= \frac{(8.314\ \text{J K}^{-1}\ \text{mol}^{-1})(77.5\ \text{K})(90.1\ \text{K}) \ln (24/290)}{(90.1\ \text{K} - 77.5\ \text{K})}$$

$$= -11.6\ \text{kJ mol}^{-1}$$

24.5 A mixture of A and B is adsorbed on a solid for which the adsorption isotherm follows the Langmuir equation. If the mole fractions in the gas at equilibrium are y_A and y_B, what is the equation for the adsorption isotherm in terms of total pressure? What is the expression for the mole fraction of A in the adsorbed gas in terms of K_A, K_B, y_A, and y_B?

SOLUTION

See Example 24.2

$$v = v_m(\Theta_A + \Theta_B) = \frac{v_m(y_AK_AP + y_BK_BP)}{1 + y_AK_AP + y_BK_BP}$$

$$= \frac{v_m(y_AK_A + y_BK_B)P}{1 + (y_AK_A + y_BK_B)P} = \frac{v_mK'P}{1 + k'P}$$

$$\text{Mole fraction adsorbed A} = x_A = \frac{\theta_A}{\Theta_A + \Theta_B}$$

$$= \frac{y_AK_AP}{y_AK_AP + y_BK_BP}$$

$$= \frac{y_AK_A}{y_AK_A + y_BK_B}$$

24.6 The following table gives the volume of nitrogen (reduced to 0 °C and 1 bar) adsorbed per gram of active carbon at 0 °C at a series of pressures.

P/Pa	524	1731	3058	4534	7497
$v/cm^3\ g^{-1}$	0.987	3.04	5.08	7.04	10.31

Plot the data according to the Langmuir isotherm, and determine the constants.

<u>SOLUTION</u>

$$\frac{1}{V} = \frac{1}{v_m} + \frac{1}{v_m KP}$$ Intercept = 0.025 g cm^{-3}

$$v_m = 1/(0.025 \text{ g cm}^{-3}) = 40 \text{ cm}^3 \text{ g}^{-1}$$

$$\text{Slope} = \frac{(1.00 - 0.025) \text{ g cm}^{-3}}{1.9 \times 10^{-3} \text{ Pa}^{-1}} = 513 \text{ g cm}^{-3} \text{ Pa}$$

$$= \frac{1}{v_m K} = \frac{1}{(40 \text{ cm}^3 \text{ g}^{-1}) K}$$

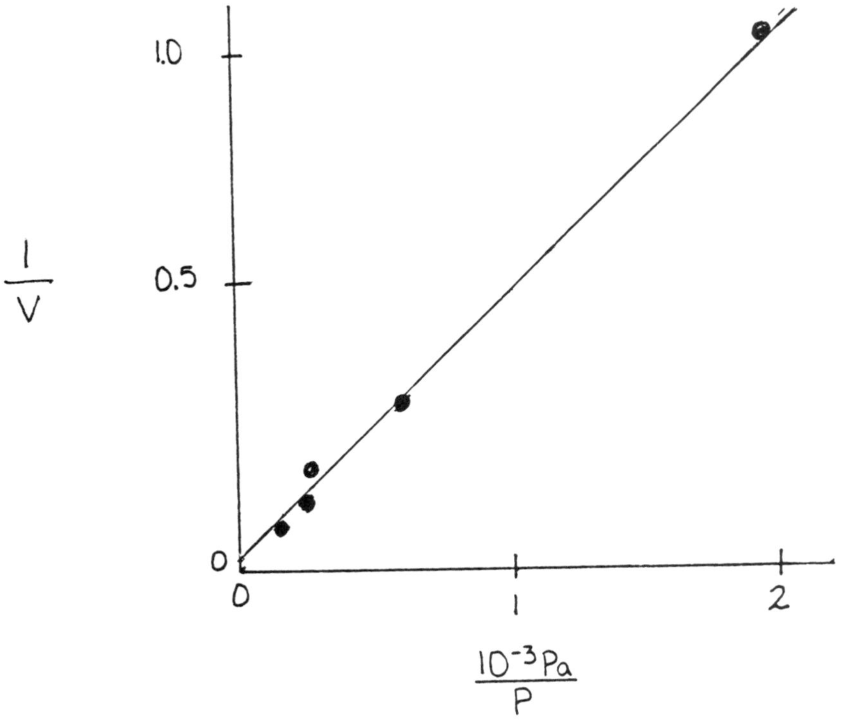

24.7 According to problem 19.10, the rate with which oxygen molecules strike a surface at 1 bar and 25 °C is 2.69×10^{23} $cm^{-2}\ s^{-1}$. If the oxygen molecules are striking a platinum surface, what are the frequencies of collisions per atom on (100), (110) and (111) planes? (See problem 23.36).

<u>SOLUTION</u>

100 planes $\dfrac{(2.69 \times 10^{23}\ cm^{-2}\ s^{-1})(10^2\ cm\ m^{-1})^2}{(1.295 \times 10^{19}\ m^{-2})} = 2.08 \times 10^8\ s^{-1}$

110 planes $\dfrac{(2.69 \times 10^{27}\ m^{-2}\ s^{-1})}{(9.157 \times 10^{18}\ m^{-2})}) = 2.94 \times 10^8\ s^{-1}$

111 planes $\dfrac{(2.69 \times 10^{27}\ m^{-2}\ s^{-1})}{(1.495 \times 10^{19}\ m^{-2})} = 1.80 \times 10^8\ s^{-1}$

24.8 One gram of activated charcoal has a surface area of 100 m^2. If complete surface coverage is assumed, as a limiting case, how much ammonia, at 25 °C and 1 bar, could be adsorbed on the surface of 45 g of activated charchoal? The diameter of the NH_3 molecule is 3×10^{-10} m, and it is assumed that the molecules just touch each other in a plane so that four adjacent spheres have their centers at the corners of a square.

SOLUTION

Area per NH_3 molecule = $(3 \times 10^{-10}\ m)^2$

Number of NH_3 molecules required

$$= \frac{45 \times 10^3\ m^2}{(3 \times 10^{-10}\ m)^2} = 5 \times 10^{23}$$

$$V = nRT/P = \left(\frac{5 \times 10^{23}}{6.022 \times 10^{23}\ mol^{-1}}\right) \frac{(0.08314\ L\ bar\ K^{-1} mol^{-1})(298 K)}{1\ bar}$$

$$= 20.6\ L$$

24.9 Calculate the surface area of a catalyst that adsorbs 103 cm^3 of nitrogen (calculated at 1.013 bar and 0 °C) per gram in order to form a monolayer. The adsorption is measured at -195 °C, and the effective area occupied by a nitrogen molecule on the surface is $16.2 \times 10^{-20}\ m^2$ at this temperature.

SOLUTION

The surface area per gram is equal to the number of molecules adsorbed per gram times the effective area per molecule. The number of molecules is PVN_A/RT.

$$A_S = \frac{(1.013\ bar)(0.103\ L)(6.022 \times 10^{23}\ mol^{-1})(16.2 \times 10^{-20}\ m^2)}{(0.083\ L\ bar\ K^{-1}\ mol^{-1})(273\ K)}$$

$$= 449\ m^2$$

24.10	2×10^{-3} L
24.11	-36.8 kJ mol^{-1}
24.12	-29.6 kJ mol^{-1}
24.13	(a) 90 Pa, (b) 30 Pa
24.14	2.70×10^{20} m^{-2} s^{-1}
24.15	560 m^2 g^{-1}
24.16	48.0 L